AF539945

ENCYCLOPEDIA OF POLYMER SCIENCE AND TECHNOLOGY

Volume - 2

ENCYCLOPEDIA OF POLYMER SCIENCE AND TECHNOLOGY

Volume - 2

by

Navid Naderpour
Bandar Imam Petrochemical Complex BIPC
Tarbiat Modares University of Tehran
Engineering Faculty-Polymer Department
Research & Development center of BIPC
Islamic Republic of Iran

Ebrahim Vasheghani Farahani
Tarbiat Modares University of Tehran
Engineering Faculty-Polymer Department
Islamic Republic of Iran

Adel Nejad Salim
National Petrochemical Company NPC
Islamic Republic of Iran

Reza Amiri
Bandar Imam Petrochemical Complex - BIPC
Islamic Republic of Iran

Simin Eydivand
Bandar Imam Petrochemical Complex BIPC
Research & Development center of BIPC
Islamic Republic of Iran

2009

SBS Publishers & Distributors Pvt. Ltd.
New Delhi

ISBN SET - 9788189741921
ISBN V-2 - 9788189741945

First Published in India in 2009

Published by:
SBS PUBLISHERS & DISTRIBUTORS PVT. LTD.
2/9, Ground Floor, Ansari Road, Darya Ganj,
New Delhi - 110002, INDIA
Tel: 23289119, 41563911
Email: mail@sbspublishers.com

Printed in India by Syndicate Binders, Noida.

Special Thanks to

Mr. KIYANOSH SHAMSEAMIRI - Manager, Basparan Complex
Mr. FATHE - Education Manager, NPC
Mr. FAKKARI - Head, Education Center, BIPC
and Specially Dr Nader Naderpour
for encouragement, help and advice.

We would also like to thank Dr HADADI ASL, Dr NASR JAFARI and Dr GHAFELEBASHI of Petrochemical Research and Technology Co. (NPC-RT)for supporting us all through these years.

We also thank
Dr M.NAVIDFAMILI,
Dr M.SEMSARZADEH,
& Dr M KOKABI
of Polymer Department of TARBIAT MODARES University

Mr. AKBARPOUR,
M.GHOLAMZADEH,
& D.MOHAMADI
of Basparan Engineering Complex

Mr. H.GODARZBOOD,
M.JOHARI,
M.HAMEDIAN,
A.MOAVI
& R.ADIBAN,
Researchers of R & D Center, BIPC
for providing us the visuals and figures for this Encyclopedia.

And finally we thank many people who have helped, often without knowing they were doing so, for, exhortation and consolation throughout the years it has taken to create this work.

Preface

This publication entitled "Encyclopedia of Polymer Science and Technology" provides the readers with key information on the said subject in an affordable three volumes condensed format. This user-friendly reference publication offers quick access to all areas of polymer science and technology. It presents the state of the art in all areas of polymer science, technology and engineering, covering new imaging and analytical techniques and new methods of controlled polymer architecture. It also talks about related information about polymers, plastics, fibers, biomaterials and polymerization processes.

It follows a distinctly new approach while discussing the interdisciplinary nature of polymer science, technology and engineering. Modern polymer science, technology and engineering are firmly rooted both in the chemistry of macromolecules and their physical chemistry. Besides, this publication provides readers with information on all the important uses of synthetic polymers. This publication fulfills the much desired need for such compact volumes on the said subject. The elaborate bibliography/reference to all applied aspects makes this book an indispensable support for both students and professionals.

This "Encyclopedia of Polymer Science and Technology" gives an introduction to polymer science and technology in addition to a basic understanding of polymers and polymer nomenclature and polymer characterization. Natural polymers, synthetic polymers and their sources are discussed briefly. Other themes which are covered include the phenomena of observing and measuring polymer/colloids properties; polymer chemistry and chemical properties of polymers; chemistry of high polymers; polymer physics and physical properties of polymers; polymer structure, tacticity and other properties; and polymer measurement devices. This publication also addresses processes like methods of polymer manufacturing; methods of polymer synthesis and modification; design and synthesis of polymers with controlled analysis; polymer blends and composites; polymer fibre composite materials; design and manufacture of composites; biopolymers and biomaterials; smart polymers and polymeric materials; and polymers strength enhancement and degradation. Other aspects which are highlighted include computational polymer science and technology; theoretical polymer science and technology; applied polymer science and technology; polymer Rheology; polymer crystallization; polymer material science and engineering; biomedical applications of polymers; macromolecules of interfaces and films; electronic and photonic molecular materials and devices; advanced polymer research; and polymer fibre applications.

The content of this publication on polymer science and technology is aimed for qualified and unqualified graduates who possess knowledge on polymer chemistry, polymer physics, polymer technology, polymer engineering and their applications. It also provides such students a skill on practical works related to polymer chemistry and processing. The objectives of this publication are to provide a broad range of knowledge in polymer science and technology as well as the skill/ability to apply this knowledge for industrial processes and modification of polymer products. This will also enhance opportunity of the graduate students for further education in polymer science and technology or related fields.

Contents

Preface *vii*

1. Methods of Polymer Manufacturing **1**

1.1 Primary Sources of Synthetic Polymers 1
1.2 Petrochemical Processing 1
1.3 Petrochemical Intermediates and Monomers 6
1.4 Petrochemical Industry 10

2. Methods of Polymer Synthesis and Modification **12**

2.1 Understanding Polymerization Process 12
2.2 Chain Growth Polymerization 13
2.3 Step Growth Polymerization 21
2.4 Co-Polymerization 23
2.5 Polymer Grades 28

3. Design and Synthesis of Polymers with Controlled Structures **34**

3.1 Design in Polymers 34
3.2 Manufacturing and Process Methods 35
3.3 Materials Selection 36
3.4 Case History: Topper Boat 39
3.5 Market Experience 50

4. Products and Stuffs Made from Polymers **52**

4.1 Polymer Products 52
4.2 Products and Services 52
4.3 Polymer Products of India 53
4.4 Products Offering 54
4.5 Advanced Polymer Products 55

4.6 Hot Cast Polyurethanes 57
4.7 Mining 59
4.8 Pipeline Pigs 60
4.9 Light Density Foam Pigs 61
4.10 Medium Density Foam Pigs 62
4.11 High Density Foam Pigs 62
4.12 Cup Pigs 63
4.13 Bi-Directional Disc Pigs 64
4.14 Social Purpose Pigs 66
4.15 Coil Pllets 66
4.16 Engineered Plastic 67
4.17 Polyurethne Foam 68
4.18 Others 70
4.19 Trulok 71
4.20 TRULOK Trommel Panel system 72
4.24 TRULOK Crossmember Wear Liner 73
4.23 Screen Cloths 74
4.24 Screen Panels 76
4.25 Wear Liners 76
4.26 Others 77

5. Polymer Processing, Testing and Analysis 78
5.1 Polymer Processing Operation 78
5.2 Single Screw Extrusion 79
5.3 Dry Spinning 79
5.4 Melt Spinning 80
5.5 Filament Winding 81
5.6 Dow's Polyurethane Systems 81
5.7 Film Blowing 83
5.8 Injection Molding 84
5.9 Batch Mixing 85
5.10 Reaction Injection Molding (RIM) 85
5.11 Spin Coating 85
5.12 Polymer Analysis 86
5.13 Polymer Lab 86
5.15 Chemical Analysis of Polymers 87
5.16 Polymers Analysis Capabilities 89
5.17 Polymers and Plastics Testing Services 90

6. Polymer Blends and Composites 94
6.1 Polymer and Composite Materials 94
6.2 Designing Polymer Blends 102

6.3 Types of Polymer Clay 113
6.4 Polymer Drying: The Basics 113
6.5 Composite 115
6.6 Polymer Composites in Construction 119
6.7 Polymer Composites Group 121

7. Polymer Fibre Composite Materials 124
7.1 Fibre-reinforced Plastic 124
7.2 Glass-reinforced Plastic 124
7.3 Carbon Fiber Reinforced Plastic 128
7.4 Polymer and Fiber Engineering 133
7.5 Bi-component Polymer Fibers Made by Rotary Process 134
7.6 Unsheathed Polymer Fibre 142
7.7 Fiber Reinforced Concrete 142
7.8 BS 4994 144

8. Design and Manufacture of Composites 145
8.1 Polymer Composites and Composite Design 145
8.2 Thermoplastic Composites Explained 147
8.3 Mold Design Inclusion in an Innovative Composites and Polymer Materials Education 154
8.4 Meaning of Composites? 160
8.5 Polymer Composites Directly Replace Metals? 161
8.6 Type of Polymer Composites Available 161
8.7 Main Factors Determining the Costs 162
8.8 Applications Currently Fabricated from Composites 162
8.9 Sports/Pleasure/Commercial Boats 162
8.10 Automotive and Rail 163
8.11 General Industry and Engineering 164
8.12 Aerospace 165
8.13 Sport and Recreation 166
8.14 Civil Engineering 167
8.16 Medical 168

9. Biopolymers and Biomaterials 169
9.1 Biopolymers 169
9.2 Biomaterials 172
9.3 NovaMatrix—Ultra-Pure Biopolymers and Biomaterials 177
9.4 Developments in Medical Polymers for Biomaterials Applications 178
9.5 Overview 187

10. Smart Polymers and Polymeric Materials **189**

10.1 Smart Polymers 189
10.3 Polymeric Medical Materials 191
10.4 Graft-and-Block 192
10.5 Polymeric Materials Novel Polymerization Technology 194
10.6 Polymeric Materials for Photo-chromic Applications 195
10.7 Molded Polymeric Material 203
10.8 Polymeric Material Properties 244
10.9 Polymeric Material Adapted for Physico-Chemical Separation of Substances 274

11. Polymers Strength Enhancement and Degradation **289**

11.1 Polymer Degradation 289
11.2 Weather Testing of Polymers 295
11.3 Degradation of Polymers 318
11.4 Design and Fabrication of Polyester-Fiber and Matrix Composites for Totally Absorbable Biomaterials 321
11.7 Torlon® Polyamide-Imide 325
11.8 Radiation Sterilization 326
11.9 Bio-Plastics 332
11.10 Chemically Assisted Degradation of Polymers 354
11.11 Thermal Degradation of Polymers 356

12. Synthetic Butadiene Rubber (SBR) **359**

12.1 Rubber Polymers 359
12.2 Synthetic Rubber 361
12.3 Synthetic Rubber Output 366
12.4 SBR (Styrene Butadiene Rubber) Sheet 366
12.5 Raw Materials 367
12.6 Synthetic Rubbers 369
12.7 Styrene-Butadiene Rubber SBR 1712 373
12.8 Styrene Butadiene Rubber (SBR) Revisited 373
12.9 SBR Rubber Molding. 373
12.10 Rubber Injection Molding 374
12.11 Rubber Compression Molding 375
12.12 Rubber Transfer Molding 376
12.13 SBR - Styrene Butadiene Rubber 377
12.14 Notification 378

Bibliography **410**

Index 413

1

Methods of Polymer Manufacturing

1.1 Primary Sources of Synthetic Polymers

The most important primary sources of synthetic polymers are crude oil, natural gas and, to a minor extent, coal. Because all are primarily fuels rather than sources of materials, the manufacture of polymers is susceptible to changes in price or supply. However, this is also true of other materials, since fuel costs are an important component of metal, ceramic and glass manufacture where very high reaction temperatures are needed for reduction of ore to metal and/or smelting. Where polymer manufacture is different is in the range of sources of the basic building blocks for the polymer repeat units. Both oil and natural gas can be used to make polyethylene for example. In parallel with the advances that have been made in polymerization and polymer structures, there have been major advances in making intermediates more efficiently using tailor-made catalysts. A wider variety of intermediate petrochemicals is also now available, particularly for speciality—materials.

1.2 Petrochemical Processing

Following distillation of petroleum into the major fractions (gasoline C_5 up to 95 °C, naphtha 75-175 °C, kerosene 175-225 °C), the naphtha cut is subjected to cracking to yield smaller, double bonded molecules. The reaction is conducted at high temperatures (400–800 °C), but under low pressure using steam for cracking. This process can yield monomers directly, such as ethylene (C_2), propylene (C_3) and butadiene (C_4), but often further reactions are required to add other elements such as oxygen and chlorine. Since cracked naphthas are complex mixtures, expensive separation procedures are needed for the co-products.

Thermal Cracking

The bulk of the major monomer and intermediate, ethylene (C_2H_4), is still produced in the UK by steam cracking without the use of catalysts. Paraffinic feedstocks are best for optimising ethylene yields, and the severity of cracking is specified by the rate of disappearance of a marker compound, usually *n*-pentane. The severity of the reaction can then be defined as follows:

$$\text{Cracking Severity} = K_5 t \qquad \text{... (1)}$$

where k_5 is the rate constant (per second) for the cracking of *n*-pentane (C_5) and t is the time in seconds. The rate of disappearance can be simply related to the degree of conversion, α, assuming first-order kinetics:

$$\alpha = \frac{\text{Amount Converted}}{\text{Amount Present Initially}}$$

$$= 1\frac{(N_A)_0 - (N_A)_t}{(N_A)_0} \qquad \text{... (2)}$$

and

$$\text{In}\ \frac{1}{(1-\alpha)} = kt \qquad \text{... (3)}$$

where $(N_A)_0$ and $(N_A)_0$ are the number of molecules (or moles) of compound A present at t=0 and t seconds respectively. Cracking severity is also dependent on temperature T through the *Arrhenius equation*

$$k = A\exp\left(\frac{-E}{RT}\right) \qquad \text{... (4)}$$

where E is the activation energy for the process, A and R are constants and k is the rate constant as in Equation (1).

The exponential dependence of rate constants on temperature for a variety of simple hydrocarbons in thermal cracking is shown in Figure 1.1 and it is clear that large molecules are easier to crack than smaller ones. Ethane for example cracks more than 20 times more slowly than *n*-hexane at 1000 K. But what products are formed? Unfortunately, ethane can break down more easily to methane and hydrogen and ultimately to carbon, particularly under the rather crude conditions of thermal cracking. Propane and higher homologues give higher yields of ethylene under similar cracking conditions but cyclic paraffins (cycloalkanes) give rather less.

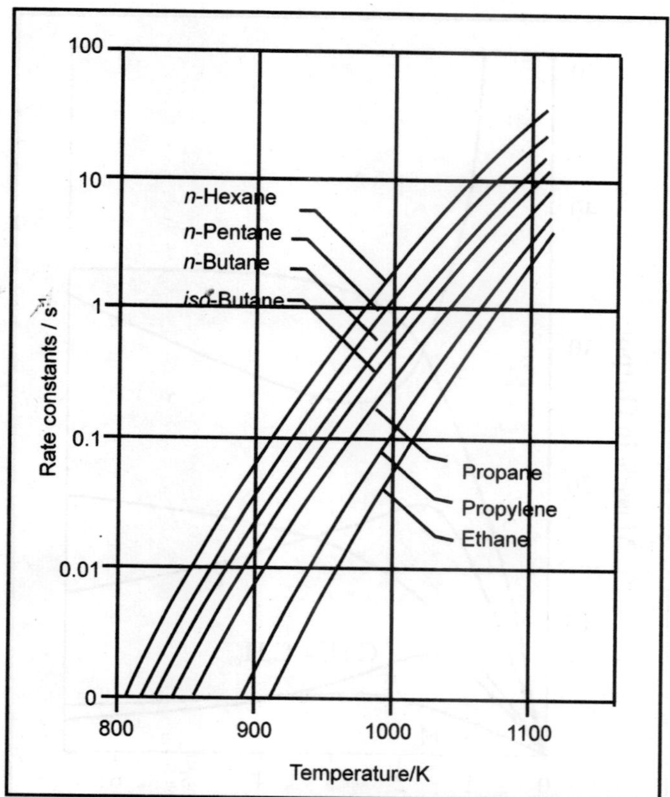

Fig. 1.1: Rate constants for the thermal cracking of selected hydrocarbons.

The fractionated naphtha is divided into two streams, one to the thermal cracker (olefin plant) and the other to the reformer (aromatics plant). The exit streams are separated and individual chemicals subjected to separate treatments depending on the intermediates, monomers or polymers required.

A highly paraffinic naphtha is thus the best cracking feedstock for high yields of ethylene, and Figure 1.2 shows the kind of product distribution from a Kuwaiti naphtha under various cracking conditions. The major co-products—ethylene, propylene, butadiene (C_4H_6), mixed butene/butane (C_4H_8, C_4H_{10}) and pyrolysis gasoline (C_{5+}) - vary in concentration depending on cracking severity. With naphthas derived from other crudes, the product distribution will be quite different (see Box 1.1). The proportion of unwanted byproducts like hydrogen, methane and carbon coke (not shown in the figure, but which gradually accumulate within the cracking tubes) increases with cracking severity. However, the product distribution may not coincide with polymer demand and consumption of ethylene for other chemicals.

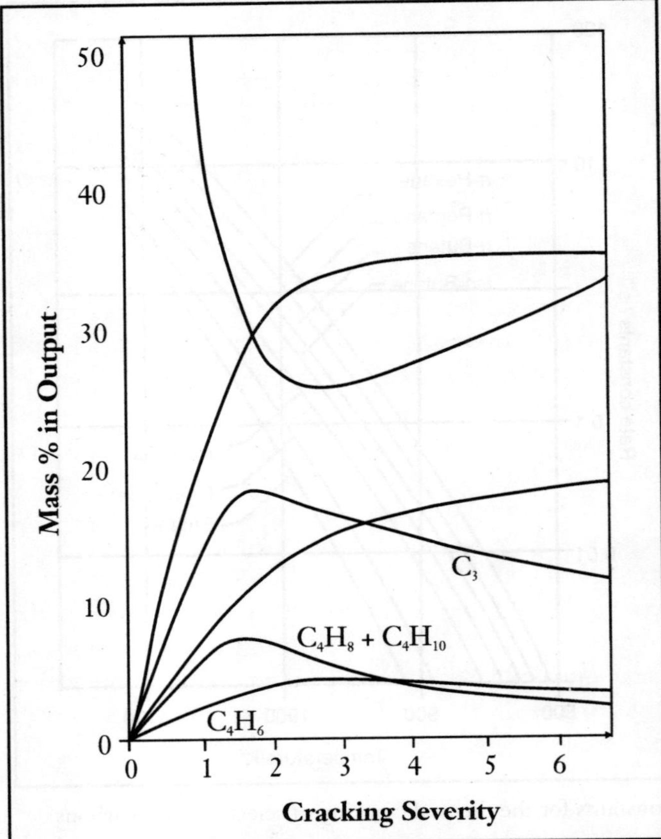

Fig. 1.2: Co-product yields for a thermal cracking of a paraffinic naphtha. Ethylene yield is maximised at severities greater than two, but the maximum propylene yield occurs at about 1.5. Severe cracking occurs between values of about 2.0 and 4.0

Ethane Cracking

Although ethane can be cracked thermally, the reaction is slow and does not necessarily yield ethylene at high severity. Careful control of reaction conditions, however, allows the reaction to occur

$$C_2H_6 \rightarrow C_2H_4 + H_2 \ (T \approx 800°C)$$

The yield of ethylene is typically nearly 50 wt per cent with the rest composed of unreacted ethane (40 per cent) and some methane and hydrogen (10 per cent). The ethane is separated and recycled to the start. This process is practised widely in the US, where 'wet' natural gas has traditionally been abundant and cheap. With supplies arriving in Scotland from the Brent complex, ethane and propane are cracked at the Shell-Esso petrochemical works at Mossmoran, Fife and the BP Grangemouth complex near Edinburgh.

Box 1.1: North Sea oil

Although we all know that Britain is effectively self-sufficient in oil and gas, and indeed exports substantial quantities, it may not be so widely appreciated that our oil is of very high quality. It attracts a high price on the Rotterdam market because it is low in the impurity sulphur and is also very light. Thus the marker crude oil. Brent, from the UK sector of the North Sea has a sulphur content of only 0.26 per cent compared with 2.5 per cent for Kuwaiti crude. Why should this be important? One of the reasons is that in much petrochemical processing (to make polymers, for example) catalysts are use to speed up process reactions; sulphur poisons such catalysts, so must be removed in expensive purification processes before catalysis. It will also burn in fuels made from crude oil (fuel oil, petrol, paraffin etc.) to pollute the environment, so low sulphur oils generally command a premium on the market.

North Sea crudes are also very light, with a density much lower than Middle East crudes. Thus Brent oil has a gravity of about 38 API degrees compared to 27 API degrees for Alaskan crude, for example. This means that it yields more light fraction liquids such as petrol (gasoline), which are more valuable than heavier fractions. Thus Brent yields about 20 weight per cent petrol compared with only 10 per cent for Alaskan crude. There is a more subtle reason why our oil is of greater value, the reason lying in its molecular composition. Although an incredibly complex mixture of organic compounds, oil composition can be expressed in terms of paraffins (alkanes), naphthenes (cyclic paraffins) and aromatics. The compositions of a typical North Sea crude and Kuwaiti crudes are shown in Figure 1.3, where it is clear that North Sea oil is much more naphthenic and aromatic than Kuwaiti. This is important for petrochemical extraction of high value aromatics for polymers like PS and PET, as well as increasing the octane rating of car petrols, especially for lead-free varieties.

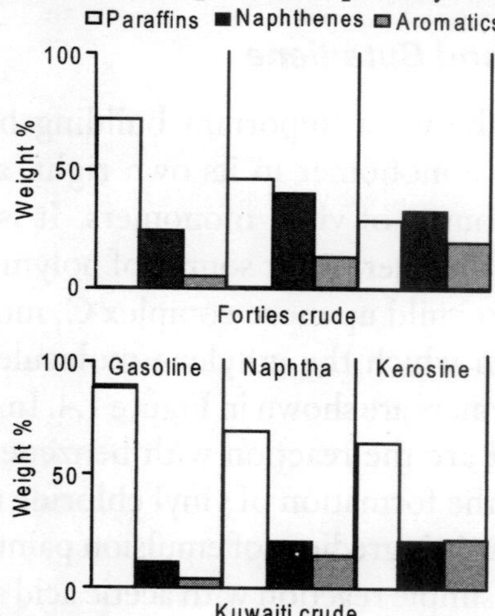

Fig. 1.3: The compositions of a typical North Sea crude and Kuwaiti crudes

In the US, however, ethane and propane are becoming more expensive, so that thermal cracking of naphtha to produce ethylene is increasing. There is no doubt that thermal processing of ethane and propane to produce monomer is more efficient because separation costs both before and after processing are much lower; it also means that dependence on a semi-fixed product distribution is much less (Figure 1.2).

In recent years, there have also been attempts to recover previously flared wet gases from the rich oilfields of the Middle East; conversion to petrochemical building blocks like ethylene clearly offers greater added-value products like polyethylene. There is now a large and growing petrochemical industry in the Middle East, and interestingly, Norway, where the enormous reserves of high-grade oil and wet gases from the North Sea assure supplies for many years to come.

1.3 Petrochemical Intermediates and Monomers

About 80 per cent of all petrochemicals end up in polymers, the most important building blocks being ethylene, propylene, butadiene and benzene. The first three can be polymerized directly but an important slice of their production is used to create more complex monomers. Ethylene is the progenitor of most vinyl monomers (Figure 1.4), so the pressure on ethylene supply is particularly strong compared for example to propylene. The C_2 and C_3 building blocks can be combined with benzene to form another set of monomers and intermediates, particularly valuable for constructing the complex repeat units noted in the last section. Other chemicals are also produced, such as plasticizers which are then added in a subsequent stage to polymers to modify their properties.

Ethylene, Propylene and Butadiene

Nowadays ethylene is the most important building block for the chemical industry, particularly as a monomer in its own right, as a co-monomer with other vinyls, and as a source of vinyl monomers. It is the prime source for ethylene oxide, which is another major source of polymers, glycols and ethers. They can also be used to build up more complex C_4 molecules and aromatics.

Some of the ways in which the ethylene molecule is modified to create other chemicals and polymers are shown in Figure 1.4. In terms of gross tonnage, the two most important are the reaction with benzene to form ethylbenzene and hence styrene, and the formation of vinyl chloride monomer by two steps with chlorine. PVA, a staple ingredient of emulsion paints and adhesives, comes from the monomer by a simple reaction with acetic acid and oxygen. In a similar way, vinyl chloride monomer is produced by a pathway which also yields chlorinated hydrocarbon solvents.

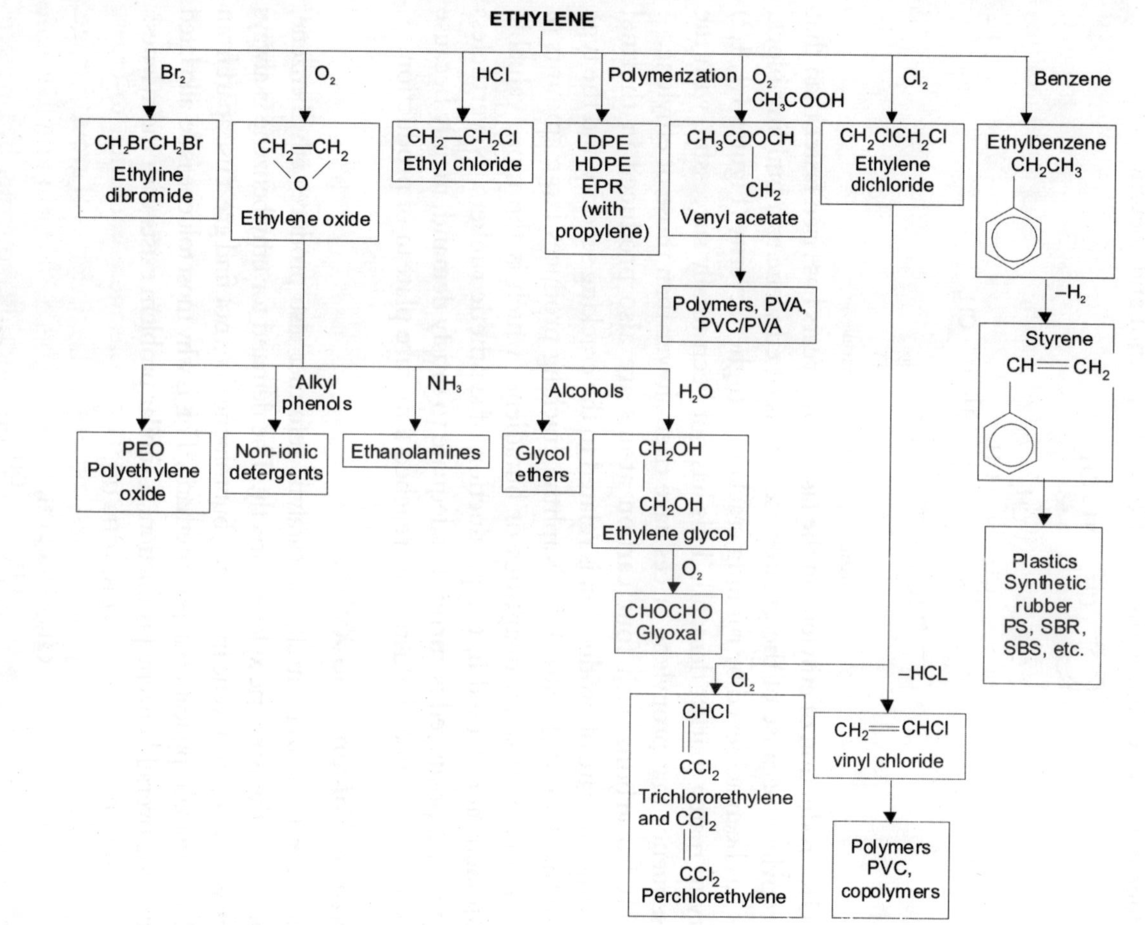

Fig. 1.4: Ethylene as a Petrochemical Intermediate and Monomer

As a co-product of ethylene from steam cracking, propylene is a useful intermediate for building C_3 monomers: propylene oxide, acrylonitrile and methyl methacrylate. Although the repeat unit of PMMA contains *five* carbon atoms, it is in fact produced from acetone and hydrocyanic acid (HCN). Acetone itself comes from an interesting reaction between benzene and propylene to produce cumene, which is then split to make phenol and acetone:

$$C_6H_6 + CH_3{-}CH = CH_2 \longrightarrow C_6H_5CH(CH_3)_2 \xrightarrow{O_2} C_6H_5OH + CH_3{-}C(=O){-}CH_3$$

cumene phenol acetone

UK production of ethylene in the last twenty years has risen faster than that of propylene because of the greater usefulness of ethylene as a building block. This has been achieved by running crackers at higher severity (Figures 1.1, 1.2) to try to match market demand. This has not been totally successful, with the consequence that propylene prices have risen slower than those of ethylene.

Hydrocarbons with four carbon atoms are also produced in thermal cracking; the rate of production is related to the cracking severity and the type of feedstock used (Figure 1.2). Naphtha cracking produces large amounts of butene and only small quantities of butadiene, which is the more valuable component and is used in the production of synthetic rubbers. Nevertheless the amount produced is currently adequate to satisfy demand, mainly because the synthetic rubber industry has reached a mature plateau of production.

Benzene, Toluene and Xylene

In addition to benzene itself, the catalytic reformer also produces ethylbenzene, toluene and the isomeric xylenes directly. The demand for ethylbenzene is always great as a source of styrene monomer, but toluene does not find great use apart from a relatively small application in polyurethane. This is why most toluene is de-alkylated to increase overall benzene production. A similar problem exists with the xylenes:

$$+ CH_3{-}CH = CH_2 \longrightarrow C_6H_5CH(CH_3)_2 \xrightarrow{O_2} C_6H_5OH + CH_3{-}C(=O){-}CH_3$$

cumene phenol acetone

Para-xylene is most widely used as a source of terephthalic acid for PET (Table 1.1), a blow moulding and fibre forming polymer. Although demand is lower, the *ortho* isomer is used to make phthalic anhydride of use in thermosetting resins and paints,

$$C_6H_4(CH_3)_2 \xrightarrow[-H_2O]{+O_2} C_6H_4(CO)_2O$$

The unused *meta*- and *ortho*-xylenes are downgraded to a mixed solvent, a step which clearly represents a loss of added value. Some attempts have been made to use *meta*-substituted groups in polymers, *Nomex* fibre for example, but the demand for this speciality material is insufficient to exploit the amount of meta-xylene available from reformed naphtha.

A much more complex procedure is necessary to make the two monomers for nylon 6,6. Benzene must first be hydrogenated back to cyclohexane which then undergoes oxidation and ring scission to create hexamethyl-enediamine and adipic acid in five and four steps respectively. It is an irony of petrochemical processing that such a long-winded procedure is needed—and it is a direct consequence of the complex mixtures of hydrocarbons produced in the first major processes of refining.

Higher Aromatics

Benzene rings can be fused in various ways to create component parts for some of the complex aromatic repeat units shown in Table 1.5. One of the most important is bisphenol A, made by fusing two phenol rings with acetone:

$$HO-C_6H_5 + (CH_3)_2C{=}O + C_6H_5-OH \xrightarrow[-H_2O]{} HO-C_6H_4-C(CH_3)_2-C_6H_4-OH$$

This intermediate is important for a number of speciality polymers, for example, polycarbonate and epoxy resins. Epoxies were developed in the early 1940s, and they found eventual use in aircraft construction for high-strength adhesive bonding. They continue in popularity as speciality adhesives and are also important in polymer composite materials, where they are used as the matrix for glass and carbon fibres. Condensation of bisphenol A and epichlorhydrin yields prepolymers of the kind:

$$CH_2{-}CH{-}CH_2{-}\left[O{-}C_6H_4{-}C(CH_3)_2{-}C_6H_4{-}O{-}CH_2{-}CH(OH){-}CH_2\right]_n{-}O{-}C_6H_4{-}C(CH_3)_2{-}C_6H_4{-}O{-}CH_2{-}CH{-}CH_2$$

Since M_R is very large (284), $n = 1$ to ca. 10 in the prepolymers of molecular masses up to 3000. Not only are there reactive epoxy groups at the chain ends but also hydroxyl groups in the repeat units which are available for crosslinking by multi-functional amines.

With bisphenol A available commercially as a relatively cheap intermediate, its potential for use in thermoplastic materials was exploited in the development of polycarbonate resins by simply reacting the material with phosgene gas, $COCl_2$:

$$HO{-}C_6H_4{-}C(CH_3)_2{-}C_6H_4{-}OH + O{=}CCl_2 \xrightarrow[-2HCl]{} \left[O{-}C_6H_4{-}C(CH_3)_2{-}C_6H_4{-}O{-}C(=O)\right]_n$$

More recent exploitation of this linked phenol has occurred in certain grades of polysulphones and polyimides.

1.4 Petrochemical Industry

The four-fold increase in the price of oil in 1973-4, together with associated political events, proved a powerful stimulus in the development and exploitation of North Sea crude oil. Increasing the price of oil does not mean that the price of the final plastic moulding increases by the same amount. For example if oil prices were doubled again then naphtha prices would typically increase by about 80 per cent, although there is no simple and fixed gearing mechanism between the two prices. The bulk of naphtha is used for gasoline with most of the balance used as petrochemical feedstock. Hence the price of naphtha at any time reflects the balance of supply with demand in *both* of the downstream markets.

An 80 per cent increase in the price of naphtha would mean that the price of ethylene would increase by about 52 per cent and polyethylene by about 28 per cent. A typical polyethylene pipe would cost about 12 per cent more. Crude oil and naphtha are traded in dollars. For the UK plastics producers each one per cent reduction in the value of sterling relative to the dollar raises the price of naphtha by about £1 per tonne.

However, the situation has changed with the availability of NGL (natural gas liquids) and the consequent direct cracking to ethylene and propylene. These processes were originally developed in Europe with Ekofisk NGL delivered to Teeside and shipped back to Norway. It enabled Norwegian companies like Norsk Hydro to develop greenfield petrochemical sites and gave them a headstart in polymer production. The same has happened in Scotland and helps to keep intermediate prices down through effective competition. Another development is the mixing of NGL with naphtha so that it can be fed into conventional thermal crackers, but this is an interim solution which does not realize the full intrinsic value of these feedstocks. Where petrochemical companies do not have easy access to feedstocks, there are problems both fundamental and political in nature. Companies like ICI, who obtain naphtha on the international market in Rotterdam, pay import levies which increase feedstock prices over and above what companies like Shell-Esso pay for their own UK supply of NGL. Some collaboration occurs between the large companies in petrochemicals—an ethylene pipeline from Shell-Esso's Mossmorran plant to BP's Grangemouth complex for example.

The production of monomers and intermediates is clearly tied to the market penetration and sales of particular polymers. Since the distribution of hydrocarbon structures in the feedstocks does not coincide closely with the repeat structures of tonnage polymers, there are clear problems of balancing supply with demand. Since vinyl polymers are in a mature stage of development, the demand for ethylene exceeds that for propylene with the result that polypropylene prices are much lower than they would be otherwise. Moreover, there are many unused co-products (*meta-xylene* is a good example) which cannot be used in quantity to make polymers. Even if new and interesting polymers based on these intermediates were developed, it would be many years before market penetration would mop up available supplies of this chemical. The trend towards polymers with high aromatic content will be helped by the high aromatic content of North Sea naphtha, although it is worth pointing out that the UK still imports considerable quantities of Middle East crude because of its high paraffin content for naphtha cracking. The difference between our own production and imports is largely exported to the USA and Germany where the higher light end content of North Sea oil is exploited to the full.

2

Methods of Polymer Synthesis and Modification

2.1 Understanding Polymerization Process

Converting monomer to long chain polymer is the final step in the polymer manufacturing sequence. Polymerization is usually highly favourable in thermodynamic terms, mainly on energetic grounds because ordering molecules into linked chains is a process where the entropy is decreased. Heat is always given out during polymerization owing to the very favourable energetics of reaction, a point you may have noticed if you have ever made GRP parts for your car, for example!

Advances in catalysis have given a high degree of control over both structure and molecular mass so that grades of a given polymer can be tailored for specific end usage. It is possible to look at polymerization in at least two different ways: the nature of the catalyst used, and the way the chains grow to form the final product. Polymerizations can be conducted in the gaseous, liquid or solid state, and now in the liquid crystal state to produce highly oriented macromolecules. An appreciation of the kind of advances that have been made is important because of the new possibilities for manufacturing finished products that are becoming available.

A basic understanding of polymerization processes is important not only because polymerization affects structure, and hence properties, but also because some processing routes can convert monomers directly to a finished shape. They offer manufacturing industry considerable benefits both in direct and indirect costs. An extra dimension to polymer structure is added by the possibilities of copolymerization, where two or more different monomers are polymerized together. In one sense it is comparable to alloying different metals to produce an appropriate balance of properties in the final product.

2.2 Chain Growth Polymerization

Chain growth polymerization is basically a three-stage process, involving initiation of active molecules, their propagation and termination of the active chain ends.

Initiation

Initiation is the mechanism which starts the polymerization process. Vinyl monomers are quite easily polymerized by a variety of activating methods. Styrene, for example, can be converted to solid polymer simply by heating, and ultraviolet light can have exactly the same effect. Usually, however, an activating agent is used. This is an unstable chemical which produces active species that attack the monomer. A good example is benzoyl peroxide which splits up when heated:

HO—C₆H₄—C(=O)—O—O—C(=O) —heat→ 2 C₆H₅—C(=O)O· —heat→ 2 C₆H₅· + $2CO_2$

benzoyl peroxide

The formulae of the products are written with a dot alongside to show that they are free radicals. A free radical is a molecule in which there is an unpaired electron. This free radical is very reactive and will attack monomer molecules when introduced into a polymerization vessel. Thus, as benzoyl peroxide is added to styrene (a reaction used with GRP), the peroxide splits to make free radicals, which react as follows:

phenyl radical + styrene (CH_2=CH—C₆H₅) → C₆H₅—CH_2—ĊH—C₆H₅

The net result is that the reactants have been linked together but the product is still a radical and so is capable of attacking further monomer molecules. In each instance the attack will lead to a larger molecule but the free radical will be preserved. The reaction is referred to as **free radical polymerization.**

Free radicals are not the only way of initiating reactions. Charged molecules can often exert the same effect. Ethyllithium, for example, is a relatively unstable molecule which can dissociate to form an ion pair:

$$C_2H_5Li \rightarrow C_2H_5- + Li^+$$

Styrene can also be polymerized by this compound:

$$\overset{Li^+}{C_2H_5} + CH_2{=}CH{-}C_6H_5 \longrightarrow C_2H_5{-}CH_2{-}\overset{Li^+}{CH}{-}C_6H_5$$

This mechanism is called **anionic polymerization**. The next two sections refer to free radical polymerization. Ionic polymerization will be discussed in Section 4.2.4.

Propagation

Once a small number of chains have been started, propagation involves successive addition of monomer units to achieve chain growth. At each step the free radical is regenerated as it reacts with the double bond. So in the case of styrene the propagation step is

$$C_6H_5{-}CH_2{-}\dot{C}H(C_6H_5) + CH_2{=}CH(C_6H_5) \longrightarrow C_6H_5{-}CH_2{-}CH(C_6H_5){-}CH_2{-}\dot{C}H(C_6H_5) \quad \text{etc.}$$

The free radical can also add on in a different way to produce

$$C_6H_5{-}CH_2{-}CH(C_6H_5){-}CH(C_6H_5){-}\dot{C}H_2$$

but this process happens only rarely since the free radical is less stable than in the first case.

The junction that is formed normally is known as a head-to-tail link while the abnormal link is head-to-head. The effect is limited to about 1 per cent of the total number of monomer links in normal polystyrene, but it is important because the head-to-head links are weaknesses in the chain. Since they are of higher energy, thermal degradation can start at these defective junctions.

Termination and Transfer

There are basically three ways in which chains terminate.

The first is known as **coupling** and occurs when two free radicals join together. This can be represented by the general equation

(polymer chain 1) – RH • + • HR – (polymer chain 2)
→ (polymer chain 1) – RH –RH – (polymer chain 2)

Such a mechanism significantly increases molecular mass, if it results in two polymer chains joining. This is the main mechanism which terminates the polymerization of styrene.

An alternative mechanism that may occur when two radicals interact is known as **disproportionation**. In this case, one molecule abstracts a hydrogen atom from the other and the other molecule forms a double bond

(polymer 1) – RH • + • HR – CH_2 (polymer 2)
→ (polymer 1) – RH2 + R = CH – (polymer 2)

Disproportionation has no effect on molecular mass. Poly (methyl methacrylate) (PMMA) terminates by a mixture of coupling and disproportionation.

The third method of termination is **chain transfer** in which a radical abstracts a hydrogen atom from a neighbouring molecule. In the case of polystyrene the effect will be as shown in Figure 2.1, where (a) shows the situation before the interaction and (b) shows the structures after chain transfer in which the radical is transferred to one of the mid-chain carbon atoms. The new radical may now attack further styrene (Figure 2.1(c)) but, because it is not on the end of the chain, side branching occurs.

A similar mechanism accounts for the side branches in LDPE where it is a more important mode of termination than in polystyrene. Transfer to monomer; initiator or solvent (if present) can also occur in free radical polymerization, and effectively increases the dispersion of the molecular mass of the final polymer.

If termination is simply by disproportionation, then

$$n = \frac{K[M]}{[I]} \quad \ldots (1)$$

where K is a constant, [M] the concentration of monomer, [I] the

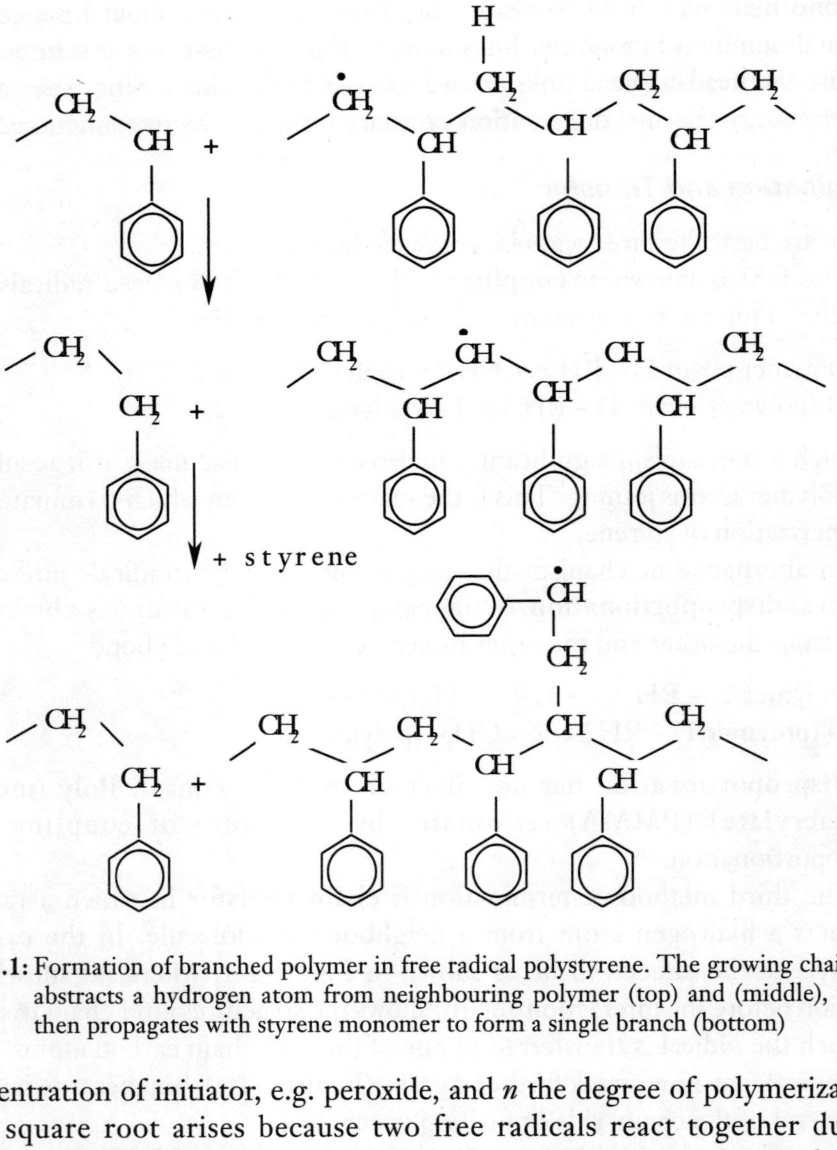

Fig. 2.1: Formation of branched polymer in free radical polystyrene. The growing chain end abstracts a hydrogen atom from neighbouring polymer (top) and (middle), which then propagates with styrene monomer to form a single branch (bottom)

concentration of initiator, e.g. peroxide, and n the degree of polymerization. The square root arises because two free radicals react together during termination. If termination is by coupling there will be an extra factor of two in the constant compared to disproportionation. So the degree of polymerization or molecular mass can be controlled by varying monomer concentration—for example, by conducting the reaction in solvent—or by varying initiator concentration.

Controlling polymerizations on an industrial scale is of critical importance for molecular mass, and hence the processability and physical properties of the polymer, and one of the most important variables is the temperature of reaction. All polymerizations are exothermic (heat is liberated due to bond formation) and the heat must be conducted away to maintain a uniform reaction temperature. This is much more easily achieved when an inert solvent is used. Another method very commonly used industrially is to emulsify the monomer with a soap and conduct the reaction in water—so-called **emulsion polymerization** (Figure 2.2). Since control of molecular mass is so vital, extra aids are used industrially in addition to varying monomer and initiator concentrations. Reactions are 'short stopped' before all monomer is consumed by adding a specific chemical which reacts with free radicals, stopping them dead. Other chemicals can be added to induce transfer reactions, so controlling molecular mass distribution.

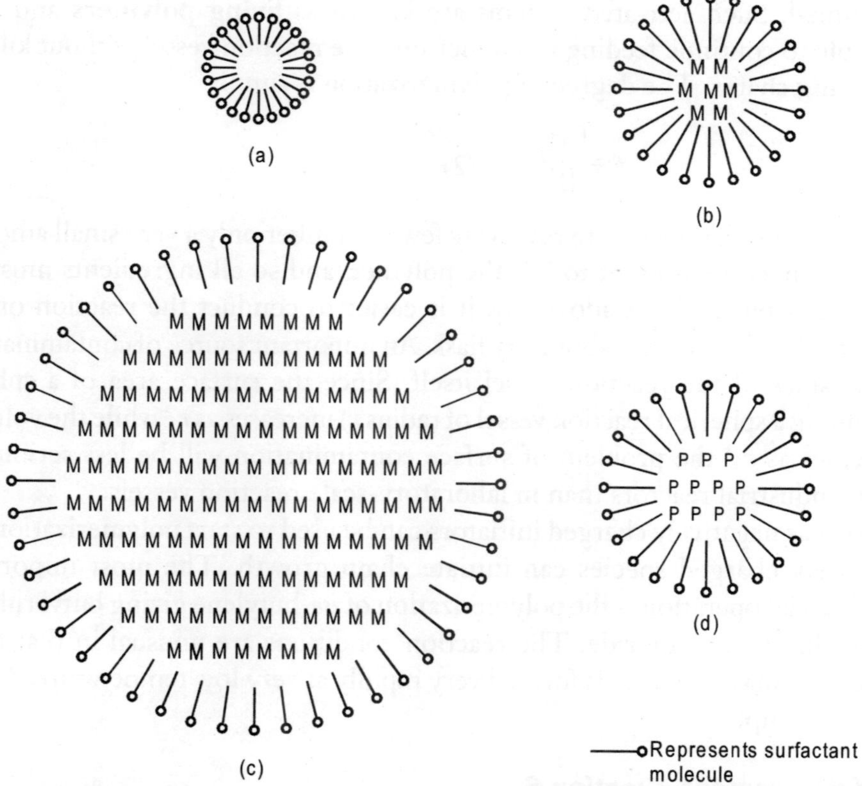

Fig. 2.2: Emulsion polymerization of vinyl monomers is conducted in water to aid heat dissipation. Soap micelles (a) swell with monomer (b) which migrates from monomer droplets stabilised by soap molecules (c). The process yields a polymer latex (d) which can be used directly or reduced to bulk polymer

Ionic Polymerization

Free radicals are indiscriminate in the compounds they attack, and their non-selective nature in polymerization reactions leads to problems such as chain branching and transfer which affect the structure of the polymer produced. Anionic polymerization overcomes many of these problems.

A typical commercial (but also see Box 2.1) anionic reaction is the polymerization of styrene using butyllithium, C_4H_9Li, in an inert solvent such as *n*-hexane. Termination does not occur by polymer-polymer interaction but by reaction with small molecules such as water:

(polymer) – Li + H2O → (polymer) – H + LiOH

This type of polymerization gives rise to very sharp molecular mass distributions because transfer processes are absent. If the solvent is extremely pure, the polymer chains will still be active after all the monomer has been consumed. Such activated systems are known as **living polymers** and it is possible to continue feeding monomer into the reaction vessel without killing the living chains. The degree of polymerization is simply

$$n = \frac{[M]}{[I]} \quad \ldots (2)$$

Since the chain ends are relatively few in number only a very small amount of water need be present to kill the polymer, and so all ingredients must be rigorously purified. Paradoxically, it is easier to conduct the reaction on an industrial scale than in a laboratory flask. An important source of contamination is the sides of the reaction vessel itself. Since the surface area of a sphere (assuming a spherical reaction vessel of radius r) increases as r^2 while the volume increases as r^3 the problem of surface contamination will be less serious in large, industrial reactors than in laboratory-scale reaction vessels.

Just as negatively charged initiators can be used to start polymerization, so positively charged species can initiate chain growth. The most important commercial operation is the polymerization of *iso*-butylene giving butyl rubber using aluminium chloride. The reaction conditions are unusual in that high molecular mass polymer is formed very rapidly at very low temperatures (-100 °C for example).

Self assessment question 6

An anionic polymerization is initiated with a solution containing 0.1 moles of *n*-butyllithium in 100 ml of *n*-hexane. Initiator (1 ml) is added to a litre of hexane solution containing 1 mole of styrene monomer. At the end of the reaction, another mole of styrene is added and the reaction is terminated with water. What is the molecular mass of the polystyrene extracted from the solution?

Now read the answer

Box 2.1: Superglue

A more familiar example of anionic polymerization occurs when you use cyanoacrylate liquid ("superglue') to stick a broken pot together. The monomer is

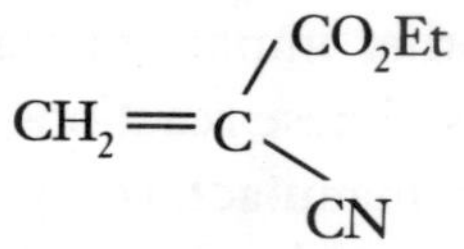

$$CH_2{=}C(CO_2Et)(CN)$$

Fig. 2.3

and being a small molecule, has a very low viscosity (Et is the abbreviation for the ethyl group -C_2H_5). This is an important property for adhesion, because it means that the liquid when applied to the broken pot will penetrate even the finest cracks in the fractured surfaces. Such surfaces will normally already be very slightly wet with water from the atmosphere (a monomolecular film is enough), and the monomer will start to polymerize anionically. The anion is supplied by the small amount of hydroxyl ions present in water:

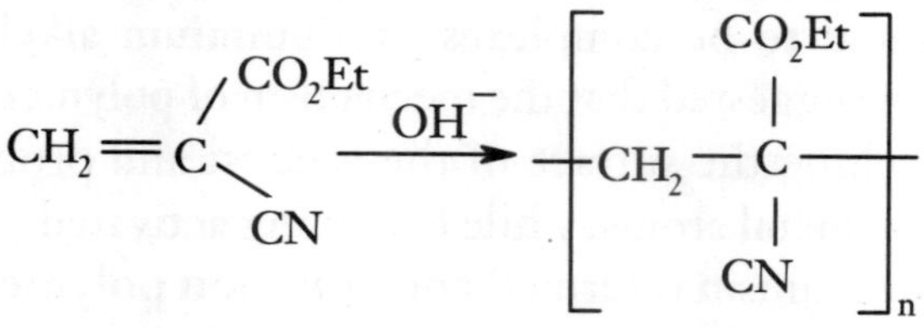

$$CH_2{=}C(CO_2Et)(CN) \xrightarrow{OH^-} \left[-CH_2-C(CO_2Et)(CN)- \right]_n$$

Fig. 2.4

Reaction is very fast, and since this is a chain growth mechanism, high molecular mass polymer is created very rapidly. You might, if unlucky, already have experienced this effect if you accidentally spilt the monomer on your fingers and they made contact! The sweat present there is more than enough to initiate polymerization. The good news is that since the polymer is thermoplastic, there are solvents available for swelling or dissolving the bond and so releasing your fingers. Termination occurs when no more monomer is present, so that all the liquid monomer present at the interface between the two parts of the pot becomes solid polymer. The polymer chains pass from one broken surface to the other, so adhesion is excellent, and strength will be maximised. A range of such cyanoacrylate monomers is available now with varying rates of polymerization and modes of initiation (e.g. thermal or pressure initiation), as well as grades which react without air or water being present, the so-called anaerobic superglues. These adhesives have slightly different substituents, so affecting the way the monomer behaves during reaction.

Co-ordination Polymerization

While most free radical and ionic polymerizations are carried out homogeneously, there is another important class of reaction which is often performed with solid catalysts. These reactions, discovered in the mid-fifties, have revolutionized polymer manufacture by permitting much less severe polymerization conditions than with other systems and by allowing a greater degree of control of polymer structure. **Ziegler-Natta catalysts**, as they are called, will convert vinyl and diene monomers to highly linear, stereoregular structures under ambient conditions.

The prototype of all vinyl polymers, polyethylene, was first discovered quite accidentally in the 1930s as the result of a very high pressure experiment; low density polyethylene is still made under such conditions. However, the thermodynamics of the polymerization reaction indicated that the reaction should be possible at a pressure of 1 atmosphere and a temperature of 300 K, if the right catalyst could be discovered. Suitable catalysts were found to be complexes of aluminium alkyls and titanium halides. It has been suggested that the mechanism of polymerization involves monomer approaching the surface of the catalyst and probably forming a bridge between the metal atoms, while becoming activated at the same time. This is why the mechanism is termed **coordination polymerization**. More monomer can approach the surface site, react with the active end and grow into chain. The chain thus grows from the surface of the catalyst. Since the activated end is probably partly charged, transfer reactions are much less likely than in free radical situations, so that branching is unlikely. As the monomer molecule must sit in a rather specific position to react, stereoregular polymers can be made.

Molecular masses in Ziegler-Natta polymerizations are often very high, sometimes too high for the polymer to be useful commercially, because too high a molecular mass makes the polymer too viscous to process easily. Molecular mass distributions are often quite broad, probably because active sites on the catalyst surface are sensitive to catalyst poisons.

Metallocene polymers are of much more recent origin (1990s). Metallocenes are sandwich compounds of cyclics and metal ions. Monomer is polymerized by the metal ion in a controlled way:

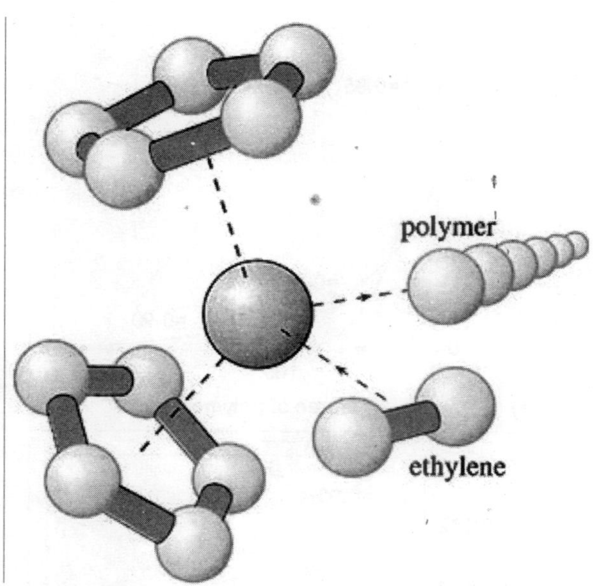

Molecular mass distributions are narrow compared to Ziegler-Natta polymers, with a typical dispersion of 2.5. Such polymers are finding application in packaging and mouldings.

2.3 Step Growth Polymerization

In contrast to chain growth reactions, where high molecular mass polymers are formed almost from the start of the reaction, a stepwise reaction results in the molecular mass of the polymer increasing slowly as the reaction progresses (Figure 2.5). Simple statistical arguments can be used to show how the distribution develops with extent of reaction a. Since a is simply the fraction of functional groups which has disappeared after time t (see Equation (13)), it can be interpreted as the probability that a functional group has reacted after time t. The probability of finding an unreacted functional group is thus (1 - a). To determine the molecular mass distribution, it is necessary to find the probability that a molecule selected at random is an *rimer*. The probability of finding a single peptide group in a nylon molecule for example will be a, and the probability of finding n - 1 of them will be a^{n-1}. The factor of unity appears because the end of the molecule will possess an unreacted carboxyl or amine group, which will have a probability of (1 - a). So the probability of finding the complete n-mer is simply $a^{n-1}(1 - a)$. This in turn is the fraction of n-mers in the entire assembly, so

$$\frac{N_n}{N_0} = (1-\alpha)\,\alpha^{n-1} \quad \dots (3)$$

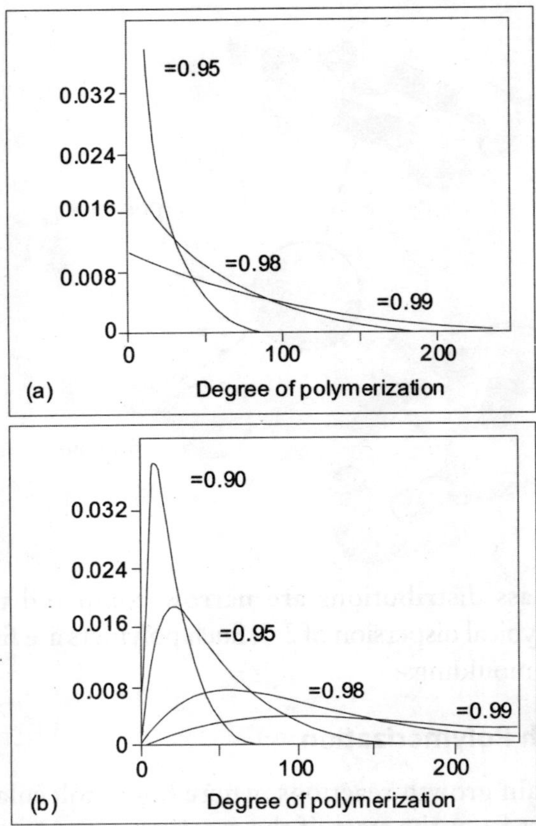

Fig. 2.5: Molecular mass distributions in step growth polymerization. The number distribution (a) is shown for three different extents of reaction, α. The mass distribution (b) shows a single narrow peak which progressively broadens and shifts to high molecular mass as α approaches unity

where N_n is the number of *n*-mers present and N the total number of oligomers present. If the total number of repeat units present is N_0, then $N = N_0(1 - \alpha)$. This is reasonable since when $\alpha = 0$, the total number of molecules present is just the number of monomer units present. As monomers combine together $\alpha \rightarrow 1$, N falls to a very small number and in the limit is just unity. Hence

$$\frac{N_n}{N_0} = (1-\alpha)^2 \alpha^{n-1} \quad \dots (4)$$

This equation represents the number distribution function for a linear stepwise polymerization [5(a)].

The mass distribution is simply

$$W_n = n(1-\alpha)^2 \alpha^{n-1} \quad ... (5)$$

and this distribution is shown in Figure 2.5(b) for several different extents of reaction. It can also be shown that the dispersion is given by the equation

$$\frac{\overline{M}_w}{\overline{M}_n} = 1 + \alpha \quad ... (6)$$

so that as $\alpha \to 1$, $\overline{M}_w / \overline{M}_n \to 2$

High molecular mass material is developed only in the very final stages of reaction ($\alpha = 0.99$ and beyond). When the reaction is 90 per cent completed, the peak degree of polymerization is only about 10. For nylon 6, this represents a molecular mass of 1130. At 95 per cent completion, it is still only about 5650. Only at 99 per cent completion does the molecular mass begin to approach the useful region, with a peak of about 11 130. This feature of step-growth reactions creates severe problems for monomer purity (see Box 2.2). In fact the peak in the mass distribution is given by the equation

$$n_p = -\frac{1}{\ln \alpha} \quad ... (7)$$

and this value is close to the number-average molecular mass $\overline{M}_n$. For thermoplastic polyesters, for example poly (ethylene terephthalate), $\overline{M}_n$ must exceed about 10 000 for film formation and about 14 000 for fibre formation. For polyamides on the other hand, $\overline{M}_n$ of commercial fibre forming polymer is 12 000-13 000, which for nylon 6,6 is equivalent to an average degree of polymerization of 53-58. The lower molecular masses needed for polyamides compared to polyesters reflects the strengthening characteristics of interchain hydrogen bonding.

2.4 Co-Polymerization

The alloying of metals to improve their properties is widespread and although many polymers used today are relatively pure (e.g. polystyrene, nylon), an increasing number are mixtures of two or more polymers. As with metals, one reason for doing this is to increase the range of properties. The major practical problem, however, is that homopolymers blend together with difficulty and even where blends are possible, as in some thermoplastics, phase separation can occur readily.

This problem is often overcome by polymerizing a mixture of monomers, a process known as **copolymerization**. It gives a much greater range of structures than is possible by mixing homopolymers because of the possibility of branching, structural isomerism within a single monomer, and the way in which the different repeat units can be added together. In addition, composition can be varied over very wide limits and, of course, molecular mass can be varied to achieve the desired balance of properties in the final product.

One central problem of copolymerization is manipulation of the order of repeat units along the length of the chain. To illustrate this, suppose that two monomers, A and B, are copolymerized. The chain could start with either a molecule of A or a molecule of B, and at each successive addition there are always two possibilities as to which monomer molecule will be attached. As shown in Table 2.1, the number of possible chain structures grows rapidly as n increases. Since the number of possible structures is proportional to 2^n, it is easy to see that even for low degrees of polymerization the number of possible copolymers is very large indeed. Some of these molecules are identical however (AB is the same as BA for example), so that the number of *real* structures will be somewhat lower, as shown in Table 2.1.

Table 2.1: Possible and real structures of linear copolymer chains created from two monomers, A and B

	Number of structures		
Degree of polymerization	**Possible**	**Real**	**Possible structures of chains**
1	2	2	A B
2	4	3	A A B B | | | | A B A B
3	8	6	A A A A B B B B | | | | | | | | A A B B A A B B | | | | | | | | A B A B A B A B
4	16	12	A A A A A A A A B B B B B B B B | | | | | | | | | | | | | | | | A A A A B B B B A A A A B B B B | | | | | | | | | | | | | | | | A A B B A A B B A A B B A A B B | | | | | | | | | | | | | | | | A B A B A B A B A B A B A B A B

Box 2.2: Making nylon.

Nylon is a familiar polymer both as a fibre (in textiles and ropes), monofilament (in fishing lines and toothbrushes) and in mouldings (such as the plugs on large screws for making attachments to brick and concrete walls). It was first made in the USA in 1936/7 by a university chemist employed by DuPont, W.H. Carothers by name. He more than any other scientist opened up the then obscure subject of polymers to commercial exploitation. He was also closely involved in the development of one of the most important synthetic rubbers, polychloroprene (*Neoprene*) during the same period. In April 1937, the first experimental nylon 6,6 stockings were made, followed rapidly by full plant production in 1938 after a good reception from the (female) consumer. The stockings, for example, proved cheaper and more durable than the silk stockings which then dominated the market. Most nylon however, was produced for parachutes, tyre cord, and rope for military use in the Second World War, which then intervened. The Germans competed by making nylon 6 in 1941-2, and the two types still compete in the market.

The industrial production of nylon (or aliphatic polyamide) is beset with problems, however. An important practical consequence of step growth behaviour, for example, is that small amounts of impurity can seriously inhibit the growth of high molecular mass polymer. To counter this problem, monomers must be purified carefully and, in the case of the nylons, the monomers must be in such a form that the numbers of different functional groups are exactly equal:

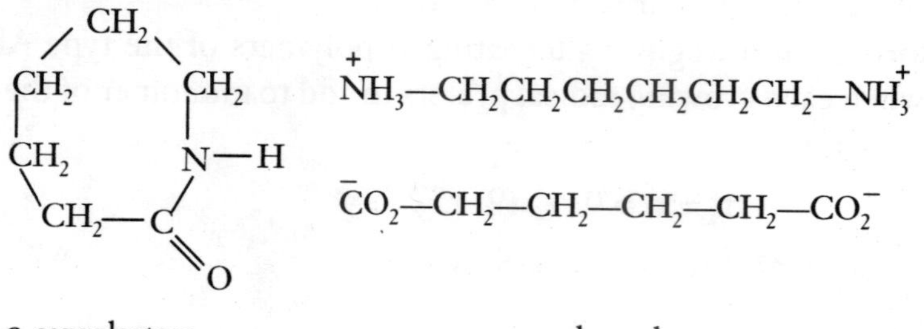

$$\overset{+}{N}H_3\text{—}CH_2CH_2CH_2CH_2CH_2CH_2\text{—}\overset{+}{N}H_3$$

$$\overset{-}{C}O_2\text{—}CH_2\text{—}CH_2\text{—}CH_2\text{—}CH_2\text{—}CO_2^-$$

ε-caprolactam (ring: CH_2, CH_2, CH_2, N—H, CH_2—C=O, CH_2) nylon salt

Fig. 2.6

Nylon 6 is prepared by breaking the ring in å-caprolactam and nylon 6,6 by heating nylon salt, when the ionic structure breaks down to form linear chains. In both cases, the amine and acid groups are of equal concentration. With other step growth polymers, there are similar problems, prepolymers often being made to overcome the stringent requirements of the process.

In fact, the structures which form are not entirely random if the reaction is started with a particular mixture of monomers. Table 2.1 shows that the composition varies from chains of only monomer A (homopolymer A) to chains containing only monomer B (homopolymer B). Between these two extremes it is possible to identify structures of the type AAA... BBB... where there are relatively long sequences of either A or B monomer units; these are known as block copolymers. There are also structures of the type ABAB... known as alternating copolymers (see section 2.2.6).

In many of the structures no regularity can be detected, although there will be short sequences of one type of unit, and the copolymer can be regarded as completely random; such copolymers are usually said to be **ideal copolymers.** These possible copolymer structures are shown schematically in Table 2.1.

The Copolymer Equation

It can be shown that the rate of change of monomer concentration in any copolymerization is given by the equation

$$\frac{d[M_1]}{d[M2]} = \frac{[M_1]}{[M_2]} \cdot \frac{r_1[M_1]+[M_2]}{r_2[M_2]+[M_1]} \quad \ldots (8)$$

where $[M_1]$ and $[M_2]$ are the concentrations of monomers 1 and 2 at any instant and r_1 and r_2., are reactivity ratios. The reactivity ratios represent the rate at which one type of growing chain end adds on to a monomer of the same structure relative to the rate at which it adds on to the alternative monomer. The **copolymer equation** can be used to predict chain structure in the three different ways, already mentioned.

The formation of regular alternating copolymers of the type ABAB... is favoured when each growing radical prefers to add to monomer of the opposite type. In this case

$$r_1 \approx r_2 \approx 0 \ldots (9)$$

and Equation (23) therefore becomes

$$\frac{d[M_1]}{d[M2]} = 1 \text{ or } d[M_1] = d\,[M_2]$$

In other words both monomers will disappear from the reaction vessel at the same rate.

An ideal copolymer will tend to form when each type of chain end shows an equal preference for adding on to either monomer. In this case,

$$r_1 = \frac{1}{r_2} \quad \ldots (10)$$

and the copolymer equation becomes

$$\frac{d[M1]}{d[M2]} = r_1 \frac{[M_1]}{[M_2]} \quad \ldots (11)$$

Hence composition depends on the relative amounts of monomer present at any time and the relative reactivities of the two monomers.

Finally, block copolymers are formed when the growing chain end has a marked preference for adding on to the same kind of monomer. In this case

$$r_1 > 1$$
$$r_2 > 2$$

As can be seen from Table 2.2, this is rarely achieved in free radical copolymerization. However, it is possible to form block structures in anionic polymerization simply by feeding different monomers to the living polymer. Step growth copolymerizations produce ideal (random) copolymers since in this special case $r_1 = r_2 = 1$.

Table 2.2: Reactivity ratios for free radical chain growth polymerization.

Monomer 1	*Monomer 2*	r_1	r_2
acrylonitrile	1,3-butadiene	0.02	0.3
	methyl methacrylate	0.15	1.22
	styrene	0.04	0.40
	vinyl acetate	4.2	0.05
	vinyl chloride	2.7	0.04
1,3-butadiene	methyl methacrylate	0.75	0.25
	styrene	1.35	0.78
	vinyl chloride	8.8	0.035
methyl methacrylate	styrene	0.46	0.52
	vinyl acetate	20	0.015
	vinyl chloride	10	0.1
styrene	vinyl acetate	55	0.01
	vinyl chloride	17	0.02
vinyl acetate	vinyl chloride	0.23	1.68

Commercial Copolymers

The main reason for copolymerizing different monomers is to adjust the physical properties of a given homopolymer to meet a specific demand. SBR elastomer, for example (Table 2.1), based on 24 wt per cent styrene monomer shows better mechanical properties and better resistance to degradation than polybutadiene alone.

By increasing the styrene content to 35 per cent, a high hysteresis (energy absorbing) material ideal for tyre treads is produced. Another example is nitrile rubber, which is produced by a free radical emulsion copolymerization of butadiene and acrylonitrile to make an oil-resistant rubber suitable for oil and petrol lines.

A second reason for copolymerization is to enhance the chemical reactivity of a polymer, particularly to aid crosslinking. Conventional vulcanization in rubbers is brought about by forming sulphur crosslinks at or near double bonds in the chain. In polyisobutylene where the main chain repeat unit is

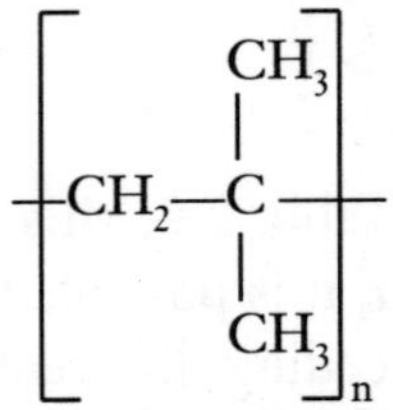

$$\left[-CH_2-\underset{CH_3}{\overset{CH_3}{\underset{|}{\overset{|}{C}}}}- \right]_n$$

there are no such bonds. So the isobutylene monomer is copolymerized with a few weight percent isoprene units to make IIR (butyl rubber) which can be vulcanized easily.

This is also the reason why EPDM rubber consists of no less than three different monomer units copolymerized together (ethylene, propylene and a diene) using Ziegler-Natta catalysts. The copolymer structure is random, so crystallinity is low and the material behaves like a rubber when vulcanized across the diene double bonds.

To show the dramatic effect of copolymer structure on physical properties, consider the change from random SBR copolymer to a block copolymer of exactly the same chemical composition but where the styrene and butadiene parts are effectively homopolymer chains linked at two points:

$$[S]_{r_1}-[B]_{r_2}-[S]_{r_3}$$

The material behaves like a vulcanized butadiene rubber without the need for chemical crosslinking since the styrene chains segregate together to form small islands or domains within the structure. Such so-called **thermoplastic elastomers** (TPEs) today form an important growth area for new polymers because of the process savings in manufacture that can be achieved with their use.

Among rigid thermoplastics, the most widely used copolymers are those of styrene and they include ABS, HIPS and SAN. Both HIPS and ABS are graft block copolymers where the elastomeric side chains are deliberately introduced to improve the toughness of the material.

2.5 Polymer Grades

Polymers synthesised by a variety of routes are available in many grades from the large polymer manufacturing companies. Naturally enough, the grades of bulk tonnage polymers, such as LDPE, PVC, HDPE and PP, run into the hundreds simply because of the multiplicity of different process routes and end functions. So what are the basic differences between grades of just one polymeric material? The most important distinguishing characteristics are structure and molecular mass.

Most suppliers of polypropylene offer grades ranging from isotactic polypropylene to crystalline ethylene propylene partial block copolymers with up to about 10 wt per cent ethylene comonomer. The copolymer grades offer greater toughness over a wider temperature range (particularly below 0 °C) at the expense of stiffness. The applications for copolymer grades are, by and large, more demanding than those for equivalent homopolymer grades. Each polymer is available in several different melt flow grades. **Melt flow index** (MFI or MFR) is a widely adopted practical way of measuring the ease of flow of a polymer grade, and so is of use in indicating the relative magnitude of process parameters for shaping the polymer granules or powder to create a finished article. It is *inversely* related to molecular mass, so that *high* MFI grades correspond to low molecular masses and vice versa. High molecular mass polymers often possess the best physical properties, which is why the most demanding uses of PP such as safety helmets and pipe fittings require low MFI grades.

Beyond the standard grades are filled and special grades which use the basic range of polymers as a matrix for other materials, such as talc, mica and glass fibre, to modify physical properties in other ways (Box 2.3). Chemical additives are also used (in smaller proportions) to improve resistance to sunlight or oxidation (AVI). Special grades have, in many cases, been developed by polymer manufacturers for very specific functions and may include, for example, added pigments and flame-retardant compounds.

Prices of Polymers

Prices of bulk and speciality polymers (Table 2.3) broadly reflect the degree of chemical processing and treatment needed to make them. Thus the polyolefins, which are directly polymerized from cracker streams, are generally the cheapest followed by vinyl derivatives of ethylene like PS and PVC. Derived polymers which require more complex treatment, such as ABS, PET and polyester thermosets are generally more expensive by factors of between two and four. Speciality engineering polymers tend to range in price (1995) from about £2000 up to £7500 per tonne or more for a material like polysulphone (PSu). These prices reflect not only more expensive feedstocks and polymerization methods but also the manufacturers' desire to recoup development costs through a premium for their special properties.

Box 2.3 Additives for Polymers

Polymer products without additives in the matrix material are rare, medical products which are in intimate contact with the human body, being the exception because of the problem of leaching by bodily fluids. But in the vast majority of products, additives are used to modify properties in a controllable way. So what are the principal types of additive? It is a surprisingly long list and includes:

- inorganic fillers
- bulking agents
- coupling agents
- crosslinking agents
- colourants (pigments and dyes)
- impact modifiers
- plasticizers
- lubricants and process aids
- stabilizers
- flame retardants and smoke suppressants
- antioxidants, antiozonants.

Fillers are added where transparency or translucency are not key design factors, and where stiffness can be enhanced. Many fillers are inorganic (such as glass fibre, talc and mica) and hence of higher inherent stiffness, but strength is usually sacrificed since particles are stress concentrators and may initiate cracks. Bulking agents (e.g. chalk, sawdust) have a much smaller effect on stiffness, and the prime motive is to reduce cost. One important factor in achieving best filler action is to ensure that there is good wetting between the polymer matrix and the filler. With silicate fillers, coupling agents are used to give a good bond between the two species. Silanes (organic monosilicates) are frequently used to bond glass fibre to polymer by reaction at the surface of the filler. The other ends then either react with the polymer or blend homogeneously to form the bonded interface.

Crosslinking agents are a vital part of thermoset formation, although modification of the backbone chain may be needed to achieve the desired effect (as in butyl rubber, or EPDM). Pigments are simply added to give the product colour, giving plastics decisive advantage over other materials since they offer extra freedom for designers. Dyeability of fibres is of fundamental commercial importance, so dye retention is an important property, especially for textile fabrics which are washed repeatedly. Many of the first block copolymers were in fact developed for fibre dyeability, since if the inserted blocks react with the dye, then it is held fast by strong chemical bonds (as in PET/polyether block polymers; ICI, 1950). The same philosophy has been used with impact modifiers, where rubber chains are permanently anchored to backbone chains, as in HIPS and ABS. Plasticizers are used extensively in PVC, producing a flexible rather than rigid product. Lubricants and stabilizers are also closely connected with PVC, improving processing and stability against degradation (like antioxidants and antiozonants which have more specific functions).

Table 2.3: Raw Material Prices.

	Price (tonne lots)/£ tonne^{-1}, 1995	*Price (tonne lots)/£m^{-3}*
Thermoplastics		
polyethylene, HDPE	450	432
polyethylene, LDPE/LLDPE	460	420
polypropylene	475	437
PVC (unplasticized)	500	530
polystyrene/HIPS	730/780	775/825
PET bottle grades	900	1220
polyester SMC/DMC	1300/1400	2270/2450
acrylonitrile-butadiene-styrene, ABS	1500-1800	1530-1835
nylon 6	2550	2855
polycarbonate, PC	2700	3270
polysulphone, PSu	7500	9300
liquid crystal polymers, LCPs	17 500	—
PEEK	45 000	5850
Metals (LME)		
lead	405	4600
zinc	680	5180
aluminium	980	2650
copper	1560	13 930
tin	3560	26 000
nickel	4360	38 800
Mild steel	400	3150
Rubbers		
standard Malaysian rubber (SMR)	790	730
SBR 1712	870	820
neoprene	1975	2430

However, it should be noted that the prices shown in the table will vary substantially depending on current supply and demand. Plant shutdowns, for example, can cause temporary price rises because supplies are often limited to a few major petrochemical plants worldwide. On the other hand, prices may slump if plant shutdowns occur (by fire damage, for example). The specific grade chosen will also affect price, those grades having many additives attracting the necessarily higher price than the raw material. The quantity purchased will influence the unit price paid by the buyer. Clearly large quantities will attract substantial discounts, and most polymer buyers will liaise with traders worldwide to achieve the best prices.

Although increases in the traded price of crude oil can push up prices of materials derived from it, the effects have been felt on *all* materials because of the consequent high energy costs in reducing ore to metal and subsequent processing. In fact, the real price of crude oil is now (1997) low in real terms compared with prices in the 1970s and 1980s. Economic recession has a much more important effect on trade prices. Thus the recession of the early 1990s caused prices to drop substantially, and polymer prices have only recently recovered to pre-recession levels.

Another factor which has helped to keep the Retail Price Index (RPI) indexed polymer prices relatively low has been the over-capacity for petrochemical production. During the 1960s, ever-larger petrochemical plants were built for the economies of scale in production. But the demand was effectively halted by the OPEC price rises, with the result that major chemical companies had been losing heavily on bulk polymers until only recently (1995). With many speciality polymers, the reverse has happened—demand has risen continuously over the years, and continues to rise at a fast rate, so contributing to the rise in polymer consumption.

When compared on a weight basis, light metals like zinc and aluminium are similar in price to engineering polymers like polyester or ABS, although the cost on a volume basis is considerably lower for polymers than for metals. Thus the cost per unit volume of polypropylene works out at about £440 m^{-3} compared with a price of over £3000 m^{-3} for mild steel. Light metals like aluminium are considerably more expensive when costed on this basis (£2650 m^{-3}). Speciality engineering polymers like polycarbonate at £3270 m^{-3} are slightly more expensive than light metals using this criterion.

In addition, most materials are used in the form of alloys, composites or mixtures which will push the alloyed price above those shown in 2.3. For example, aluminium is frequently alloyed with copper to improve its stiffness and strength, and additives such as expensive antioxidants or pigments are often mixed with polyolefins. On the other hand, fillers like chalk dust, mica and carbon black can reduce the cost of the blended product while often enhancing the valuable properties of the end product. So the prices of various grades of polypropylene will vary according to the fillers and other additives incorporated. However, it is also important to be aware of the fluctuations which occur in raw materials prices—this is particularly important for general-purpose materials subject to the market forces of supply and demand.

Material Costs in Manufacturing

For high added-value products like boats and cars, material costs form a relatively small proportion of total costs. For directly manufactured products, however, which are sold without much assembly or finishing, material costs do form a relatively large proportion of the total production cost. This applies particularly to polymeric containers for foods and drinks but not, for example, to containers for more sophisticated products like electronic or electrical goods. What is much more important in high added-value products is that the polymer container protect the contents from the environment for which that product is destined. Equally relevant is the way that the container is produced, because different routes of production have significantly different costings.

3

Design and Synthesis of Polymers with Controlled Structures

3.1 Design in Polymers

A Fresh Approach?

Polymeric materials offer substantial benefits over conventional materials in terms of their low density, relative freedom from corrosion, transparency or translucency, and a range of physical properties which cannot be achieved with metals, glasses or ceramics. Such unique properties include low coefficients of friction (e.g. PTFE), resistance to extreme environments (e.g. PTFE, silicones) as well as the ability to absorb and modulate damaging vibrations (e.g. most rubbery polymers). It is these properties which have excited designers because of their potential in manufactured products, despite the generally low short-term moduli of these materials compared to metals.

But an additional bonus is offered by the great and ever-expanding range of routes by which they can be processed into shape and assembled into relatively complex artifacts. A company manufacturing with traditional materials today is often presented with the problem of developing new polymer products. This may arise in several ways:

- competitor companies develop products which are lighter, or more efficient functionally, or more aesthetically acceptable;
- the company may decide to redesign its products to fulfil new consumer needs in response to market demand;
- the company may see possibilities for entirely new products which fit into its existing range, or exploit its existing expertise.

Such pressures require a fresh approach using the newer materials, and more rapid or more flexible production methods. The response will vary according to the breadth of expertise and equipment already present in-house. Thus companies with much moulding machinery should be better placed to exploit new opportunities. If the equipment is rather old, however, they might fail to gain all the benefits that are currently available using entirely new machines. The rate of obsolescence in plastics processing equipment has been very high in the last two decades or so, for a number of reasons, especially the introduction of computer control systems, both in the machine itself and between different machines and the staff. This might give a competitive advantage to companies entering the field for the first time, provided, of course, that they are aware of the pros and cons of the various processes available.

Alternatively, it may be preferable to hire the specialist expertise and equipment needed, by using so-called 'trade moulders' for example. This is a very active sector of SMEs, where entrepreneurial skills are applied to generate new business. Sometimes, such enterprise is exceptional in producing new designs and new ways of using existing machines. Such is the story of the Topper boat.

3.2 Manufacturing and Process Methods

Different production routes entail significantly different costings, and the selection of the manufacturing method is therefore a key step in the development of a product. For example a simple closed box in a thermoplastic could in theory be made in several different ways:

- fabricating from cut sheet, e.g. by welding
- rotationally moulding from polymer powder
- vacuum forming from sheet material
- blow moulding from molten polymer
- injection moulding two subcomponents which are joined in a subsequent welding operation.

The first option, fabrication, involves a small capital investment but, like GRP hand lay-up, is very labour intensive. In all the other process routes, some form of metal mould is required to reproduce the final shape of the container. In the case of rotational casting this might involve a mould welded from sheet metal. The required weight of polymer powder is placed in the mould which is closed and rotated in an oven. The powder spreads uniformly over the mould inner surface and melts together. The mould is designed to be easily split so that the object can be removed after cooling and solidifying. Rotational moulding (or rotomoulding) is widely used for completely closed objects like footballs or for containers of relatively simple shape. The time required to produce such castings can be up to 30 minutes or more.

A somewhat more advanced way of using a female mould to produce open objects like domestic baths is vacuum forming. Sheet of the required size is heated to a softening point, well below its melt temperature, so that it is in a flexible rubbery state. It can then be literally sucked into the polished metal mould by an applied vacuum so that it assumes the desired shape. Vacuum forming can also use male moulds which are pushed into the softened sheet before the vacuum is applied between the sheet and the mould. Although faster than rotocasting, the method is severely limited by the shapes that can be formed. Capital costs are modest.

Blow moulding is an entirely different method of achieving the final product shape. A tube of molten polymer is blown so that it expands as a bubble until it meets the surface of the surrounding cool metal mould, which can then be split to extract the solidified moulding. The method is ideal for long production runs of relatively simple hollow shapes such as containers. The tubes which are to be blown can be produced by high-speed production techniques based on extrusion or injection moulding. Capital costs are quite high. As with most of the other methods already discussed there are quite severe restrictions on component shape.

Injection moulding overcomes most of the previous restrictions on product shape but the capital cost for machinery and moulds is the highest of all the process methods. High production rates are possible with moulding cycle times measured in tens of seconds. Long production runs are normally essential to recoup the initial investment. This is reflected in Figure 3.1 which shows the unit cost for a simple 200 g box container as a function of the annual production rate. Comparative data are shown for the other production methods. With the exception of the labour intensive simple fabrication, all reflect the economy of longer production runs offsetting the higher capital costs. However, the relative positions and the points at which the curves flatten out differ significantly owing to the different investment and labour cost of each technique.

A consequence of the heavy investment in machinery and tools is extra pressure on management to maximise plant potential. In operational terms this effectively means using the full capability of the moulding machines whilst minimising the time it takes each machine to produce the desired product.

3.3 Materials Selection

Good design fulfils the product specification under the required service conditions as well as contributing to the cost effectiveness of its manufacture and maintenance. The product specification itself must be an interpretation of the market needs. Hence good design means giving product appeal at the point of sale. Selecting the polymer is just one stage in this design exercise, both in terms of information on various properties of materials, as well as the detailed evaluation and selection of the best material for the product in question (Figure 3.2). It is obviously an important stage, however, and the variety of

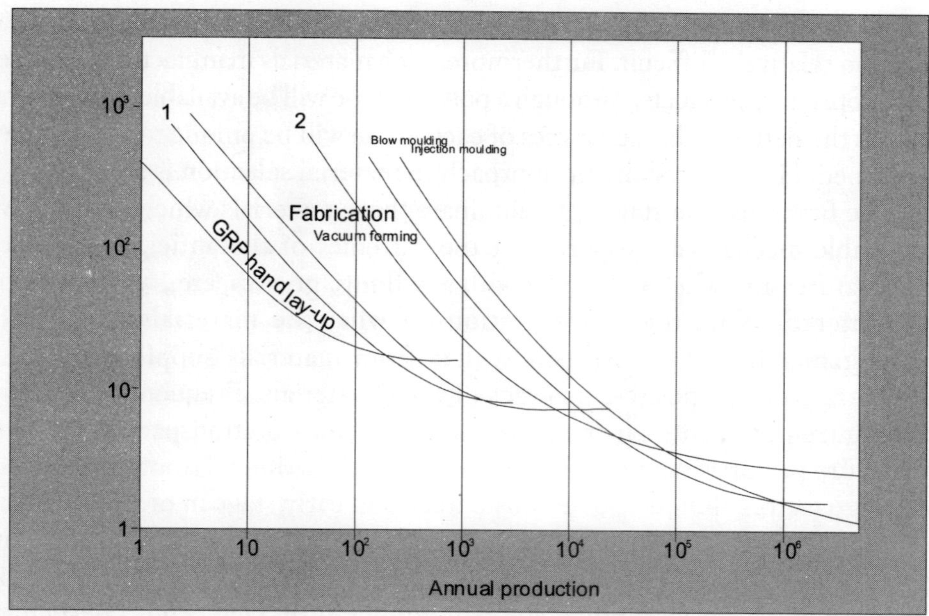

Fig. 3.1: Comparative costs for making a 200 g polyner box by different process routes. 1. Rotational moulding using simple equipment. 2. More sophisticated rotational moulding.

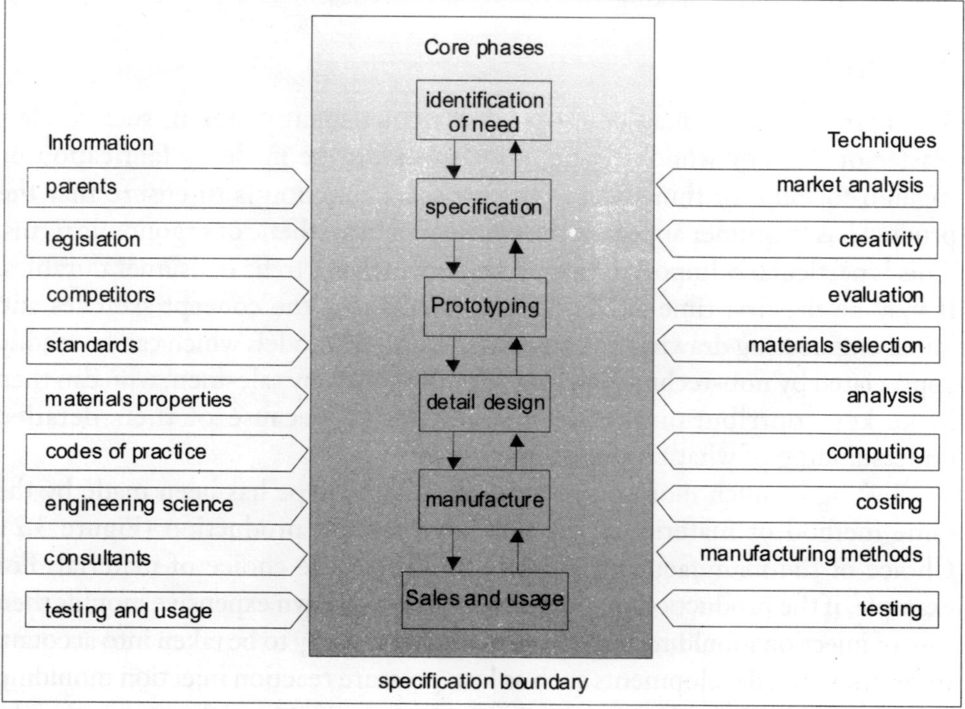

Fig. 3.2: Design process model showing sources of information and techniques used in progressing through the core phases.

polymer types plus the many different grades within each type make that selection relatively difficult. Furthermore, each materials manufacturer supplies only proprietary products. Although a polymer type will be available from several sources the detailed characteristics of each grade will be unique to the supplier concerned. Hence a systematic approach to material selection is sensible.

The first selection stage is to eliminate those materials which do not have reasonable mechanical properties at the extremes of the anticipated service temperatures and when in contact with any fluids, greases, etc., which may be encountered. Note the concentration on what the materials cannot do. Understandably, but unfortunately, data from materials suppliers tends to concentrate on the positive qualities of their materials. Frequently there are clear-cut requirements for the material, e.g. it must be transparent, or must meet a fire performance specification of the Underwriters' Laboratory. These go/no-go factors should also be considered at an early stage in order to reduce further the number of candidate materials. Some of the most important properties used for such initial screening include density, short-term tensile modulus and tensile strength at ambient temperature. Detailed property data for design must be obtained from the suppliers of the materials and may need to be supplemented by in-house testing. However, materials selection cannot be finalised without making and testing prototypes.

Prototyping

The first prototypes may be made of any convenient material, such as clay, plaster-of-Paris or wood. Plastic models can also be made by fabrication or vacuum forming of thin sheet. Their primary function is to ensure that the product has 'customer appeal' when considered in aesthetic or ergonomic terms. This is particularly important for products which are to be consumer durables. It may be the first time in the design process that the concept sketches and initial engineering drawings are translated into 3-D models which can be readily appreciated by non-technical people, e.g. the company salesmen, who can then make key contributions to the further design because of their detailed understanding of what their customers want.

Testing is much more meaningful if the prototype has been made by the same method or material as that intended for full production (Figure 3.2). Choice of the manufacturing process influences the choice of material. For example, if the production quantities justify laying down expensive moulds then ease of injection moulding is clearly a material property to be taken into account. In recent years, developments such as low-pressure reaction injection moulding have introduced further options in manufacturing and hence materials. Companies are also influenced in their materials selection by the in-house availability of plant. For example Western Electric in the USA have used their existing metal-forming presses to produce large cable connector cases by warm stamping polypropylene sheet coupled to, and sandwiching glass fibre mat.

Choice of the processing route can be important for its effect on properties and performance of the final product, through polymer chain orientation for example. Obviously the later prototypes should also be made from the shortlisted materials so that a final choice between them can be made. To facilitate this final stage much effort has been put into the development of cheap moulds particularly for injection moulding. They can be machined out of aluminium or cast in aluminium using established foundry practices. A well-made cast aluminium mould can be used to produce up to 10 000 injection mouldings in glass fibre reinforced nylon without unacceptable wear. Such a tool typically costs under £3000 compared with many times that cost for a hardened steel production tool. Moulds can alternatively be made out of low melting point metal alloys or metal-filled epoxy resins or by metal spraying a pattern of the product. In the latter case the metal coating is stripped from the pattern and then stiffened with an appropriate backing material. At the end of the prototyping and prototype testing stage, the design engineer should be in a position to design the metal mould for mass production, a critical phase of the operation owing to the large costs incurred in mould manufacture.

3.4 Case History: Topper Boat

Replacement of one polymeric material by another may be undertaken entirely for manufacturing reasons, and this is what happened in the redesign of the Topper dinghy for thermoplastic polymer. The dinghy was originally designed for hand lay-up GRP in 1969 by Ian Proctor, a well known designer of small boats and yachts (Figure 3.3). but sales had been restricted by the low output rates. The potential production rate if the boat could be injection moulded was very much greater, provided that a suitable thermoplastic could be found.

It was thought at the time of the design exercise in the late 1970s, that there was market demand for a small lightweight sailing boat which could be easily lifted on to a car roof-top for transport. Leisure activities had increased in the seventies, and the membership of sailing schools had increased during this period (Table 3.). In addition, the existence of the GRP boat meant that there was already substantial experience of the craft in terms of its sailing characteristics, so that prototyping might not be necessary. There were clear cost savings to be made by jumping this step in the design sequence (Figure 3.2).

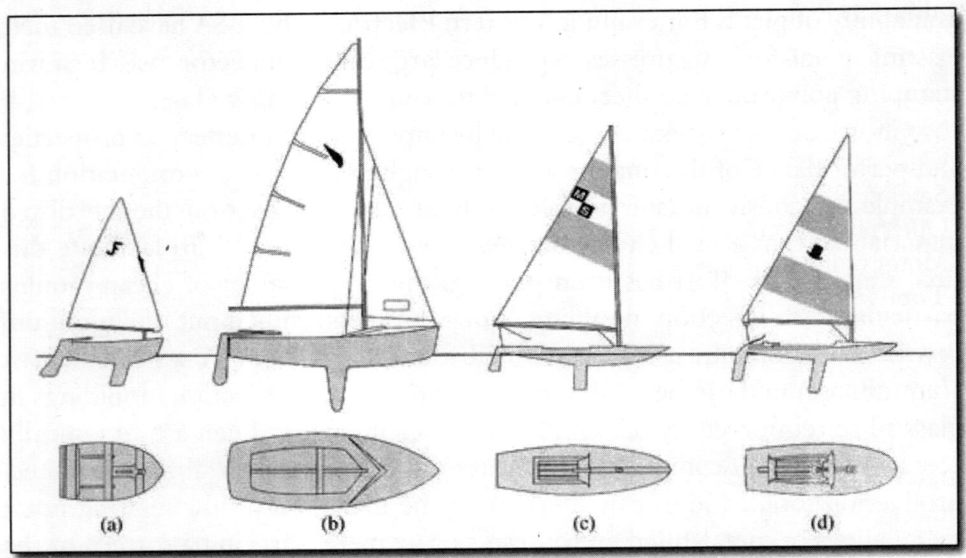

Fig. 3.3: Evolution of the Topper from earlier Proctor designs: (a) Jiffy dinghy for children; (b) Kestrel family sailboat (1960); (c) Minisail sailing surfboard (1964); (d) Topper dinghy (1969). All were designed for GRP except the early versions of the Minisail

Table 3.1: Some Data for Various Popular Small Dinghies

Class	*Designer*	*Overall Length/m*	*Approx. Sailing Mass/kg*	*Sail Area/m^2*	*Hull Material*	*Crew*
Jiffy	Ian Proctor	2.29	36.2	3.34	GRP	1-2
Optimist	Clark Mills	2.31	35.0	3.25	marine plywood	1-2
Durafloat	Colin Mudie	2.29	27.2	3.25	ABS (acrylonitrile-butadiene-styrene) plastic	1-2
Cadet	Jack Holt	3.22	65.6	5.20	GRP or marine plywood	2
Mirror	Jack Holt, Barry Bucknall	3.30	61.3	6.41	marine plywood	1-2
Gull	Ian Proctor	3.35	93.0	6.50	GRP or marine plywood	1-3
Moth	various	3.35	65.6	6.26	mainly GRP or marine plywood	1
Topper	Ian Proctor	3.35	54.1	5.20	GRP	1-2

Table 3.2: Estimated UK 'Boat Park' by Boat Type in 1978

Boat Type	*Number of Craft*	*Share (%)*
Sea yachts	40 000	7
Class sailboats	148 000	24
Small funboats	235 000	39
Open power craft	120 000	20
Motor cruisers	67 000	10
Totals	**610 000**	**100**

Source: O'Connor, M., 1978, *Management and Design in the British Small Boat Industry*, Report to the Design Council.

In terms of physical properties alone, GRP is an excellent boat-building material because of its high stiffness and high specific strength. Chopped strand mat (CSM), where randomly oriented glass fibres ca. 5 cm long are embedded in a polyester matrix, is the normal form of GRP used for small boats; it possesses an average tensile modulus of 8500 MN m^{-2} and a tensile strength of about 170 MN m^{-2}. One way of assessing the relative merit of different materials for a particular application is by calculating so-called **merit indices**. One of the simplest merit indices is for a tensile application, where the product is subjected to a pulling force along one axis, viz, a tensile force. In this case, the merit index is just

$$E/\rho \quad \ldots (1)$$

where E is the tensile modulus of the material, and ρ its density. The merit index, the **specific stiffness**, provides a measure of the mass of material required to achieve a given level of stiffness in a product. Such an index is clearly of interest to a designer when weight-saving is an important design criterion. Thus this index will be of great importance when a product is used primarily in tension, such as a rope or textile fibre which must be lifted or handled. Since CSM has a density of 1.5 Mg m^{-3}, this gives a specific stiffness of 5.6 and specific strength of 0.11. Clearly, any material chosen to replace GRP should have a greater merit index.

A more important design criterion for small boat *hulls* is the flexural stiffness. It can be shown from simple beam theory that the **specific flexural stiffness** is $E^{1/3}/\rho$. Direct comparison of merit indices of different materials shows which represent the best value in terms of flexural stiffness per unit mass: materials with a high merit index are lighter than those with a low merit index for an identical stiffness in bending. Another common hull material is marine ply (E=6500 MN m^{-2}, ρ = 0.5 Mg m^{-3}) with a merit index of $6500^{1/3}/0.5$ = 37.3. The merit index for GRP is 13.6, so that a GRP hull will be 37.3/13.6 or 2.75 times as heavy as a marine ply hull of the same flexibility. Unfortunately, marine ply cannot be shaped easily unless cut and bonded together. Monocoques of uniform structure are thus difficult, if not impossible, in marine ply.

It can also be shown that the hull thickness t is related to modulus by the expression

$$\frac{t_1}{t_2}=\left(\frac{E_2}{E_1}\right)^{1/3} \quad \ldots (2)$$

A marine ply hull will therefore be $(8500/6500)^{1/3}$ or 1.09 times as thick as an equivalent GRP hull.

Materials Selection

Among the common thermoplastics available in the mid-1970s, polypropylene appeared as a front runner on grounds of toughness, density and cost. However, it is subject to creep (being uncrosslinked) and possesses a low tensile modulus of ca. 1500 MN m^{-2}. Its merit index is 12.7 due to the low density of 0.9 Mg m^{-3}, making it comparable to GRP. If it replaced GRP, here would be a weight penalty of 3.6/12.7 = 1.07 or about 7 per cent. On the other hand, the hull would have to be considerably thicker:

$$t_{pp}=\left(\frac{8500}{1500}\right)^{1/3} t_{GRP=1.78} \; t_{GRP}$$

the original boat the GRP hull was 2.4 mm thick, so that the polypropylene boat needed to be 2.4 × 1.78 or 4.3 mm thick to give the same flexural stiffness. The assembled hull (Figure 3.4) possessed considerable intrinsic stiffness owing to the monocoque construction, with the deck and hull joined at the gunwale.

$$\theta=\frac{TL}{2\pi r^3 tG}$$

G is the shear modulus and may be taken as $E/2.6$ for both GRP and polypropylene. Use Figure 3.4 for approximate dimensions and calculate the angular distortion created by a maximum torque of 100 N m. (Figure 3.5)

Now read the answer

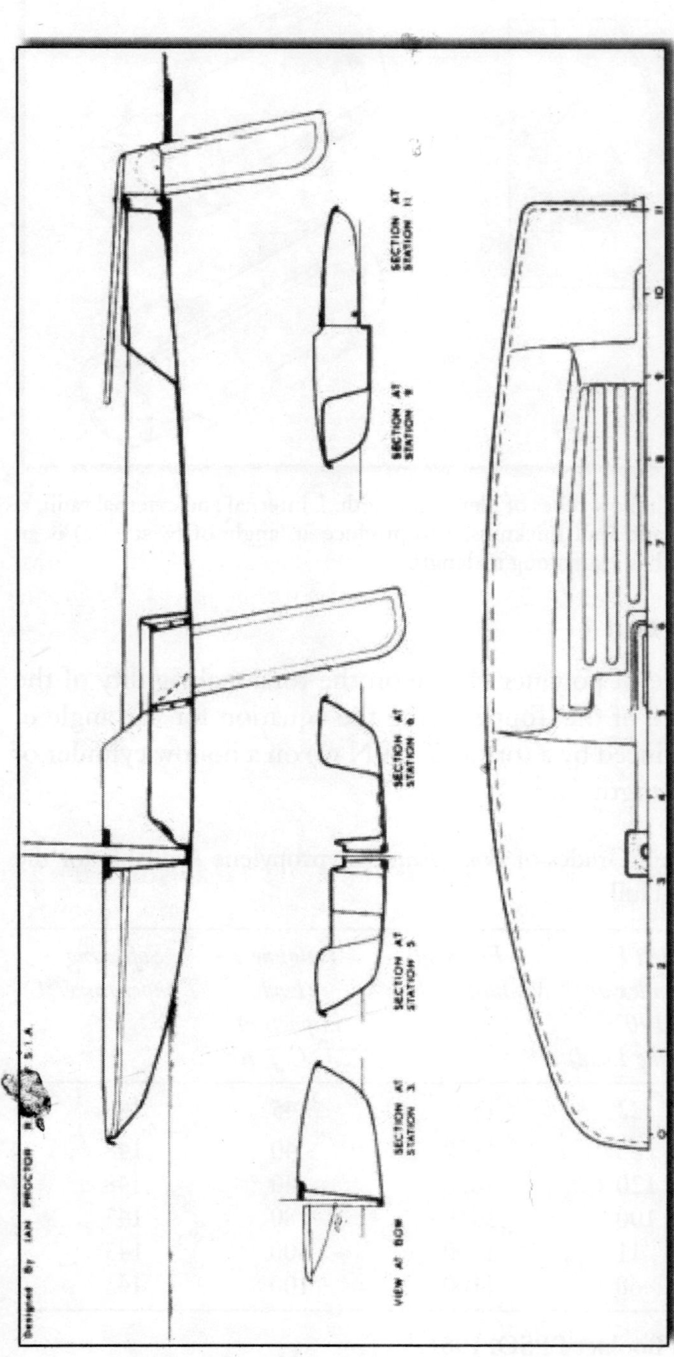

Fig. 3.4: Design of the Topper boat showing monocoque construction for the two halves (deck and hull) which are welded together at the gunwale. This is a scale drawing with dimensions in feet

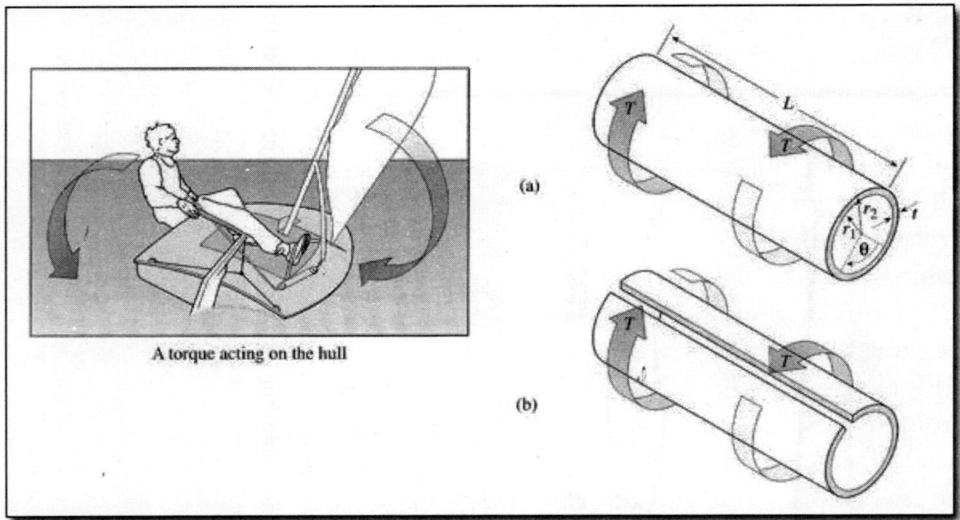

Fig. 3.5: Torque, T, acting on hollow tubes of identical length, L internal and external radii, r_1 and r_2, respectively, and wall thickness, t to produce an angle of twist è. (a) is an integral tube and (b) has a slit along its length

Self Assessment Question 11

Assess the importance of polymer choice on the torsional rigidity of the monocoque hull structure of the Topper using the equation for the angle of twisting, θ (radians), produced by a torque T (in N m) on a hollow cylinder of thickness t, radius r and length L.

Table 3.3: Some of the Main Grades of *Propathane* Polypropylene Available for the Topper Dinghy Hull

Polymer Grade	*Melt Flow Index at 190°C (10kg Load)*	*Flexural Modulus/MNm$^{?2}$*	*Toughness (Izod impact) at 23 °C/J m$^{?1}$*	*Softening Temperature/°C*
GW522M	22	1800	45	148
GX543M	60	1800	40	148
GY545M	120	1800	40	148
GY621M (copolymer)	100	1410	80	147
GY703M (copolymer)	11	1000	500	143
GY702M (copolymer)	60	1100	100	145

Source: ICI Technical Data Booklet PPSO, 1981.

The design team at Rolinx evaluated the torsional stiffness of the new Topper by covering the forward part of the deck with a grid of lines to give a visual check on distortion during sailing. The distortion was certainly visible, but did not apparently affect sailing characteristics too seriously.

Similarly, the problem of creep was thought not to be important, partly because of the intermittent nature of sailing stresses, both in terms of magnitude and direction, and partly because of the short-term expected loadings. Small sail boats are used perhaps a few hours a day before disassembly and storage. Creep, and distortion resulting from creep, increase with time and severity of loading.

The main grades of ***Propathane*** polypropylene available to the moulders are shown in 3.4; they encompass both homo- and copolymer grades of varying MFI, flexural modulus, impact strength and softening points. The impact strength in the Izod test is measured on a standard centrally notched bar of material, which is struck with a falling pendulum. The impact strength is simply the energy absorbed per metre of notch when the specimen breaks in a brittle fashion. The variation in impact strength at ambient temperature is much greater than the variation in modulus, and it was this parameter which assumed greater importance as the project developed. Sailing hulls, particularly those for small, hand launched boats, suffer a considerable number of impacts in service and it could prove catastrophic if a brittle crack were to propagate through the hull!

The upper softening point—the temperature when the polymer starts to deform, which is some distance below the true melting point—was not an important design parameter. The greatest temperatures to which the boat could possibly be exposed did not exceed 70 °C (hull surface temperature in bright sunshine in Middle Eastern climates), well below possible softening points. The lower limit was felt to be important however; the hull could be exposed to sub-zero temperatures when sailed in freezing sea water. The copolymer grades possessed much greater strength below 0 °C owing to the depressant effect of the ethylene component on the glass transition temperature.

Despite the greater flexibility of the toughest grade (GY703M), it represented the best choice on the grounds of toughness over a wide temperature range. However, the constraints of manufacture meant that a compromise with a lower molecular mass grade was necessary.

Manufacturing the Boat

Filling the mould is a serious problem in injection moulding: the lower the MFI, the more difficult it is to push molten polymer down narrow tubes into a metal mould. The engineers at Rolinx, the trade moulders who initiated the thermoplastic project, were working at the limits of their machinery in moulding such large objects in one operation. In the event, they were forced to blend the low MFI copolymer grade with a higher MFI grade material (GY702M) in order to achieve their objective (Figure 3.6). Following injection of the hull and deck, the two parts are fused together near the waterline by incorporating a tape made from metal wire and PP fibre at the joint. When a large electrical current is passed in the wire, it heats up and the two parts fuse together, a process aided by the application of an external load (Figure 3.7). Apart from several fixed parts such as protective plates for the rudder zone, the accessories (mast, rudder, daggerboard, etc.) can then be added by hand assembly, a particularly easy task since the boat is designed for ease of disassembly for the consumer or user. Indeed, these items are usually packed in kit form for ease of transport to the point of sale.

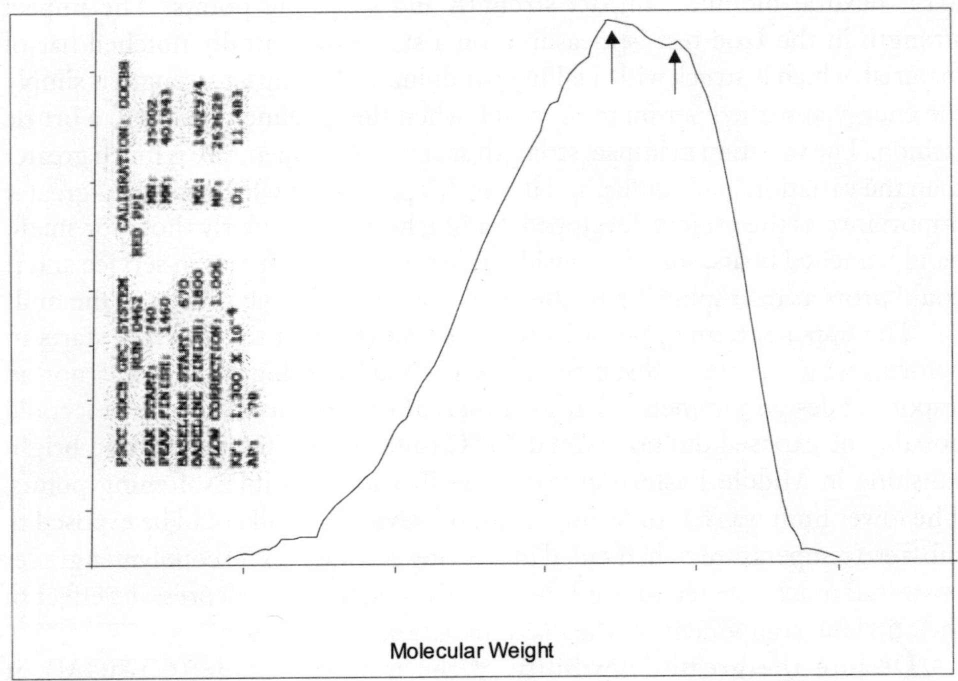

Fig. 3.6: Molecular mass distribution of propylene copolymer determined by high temperature GPC. The curve shows two peaks since it is a blend of low and high MFI materials

The sailing, if not commercial, success (total sales about 2000) of the original GRP Topper provided the Rolinx engineers with design guidelines which were explored in depth with the original designer, Ian Proctor. In one sense, the early GRP boat replaced a conventional prototype, but with the benefit of market feedback. Switching to thermoplastic produced many design changes, some such as a higher rake angle for injection moulding (Figure 3.8). In addition, most accessories were replaced by thermoplastic equivalents which have proved successful in the new version. At the same time, the redesigned hull incorporated several features which enhanced its ease of assembly, performance and integrity:

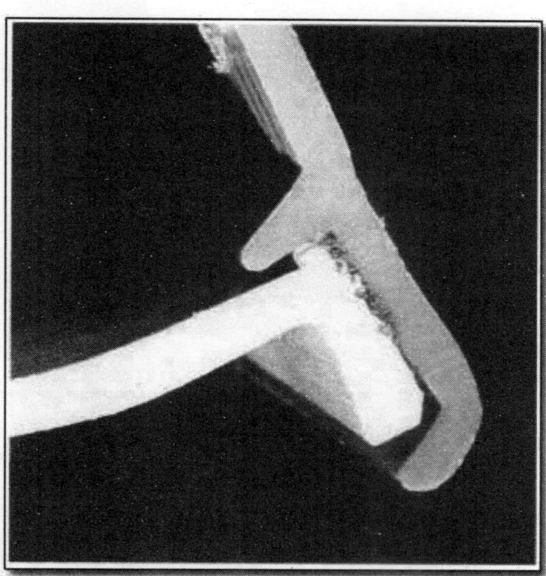

Fig. 3.7: A section through the hull-deck weld showing how the flange on the hull section fits into a recess in the deck section, and the joint fusion welded using sacrificial metal tape

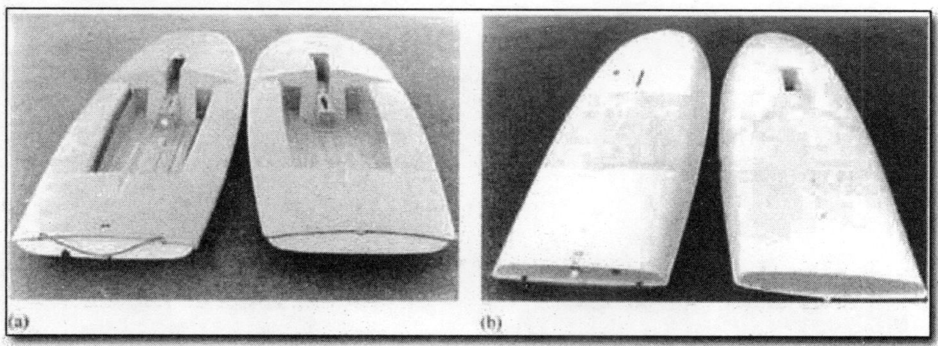

Fig. 3.8: The GRP (left-hand side) and (right-hand side) Topper boats compared: (a) deck sections, (b) hull sections. The redesigned boat incorporated a higher rake angle on the mast access slot and a daggerboard box with failsafe plate

for example, the sacrificial daggerboard box, which provides failsafe protection for the hull against grounding forces (Figure 3.9), and the mast locking gate which enables rapid insertion of the sail and mast on launching (Figure 3.10). The rudder and daggerboard are parts which must be able to withstand severe impact blows without failure, and both are made from glass-reinforced polypropylene. To reduce the weight as well as improve manufacturing efficiency, both are made by foam moulding, where a foaming agent is added to the solid polymer before injection, so that it expands on heating the tool to create a foamed interior.

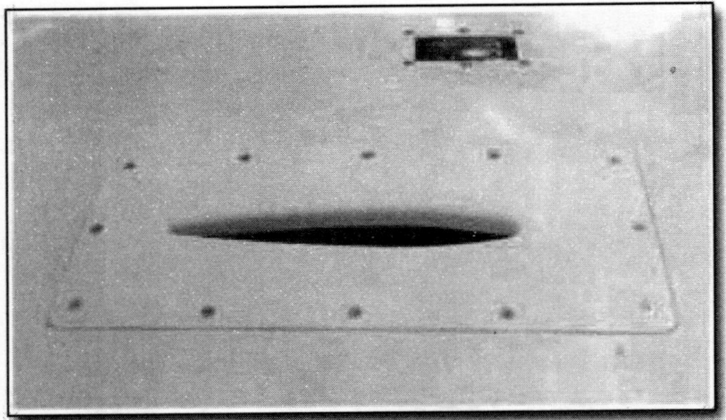

Fig. 3.9: Sacrificial dagger board plate attached to base of hull—grounding forces will only damage the plate and cracks cannot propagate into the hull itself

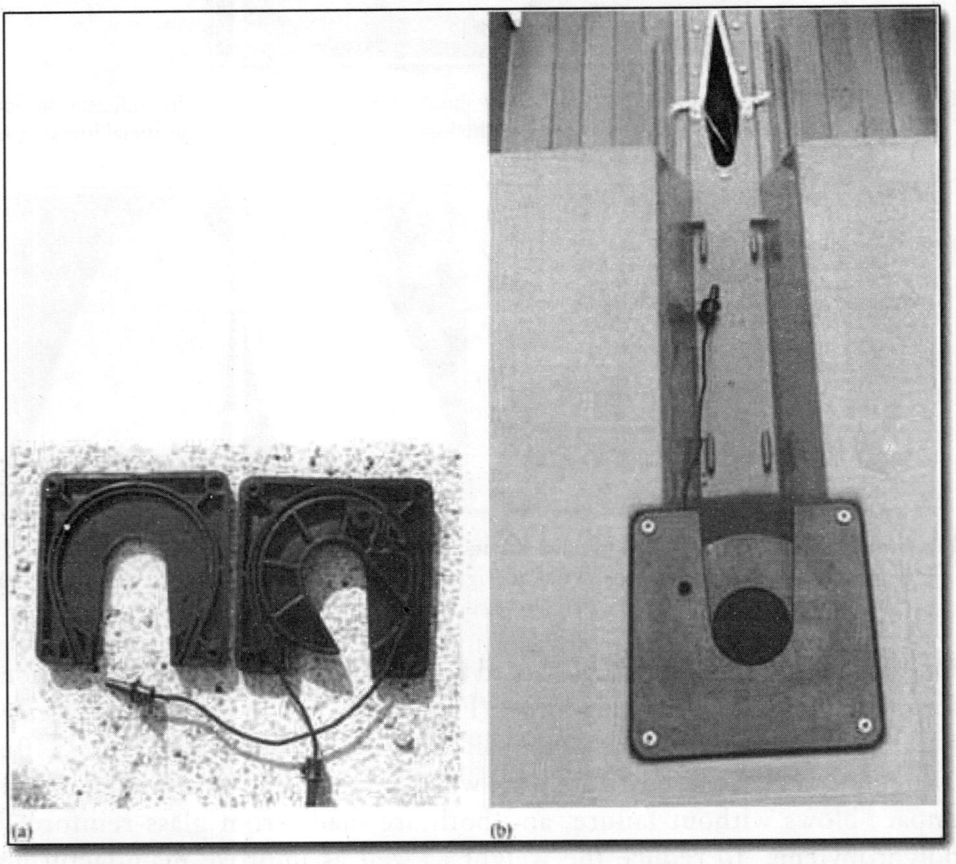

(a) (b)

Fig. 3.10: Mast locking gate for the Topper dinghy redesigned in glass-filled polypropylene: (a) split apart to show component parts. Upper and lower plates and rotating cam activate by drawcord; (b) shown *in situ* in closed position without mast. The composite moulding replaced a single marine ply board drilled with a single hole for the mast

By 1984, sales had exceeded 23 000 and the boat was still selling over 2000 per annum, so the high capital investment in the tools and machinery had been well justified. The story of the Topper is one of relatively small businesses and companies spotting the potential for efficient mass production of a high added-value consumer product. Design in polymeric materials demands a bipartisan approach—design for function *and* design for manufacture (but see Box 3.1).

Box 3.1: The Toucan Boat

Following the successful development of the Topper boat, attempts were made to design a larger boat using injection moulding. It was thought that a large rowing boat could be designed in the same way, having a significant competitive edge over wooden equivalents owing to the rot-resistant properties of polypropylene. The resultant prototype, the Toucan, is shown below in Figure A10, compared with the Topper at left.

Although the lower hull was moulded in one piece, the structure was by its very nature, an open-sided shell. The deck was positioned below the top of the hull, and was itself composed of several subsidiary moulded parts. So the project involved more than the two tools needed for the Topper, in itself increasing the development costs. The stiffening element of the monocoque structure of the Topper was much less significant with the Toucan, making the boat more flexible (especially when loaded in torsion). Finally, the design did not exploit the lightweight properties of the plastic to its greatest advantage. The Topper is designed to hydroplane over the surface of the water, giving the user exciting sailing at speed when under wind. The same couldn't be said about the Toucan, a boat with a deep hull designed more to plough through the water rather than skim along the top. The project was eventually abandoned following poor performance in trials, showing the limits of plastic materials when simply used to copy existing conventional designs without allowing for the problem of high flexibility. Existing wooden dinghies exploit the properties of wood, a natural composite, to its greatest advantage. Thus the side timbers of the hull are planks following the grain, so are intrinsically stiff and resistant to imposed bending and torsion loads. This gives a stiff structure capable of resisting sailing stresses, and so giving safe and predictable sailing.

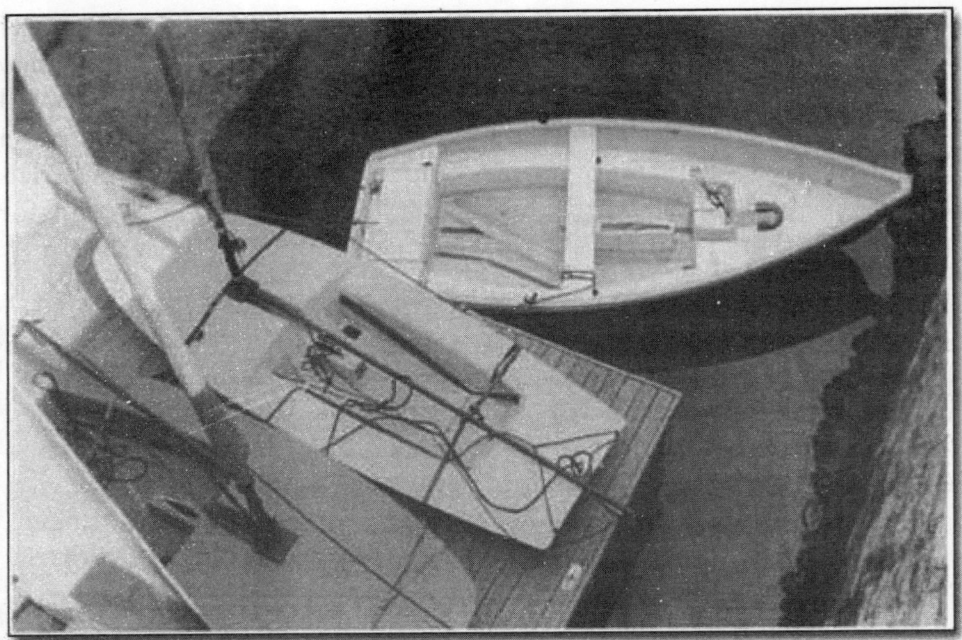

Fig. 3.11

3.5 Market Experience

It is some 20 years since the Topper project was conceived by Peter Bean, Technical Director of Rolinx and Ian Proctor, the designer of the original GRP boat. Sales initially were excellent, especially to sailing schools and clubs where there was much demand for a small, light and very safe sailing boat for children. But after that, the market became saturated, sales were heavily dependant on individuals and families, so decreased despite attempts to export the boat to the USA and Israel, for example. The sales company moved from its location in Torpoint to Kent, and the tools were moved to Germany for injection moulding of the hull and deck.

However, sales have picked up again with the expansion of leisure activities, and the personal market remains active. The boat has clearly filled a particular niche in the market, where larger boats such as the Laser (a GRP boat) are too difficult to handle for children, and alternatives such as the Mirror (self-assembled in kit form marine ply) cannot offer the same intrinsic safety as the all polypropylene boat. The design, in other words, has proven right for the market it is aimed at, and the consumer or user have voted with their pockets for the product. Sales now (1997) exceed 41 000, the boat has official International Status and has a flourishing Class Association. Large moulding machines capable of making the boat are also available commercially now, so proving the validity of the original exercise. But have any changes in the design been necessary? One important change is related the flexibility of the construction material, polypropylene. The monocoque construction, where hull and deck are fused together thermally, ensures that the boat is much stiffer as a product than either hull or deck alone. However, there are areas such as the front part of the deck ahead of the mast, where point loads may deflect it unduly (such as standing on it!). This has now been solved by incorporating a foam block of polystyrene to resist such loads (Figure 3.12). Such a solution harks back to the use of foam-core sailboards and windsurfers, and is a very common way in which hollow plastic structures can be reinforced.

Fig. 3.12: Foam interior to front deck to improve compressive stiffness against point loads.

4

Products and Stuffs Made from Polymers

4.1 Polymer Products

In 1966, Jack Crudup and Bryant Reed began a long standing business relationship while working together in the cast polyurethane department of The Acushnet Company. Five years later, with Acushnet closing their southwest division, Jack and Bryant founded a custom cast polyurethane manufacturer, servicing both the oil and gas and material handling industries.

Polymer Products began casting parts in 1971 based in a small building in Grand Prairie, Texas. Unable to sufficiently accommodate the growing need for cast polyurethane, PPI moved into a new 15,000 square foot building in 1977.

While the early 1980's saw a significant downturn in the oil and gas industry, Polymer Products continued its growth and success by finding new industries and venues for polyurethane products. This early opportunity to persevere proved to be beneficial when in 1991, Polymer Products overcame a plant fire. Polymer then relocated into its current 40,000 square foot facility.

Jack Crudup became sole owner of Polymer Products in the early 90's. Deciding to retire to Arizona in 1993, Jack ensured PPI would be left in capable hands. His son, Steve Crudup, already successful in the rubber and electronics industries, became owner and president of PPI. Throughout his fourteen years in charge, Steve has continued the legacy of growth and advancement left by his father. To this day, PPI continues to be a pioneer in the cast urethane business and offers custom urethane products to all fields and markets.

Jack Crudup passed away in January of 2004. He is fondly remembered by his friends and co-workers at Polymer Products and with his son and, now, grandson involved; his dedication to his employees and customers will never be forgotten.

4.2 Products and Services

From prototype samples to high volume customer molded polyurethane products, we pride ourselves on making high quality parts for each customer.

- Valve Inserts
- Bonded Valves
- Pistons
- Hydrocyclones
- Wipers
- Scrapers
- Sheets
- Rollers
- Rods
- Tubes
- Rings And much more...

For over 30 years Polymer has built long lasting relationships with customers, and as a result, business has continued to grow. We pride ourselves on making high quality urethane parts for each customer. Polymer offers a wide range of services:

- Material and Design Assistance
- Mold Design and Creation
- Prototypes
- High Temperature & Specialty Materials
- Open Casting, Compression, & Low Pressure Hose Fill
- Metal Bonding
- Turn-Key Production with Machining Capabilities

4.3 Polymer Products of India

Company Profile

Polymer Products of India was established in 1991 as a leading rubber products manufacturing company to address the challenging requirements of the niche rubber manufacturing field.

Polymer Products is headquartered in Bangalore, India and is addressing India as its primary market. The company is staffed with leading professionals who support the manufacturing, marketing, sales and service offerings of the company.

Vision Statement

To be a Transnational company offering the best products and services in our chosen areas of competence

Mission Statement

We will add significant value to our customers business through high quality products and services, by continuously building our technological capabilities and delivery capabilities guided by our ethical values.

Executive Profiles

Dr. N.P. Surya Narayana B.Sc, M.Sc, PhD, FPRI London—Chief Rubber Technologist

Dr. Surya Narayana is one of India's renowned rubber technologist who has more than 35 years of experience in the field of Rubber Technology & R&D.

Vishnu N.M.—Director Quality and Production

An engineering graduate from the Regional Engineering College, Srinagar (one of the leading engineering colleges in India) has more than 15 years of rich and varied experience in the field of Rubber Technology, Production and Quality Assurance

Venkat Subramanyam—Director Management and Finance

An engineering graduate from Dharwar University, one of the foremost universities in Karnataka has more than 15 years of expertise in the field of rubber manufacturing, Finance & management. Venkat also holds the position of Director—Sales and Marketing portfolio in the company.

4.4 Products Offering

Engine Mounts

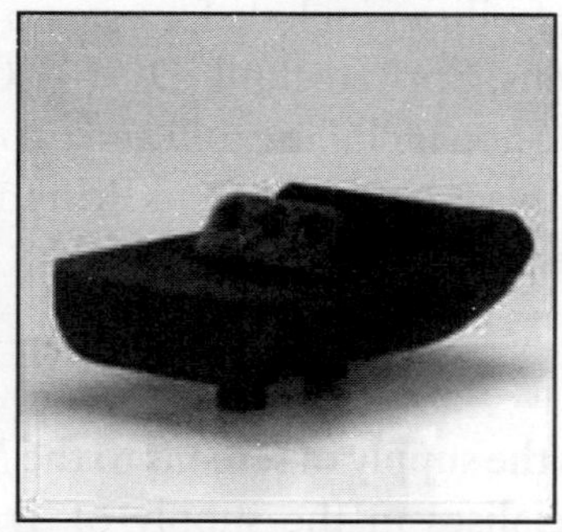

Rock Fenders

4.5 Advanced Polymer Products

Elastomers

Advanced Polymer Products users a range of different elastomers, selected specifically for the application. Experience encompasses TDI Polyethers and Polyesters, MDI Polyesters and Polycaprolactones. Additionally APP has developed a number of composite materials incorporating high slip additives and ceramic additives.

Processing

Advanced Polymer Products manufacturing facilities include four steam heated casting tables (max 4000 x 1650), seven ovens, Spritztechnik DG103 Casting Machine, Moca dispensing system, Machine Shop including mill and two lathes, Fabrication Shop, De-greasing facility, Gritblasting facility and Priming area.

All Product design is completed in-house utilising CAD facilities.

Products

Advanced Polymer Products core business is the supply of screens to the Mining and Mineral Processing industries. It specialises in the supply of Trommel Screens, Cross Tensioned Screens and Modular Screens with an emphasis on engineering for difficult applications and unusual circumstance.

Customers include Placer Dome, BHP, Boral, Zinifex, Readymix, Iluka, Pioneer and KCGM.

In recent years, the business has developed a modular screening system marketed under the name of TRULOK. A number of installations have been completed that have established the system as a cost effective, easy installation screening system.

A range of other products are supplied to the mining industry including wear linings, sieve bend screens, chute linings, rollers, bucket wheel excavator bucket linings, slurry pipe elbows and various other polyurethane mouldings.

The company also supplies a range of polyurethane mouldings to general industry in Australia. Customers include Holden, Mitsubishi, Clyde Apac, Kempe and Scholle Industries. Products vary from large conveyor head drum rollers and liners for bucket wheel excavators through to agricultural industry components and suction cups for robotic assembly applications.

Experienced in developing solutions in a wide range of industries including wine industry, timber industry, automotive, materials handling, bottling lines—the opportunities are endless.

The company also offers a range of engineering plastics including Polyethylene (HDPE & UHMWPE), Nylon, Acetal, Polypropylene, Polycarbonate, Acrylic, PETP and PTFE.

Tooling

Advanced Polymer Products owns a significant amount of tooling developed over 20 years. Cast polyurethanes as used by Advanced Polymer Products do not require the use of expensive tooling like that required for injection or compression moulding processes.

Advanced Polymer Products is able to design and manufacture all tooling in-house or in close consultation with local toolmakers and fabricators. We are also happy to provide advice on tooling design should customers wish to manufacture the tooling themselves.

History

Advanced Polymer Products commenced business in its own right on 1st March 1991. Prior to this date the business operated as a division of another company for about eight years, and the founding technology was procured from a German company, in the early 1980's.

In 1991 the division was sold and the business rebadged as Advanced Polymer Products.

The business was a three-person operation in 1991, and was located in Mile End, Adelaide. In 1994, the business was relocated to Regency Park, Adelaide. In 2005 the business acquired Hicks Trading and integrated that operation into the current business. It now employs 18 people and services a diversified range of industries across Australia with export to a number of countries.

4.6 Hot Cast Polyurethanes

Polyurethanes play a vital role in many industries- from shipbuilding to footwear, construction to cars.

The biggest demand for polyurethane products is in the form of foams and these constitute some 90per cent by weight of the total market for polyurethanes.

Advanced Polymer Products specialises in solid polyurethane elastomers, tailored to meet the needs of specific applications. In mining applications where abrasion resistance, impact and tear strength are so vitally important, the integrity of the polyurethane is critical. The name polyurethane has been used to describe a great number of polymers. In order to simplify the detail of polyurethanes we will provide a background to the principle components and describe the formation of the polyurethane molecule.

A Polyurethane Molecule is Comprised of Three Principle Components

- ❖ **A Resin** This forms the backbone of the polyurethane molecule and is more correctly referred to as the polyol. APP uses three main types of polyol- polyesters, polyethers and polycaprolactones. An Isocyanate—in our case more specifically a diisocyanate. APP predominately uses various Toluene Diisocyanates (TDI) and Methylene Diisocyanates (MDI).
- ❖ The resin or polyol and the diisocyanate are chemically reacted together to

form a prepolymer. These two components can be greatly varied to give maximisation of different properties to the polyurethane molecule and the final polyurethane product.

- **A Curative** This is the final ingredient in the process and in chemical terms is known more correctly as a chain extender. This is used in a further chemical reaction to convert the liquid prepolymer into a solid elastomeric product. This process is also called crosslinking and it is the binding together of the various atoms to form the polyurethane molecule. Again different curatives give different properties to the final polyurethane product because of the way they crosslink the various atoms in the polyurethane molecule. The curative is a critical factor in a high performance polyurethane.

Cast Polyurethane Elastomers (CPU)

As prefaced previously, cast polyurethane elastomers are made by mixing a reactive liquid prepolymer with a curative and pouring the mix into an open mould. There are a large number and variety of cast polyurethanes with a wide range of physical properties and cost. We will limit this detail to the high performance polyurethanes used by APP in mining products.

The best known and most commonly used high performance CPU are Polyether/TDI prepolymers reacted with a curative 3,3'-dichloro-4,4'-diaminodiphenylmethane which is commonly known as MOCA. This curative or diamine chain extender gives the highest physical crosslink density to the final poyurethane molecule of any curative on the market today.

The above mixture is degassed and cast into moulds and this process is undertaken at a temperature of about 100 C. In order to ensure the highest performing product the casting and curing conditions are very important. All of our products are cast under critical temperature control and after an initial mould cure there is a final post cure to ensure that the chain extender fully cross links and ensures optimum properties in the finished product.

Injection Moulded Thermoplastic Polyurethane Elastomers (TPU)

Thermoplastic polyurethane elastomers are supplied as granules or pellets for processing by techniques such as injection moulding or extrusion. TPU's are generally classified by their hardness and are supplied by a number of major chemical companies.

As such the processor accepts what he is supplied and cannot vary the chemistry of the product.

TPU is usually made from a MDI (diioscyanate) reacted with a polyether or polyester and also a chain extender usually 1,4-butanediol. These ingredients are set in the supplied pellets.

In the injection moulding process the components are constituted in the pellet and they are then melted together and injected into a mould. This process takes around five minutes. The TPU is constituted so that it can be melted and reformed. In other words it has a different molecular structure to CPU and does not have the crosslink molecular strength that a CPU has. The injection moulding process also limits product applications of TPU.

The Difference?

Well its all in the formation of the polyurethane molecule and how it is accomplished.

Firstly, the ingredients in the prepolymers used by APP are varied to provide the best properties in the polyurethane that forms the finished product. They are not an off the shelf generic product. The science of this variation is called stoichometry and APP's technical staff have developed formulae variations to fit a range of applications. The use of polyols such as polycaprolactones provides a much wider range of application, life and performance.

Secondly, the curative or chain extender used by APP which is MOCA provides a crosslinked polyurethane molecule that is superior to any other in either a cast or thermoplastic process. MOCA is not available as a chain extender in thermoplastic systems.

Thirdly, the hot cast process provides a long slow cure and post cure that enhances the crosslinking of the polyurethane molecule and ensures that it is complete and that there is no unreacted components. This cannot be accomplished in the relatively short processing time in injection moulding.

The Result?

Advanced Polymer Products cast polyurethane products are manufactured from only the best materials available using techniques that ensure both quality and technical superiority. Head to head trials have regularly demonstrated that a cast polyurethane will outperform an injection moulded polyurethane in a wide range of mining applications—from Granny Smith Gold Mine, Ok Tedi, Boral Quarrying Operations and Hanson, the proof from the field categorically supports the chemistry of cast polyurethane technology.

4.7 Mining

The mining, quarrying and mineral processing industries have always demanded the best in wear and impact resistant materials.

In conjunction with innovative and flexible design, polyurethane materials have been able to provide tailored solutions for problem areas within the mining industry for many years.

Why be forced to conform to your supplier's existing product line?

Products commonly supplied but not limited to this industry include:

- Screen cloths
- Screen panels
- Trommel panels
- Screen deck module systems
- Sieve bend screens
- Dewatering screens
- DSM Screens
- Clamp bars
- Side tension bars
- Wear liners
- Bin liners
- Lip liners
- Hopper liners
- Impact pads
- Elbows
- Cyclone liners
- Rollers
- Scraper blades
- Classifier flights

4.8 Pipeline Pigs

Advanced Polymer Products provides a complete range of Pipeline Pigs.

Pipeline Pigs are used in pipelines throughout all industries for a variety of reasons such as:

- Pipeline Cleaning
- Pipeline Clearing and Rehabilitation
- Removing debris from pipelines
- Product Separation
- Pre-Inspection Cleaning
- Applying internal coating and inhibitors
- Removal of rust, scale and other internal formations
- Commissioning of pipelines
- Testing of pipelines
- Filling and Dewatering of pipelines
- Decommissioning of pipelines
- Drying of pipelines

APP can advise on the selection of pipeline pigs to suit most applications. This phase of an intended pigging operation is very important as choosing the wrong pig can not only affect the success of the pigging operation, but may also cause damage to the pipeline, the pig and could seriously affect the ongoing operation and profitability of the pipeline in service.

4.9 Light Density Foam Pigs

This type of pig is usually manufactured from an open cell polyurethane foam material with an average density of approximately 2 lbs./cu.ft. Also known as "Swabs" these pigs are typically used in the following applications:

Proving

When cleaning or clearing pipelines these pigs are often used as a first-pass" pig to prove that the line is piggable and also to check for any major obstructions or restrictions in the pipeline. If the pig encounters a major restriction in the pipeline it may be able to deform itself around the object and/or it will disintegrate in the line.

Light Duty Pigging

Swabs are used for light duty Pigging applications where the customer is seeking to either protect the internal coating of a pipeline and/or leave an internal layer or coating on the inner wall of the pipeline to help prevent against breakage or leakage. Pigging of DICL and AC pipelines in the water industry often use this type of pig.

Drying

Light density foam pigs offer good absorption qualities due to the open cell foam structure and are used by pipeline construction contractors to dry pipelines after they have been hydrostatically tested.

4.10 Medium Density Foam Pigs

This type of pig is also manufactured from an open cell polyurethane foam material with an average density of approximately 5-7 lbs./cu.ft. These pigs are manufactured with a range of coatings and/or attachments such as:

- Plain Foam
- Polyurethane Coated Foam
- Polyurethane Coated Foam with Jetting Nozzles
- Polyurethane Coated Foam with Carbide Grain Coating
- Fitted with Rope Loops for easy handling
- Fitted with Cavity for housing Pig Tracking and Locating unit

Medium density foam pigs are typically used in the following applications:

Pipeline Cleaning or Clearing

The stronger more durable medium density foam offers effective cleaning power as the pig does not deform as much as Swab" type pigs. Generally this type of pig is suitable for cleaning or clearing pipelines up to 40 Kms in length, however, longer runs can be achieved in internally lined pipelines, HDPE, PVC, FRP, GRP or plastic pipelines.

Use the PU coated foam accessory type pigs for enhanced cleaning or clearing power over the standard pig. Note, the use of wire brush type pigs or carbide grain coated pigs in HDPE, PVC, FRP, GRP or internally lined pipelines is generally not recommended.

In pipelines where there is a significant build up of debris it is recommended that progressive pigging be utilized. In pipelines where there is a likelihood of debris becoming compacted ahead of the pig, we recommend the use of Jetting Pigs such as the RX4-JN to keep the debris in suspension ahead of the pig as it traverses the pipeline.

Filling, Dewatering or Drying

Medium density foam pigs are used for filling and dewatering or specialised drying of Pipelines. They also offer good absorption qualities due to the open cell foam structure and are used by pipeline construction contractors to dry pipelines after they have been hydrostatically tested.

4.11 High Density Foam Pigs

This type of pig is also manufactured from an open cell polyurethane foam material with a higher average density of approximately 8-10 lbs./cu.ft. These pigs are manufactured with a range of coatings and/or attachments such as:

- Plain Foam
- Polyurethane Coated Foam
- Polyurethane Foam with Jetting Nozzles
- Polyurethane Foam with Wire Brush
- Polyurethane Foam with Carbide Grain Coating
- Heavy Duty Wire Brush Pig
- Polyurethane Coated Foam with Plastic Bristles
- Polyurethane Foam with High Tensile Steel Studs
- Fitted with Rope Loops for easy handling
- Fitted with Cavity for housing Pig Tracking and Locating unit

High density foam pigs are typically used in the following applications:

Pipeline Cleaning or Clearing

Generally this type of pig is suitable for pipelines above 40Kms in length or, where effective cleaning or clearing is required.

Use the Polyurethane Coated type pigs for enhanced cleaning or clearing power over the standard pig. Note, the use of wire brush type pigs or carbide grain coated pigs in HDPE, PVC, FRP, GRP or internally lined pipelines is generally not recommended. The Plastic Brush Pig is well suited to cleaning pipelines manufactured from sensitive materials and/or internally lined pipelines.

In pipelines where there is a significant build up of debris it is recommended that progressive pigging be utilized. In pipelines where there is a likelihood of debris becoming compacted ahead of the pig, we recommend the use of Jetting Pigs to keep the debris in suspension ahead of the pig as it traverses the pipeline.

Filling, Dewatering or Drying

High density foam pigs are used for filling and dewatering pipelines, they also offer good absorption qualities due to the open cell foam structure and are used by pipeline construction contractors to dry pipelines after they have been hydrostatically tested.

4.12 Cup Pigs

Advanced Polymer Products provides cup pigs in many configurations. These pigs are produced from a number of components and are usually screwed or bolted together. The individual components are replaceable allowing the pigs to be fully maintained and therefore more economical for ongoing pigging operations than solid, foam pigs or pipeline spheres.

Cup pigs are used in a variety of applications such as:

Batching/Product Separation

To allow two or more different products or batches to be pumped through the same pipeline with minimal cross contamination. This negates the need for dedicated pipelines and saves significantly on capital expenditure and ongoing pipeline maintenance.

Gauging

Cup pigs are often used in the construction of pipelines to gauge the internal bore of the pipe after it has been laid. The pig is fitted with one or more Aluminium or Steel plates which are normally sized to approximately 95per cent of the internal diameter. This process is designed to confirm that the contractor has installed the pipeline free from defects and also to confirm that the internal bore of the pipeline is free of dents, buckles or other possible deformations which may affect its performance once commissioned.

Cleaning

When fitted with circular or spring loaded wire brushes, cup pigs can provide an effective cleaning tool for most pipelines. Plough Blades can also be fitted for wax removal in oil pipelines. Jetting nozzles prevent compacting of debris such as sand, pipe scale and slurries. Double length cup pigs are fitted with a universal joint for enhanced cleaning power, whilst still retaining the ability to negotiate tight radius bends.

Removal of Ferris Debris

Magnetic cup pigs are used for removal of ferris debris such as pipe scale, used welding rods and other metallic objects.

Multi-Diameter Pigging

Cup pigs can negotiate changes in pipeline diameters of up to 15per cent for Standard Cup pigs, and up to 25 per cent for Conical Cup type pigs. In addition special cup designs can also be supplied to met specific applications.

4.13 Bi-Directional Disc Pigs

APP provides bi-directional disc pigs.

Like cup pigs, they are produced from a number of components and are usually screwed or bolted together. The individual components are replaceable allowing the pigs to be fully maintained and therefore more economical for ongoing pigging operations than other types such as solid polyurethane, foam pigs or pipeline spheres. When properly sized and configured for a particular application, bi-directional disc pigs will in most cases offer more efficient cleaning, product separation and clearing of pipelines due to their ability to maintain a more effective seal against the internal bore.

In addition bi-directional pigs offer the advantage of being able to traverse a pipeline in either direction. This feature is particular effective in pipeline construction when filling and dewatering a pipeline during hydrostatic testing. For applications up to 20 Kilometres in length we recommend the Six (6) Disc Bi-Directional type pig and for applications above 20 Kilometres we recommend the Eight (8) Disc Bi-Directional type pig which provides increased stability via the extra Guide/Scraper discs.

Bi-directional disc pigs are used in a variety of applications such as:

Batching/Product Separation

To allow two or more different products or batches to be pumped through the same pipeline with minimal cross contamination. This negates the need for dedicated pipelines and saves significantly on capital expenditure and ongoing pipeline maintenance.

Gauging

Bi-directional pigs are often used in the construction of pipelines to gauge the internal bore of the pipe after it has been laid. The pig is fitted with one or more Aluminium or Steel plates which are normally sized to approximately 95 per cent of the internal diameter. This process is designed to confirm that the contractor has installed the pipeline free from defects and also to confirm that the internal bore of the pipeline is free of dents, buckles or other possible deformations which may affect its performance once commissioned.

Cleaning

When fitted with circular or spring loaded wire brushes, bi-directional pigs are an effective cleaning tool for most pipelines. Plough Blades can also be fitted for wax removal in oil pipelines. Jetting nozzles will prevent compacting of debris in front of the pig, such as sand, pipe scale and slurries.

Removal of Ferris Debris

Magnetic bi-directional pigs are used for removal of ferris debris such as pipe scale, used welding rods and other metallic objects.

Multi-Diameter Pigging

When fitted with special Guide/Scraper and Sealing discs bi-directional disc pigs can negotiate changes in pipeline diameters of up 25 per cent. In addition special designs can also be supplied to accommodate specific applications.

4.14 Social Purpose Pigs

APP can also provide special purpose pigs

- ❖ Solid Sphere
- ❖ Inflatable Spheres
- ❖ Scraper Disc/Cup Pigs
- ❖ High Temperature Pigs
- ❖ Multi Diameter Scraper Pigs
- ❖ High Efficiency Cleaning Pigs
- ❖ Solid Polyurethane or Rubber Pigs
- ❖ Food Grade Polyurethane Pigs
- ❖ Descaling/Decoking Pigs
- ❖ Combination Pigs
- ❖ Multi Cup Pigs
- ❖ By-Pass Pigs
- ❖ Hi-Seal Pigs
- ❖ Stud Pigs

4.15 Coil Pllets

Advanced Polymer Products has worked with Holden, Bluescope and Stratco to develop a polyurethane coil pallet that offers the following benefits:

- ❖ Reduce damage to coils
- ❖ Minimise wastage
- ❖ Improve handling and storage
- ❖ Suits a wide range of coil sizes
- ❖ Made from high performance polyurethane
- ❖ Overall size 890 x 455x50
- ❖ Lightweight and easy to handle (as light as 10 kg)

4.16 Engineered Plastic

Advanced Polymer Products is able to offer an extensive range of Engineering Plastics for wear resistant and abrasion applications as stock rods, sheets etc and also provide finish machined or cast components.

Materials Include

Nylons, Acetals, Acrylics, P.E.T.P. (Ertalyte), P.V.C., P.T.F.E. (Teflon), Polyethylene (UHMWPE, HDPE), Austlon Nylons, Polycarbonate, Polypropylene.

We can also source other Engineering Plastic materials not included above.

Material	*General Properties*
Nylon	● Excellent bearing properties ● High impact strength and toughness ● Low power factor requirements ● High wear resistance ● Very high sliding properties
Acetal	● High hardness and stiffness ● Very high mechanical properties ● Excellent creep resistance and dimensional stability ● Excellent dielectrical properties ● Low co-efficient of friction
Acrylic	● Excellent optical clarity ● 10-17 times greater impact than glass ● Excellent weatherability ● Easily fabricated and glued ● High surface hardness
Polyester P.E.T.P. (Ertalyte)	● Similar properties to Acetal with improved tensile strength, dimensional stability, operating temperature and physiologically inert, reduced notch impact resistance.
P.V.C.	● Excellent electrical insulating properties ● Very high chemical resistance ● Thermo formable ● Moderate impact resistance and service temperature ● Very good moisture resistance ● Good dimensional stability ● Bondable ● Self extinguishing

(*Contd.*)

Material	*General Properties*
P.T.F.E. (Teflon)	● Excellent chemical resistance ● Lowest co-efficient of friction ● High operating temperature stability ● Physiologically inert (virgin) ● Excellent electrical properties
Polyethylene HDPE	● Very low co-efficient of friction ● Very high surface release properties ● High chemical resistance ● Excellent impact resistance ● Very good damping properties
Polyethylene UHMWPE	● Similar properties to HDPE with excellent impact strength, wear and abrasion resistance.
Polycarbonate	● Excellent impact resistance (250 times stronger than glass) ● Very good optical properties ● Ability to be cold formed in thinner gauges ● Moderate chemical and scratch resistance ● Self extinguishing ● Excellent acoustic properties
Polypropylene	● Very high chemical resistance ● Excellent impact resistance ● Higher scratch resistance than HDPE ● Thermoformable ● Excellent moisture resistance ● Food grades

4.17 Polyurethne Foam

Advanced Polymer Products has been appointed as the South Australian and Northern Territory distributor for the Fomo range of one- and two-component polyurethane (PU) foam systems, adhesives and sealants in pressurised and disposable packaging.

HANDI-FOAM One Component Foam

One component Handi-Foam products come in various sizes and packaging.

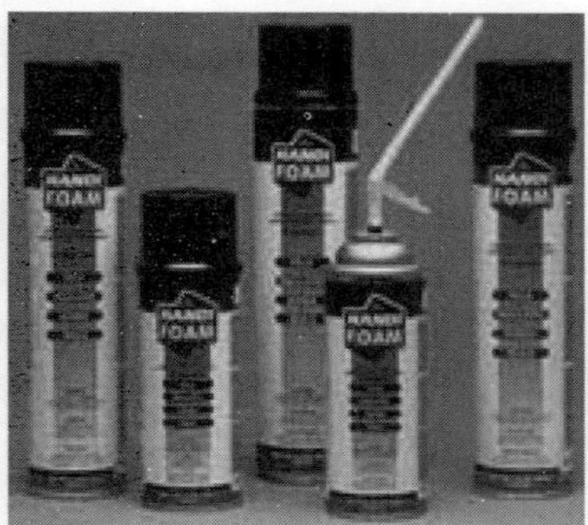

These products are easily dispensed and have various application.

HANDI-FOAM Two Component Foam

Two component foams come in a dispensable system as pictured

These products are applied with a dispensing gun.

If you are interested in the Handi-Foam two component dispensable system, please view the Technical Datasheet.

HANDI-STICK Adhesives

Handi-Stick adhesive is available as a multi-purpose construction adhesive and a subfloor adhesive.

Pictured is the subfloor adhesive being applied.

HANDI-SEAL Window and Door Sealants

Handi-Seal window and door sealant is available in a one-component can as pictured.

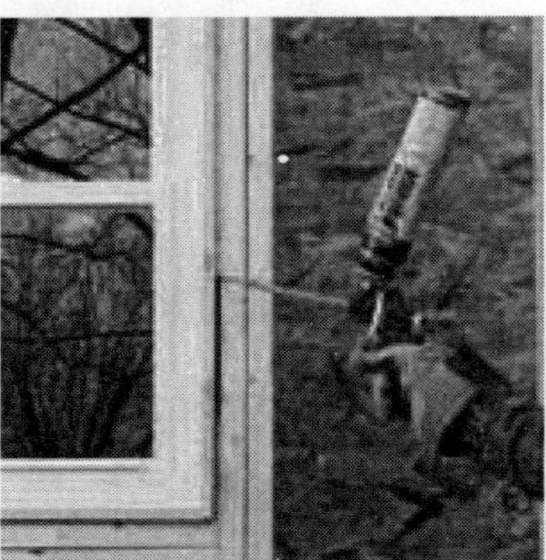

4.18 Others

Advanced Polymer Products is able to design and manufacture polyurethane products specific to your requirements. With the large range of polyurethane materials available, a design and material specification can be put together to best suit your application.

The polyurethane production method enables the use of inexpensive tooling whereas injection moulding methods often necessitate a substantial investment in tooling. Large polyurethane parts can be produced in a premium material without the need to spend tens of thousands of dollars on a mould.

We also have a selection of stock Red 90 shore A polyurethane sheet and rod available either off the shelf or within one or two days. Other material hardnesses and colours are also available on request.

Rod

Polyurethane rods are available from Ø12mm to Ø200mm at 300mm lengths and from Ø12mm to Ø125mm at 1000mm lengths.

Sheet

Spun cast polyurethane sheets are available from 1.5mm to 12mm in thickness at 600mm x 3000mm sheet sizes.

Flat cast sheets are available from 5mm in thickness and our standard sheet size is 1000mm x 1000mm. Flat cast sheets can be cast at larger sizes upon request.

4.19 Trulok

Advanced Polymer Products has developed a clip on module system for use on screen decks that includes the following features;

- Unique design featuring both rail and locking strip integral fixing
- Locking system ensures modules stay in place by eliminating stress on the fixing rail
- Easy replacement of screen modules
- System is adaptable to suit various machines
- Supplied as a fully cast panel in a range of materials and hardnesses
- Australian designed and manufactured

The design of the system allows a certain level of design flexibility to enable us to engineer a solution that fits your application.

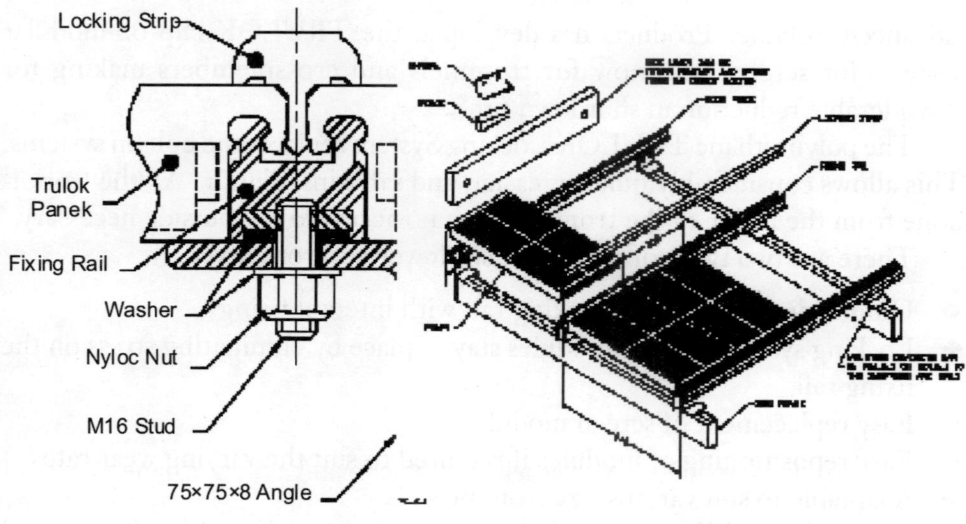

Trulok fixing detail **Trulok deck assembly detail**

Installation of Trulok System

Complete Trulok Deck

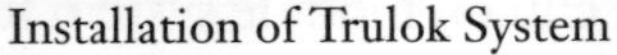

4.20 TRULOK Trommel Panel system

Advanced Polymer Products has developed the TRULOK clip-on modular systems for screens and now for trommels and crossmembers making for considerable reduction in shutdown time.

The polyurethane TRULOK Locking System replaces the bolt on systems. This allows considerably quicker, easier, and safer installation. All the work is done from the inside of the trommel with no access to the outside necessary.

There are two types of modules, flat (lower cost) or curved:

- ❖ Unique design featuring locking rail with integral fixing
- ❖ Locking system ensures modules stay in place by eliminating stress on the fixing rail
- ❖ Easy replacement of screen modules
- ❖ Easy repositioning of modules if required to suit the varying wear rates
- ❖ Adaptable to suit various size trommels
- ❖ Supplied as a fully cast panel in a range of materials and hardnesses and aperture sizes
- ❖ Options can be fitted-Impact areas, Lip Liners, Flights, Lifter Bars, Deflectors
- ❖ Australian designed and manufactured

The design of the system allows a certain level of design flexibility to enable a solution to be engineered for your application.

Installation

The engineering staff at Advanced Polymer Products work with the customer to design the fixing rails and panels to suit the trommel frame (existing or new).

Once these fixing rails are all bolted in place, starting from one end, simply drive the TRULOK Trommel Panels onto the TRULOK Fixing Rails using a rubber mallet. This results in a good positive lock that holds the panels in place.

To remove the panels, use a lever between the panels, starting at the corner, lift the panel off of the TRULOK fixing rail. The fixing rail does not normally need replacing as the panels protect it.

The Century Zinc in Queensland has reduced the shut-down time to change the trommel panels by two thirds. What used to take three shifts now takes only one.

4.24 TRULOK Crossmember Wear Liner

Advanced Polymer Products has developed the TRULOK clip-on modular systems for screens and trommels now also crossmembers making for considerable reduction in shutdown time.

Designed to protect crossmembers on screening machines, the TRULOK Wearliners are made in parts that clip together forming a positive lock and seal.

The polyurethane TRULOK Locking System simplifies the installation by removing the problems with adhering rubber onto the crossmembers or the need for bolt on type liners.

- ❖ Unique design integral positive locking fixing.
- ❖ Easy to fit, easy to handle.
- ❖ Available 3000mm long and in a range of materials to suit specific applications
- ❖ Customers can specify the specific length required to minimize the time spent on site.
- ❖ Standard Sizes suit 50x50 up to 350x350mm SHS and RHS.
- ❖ Other sizes and shapes, including CHS, can be manufactured as required
- ❖ Australian designed and manufactured

The design of the system allows a certain level of design flexibility to enable a solution to be engineered for your application.

Trommel Panel

Advanced Polymer Products is a specialist manufacturer of polyurethane trommel panels used within the mining industry.

From small light duty frames with two panel sections to over Ø3m x 5m long heavy duty frames with over one hundred panels. Every trommel frame is unique and we are able to design a trommel panel to suit.

In most cases, we are either able to design a trommel panel to suit the existing fixing methodology or provide an alternate simple locking pin or fixing rail system to allow quick and easy change out of panels.

We have a large range of existing aperture sizes available and new sizes are easily created where required.

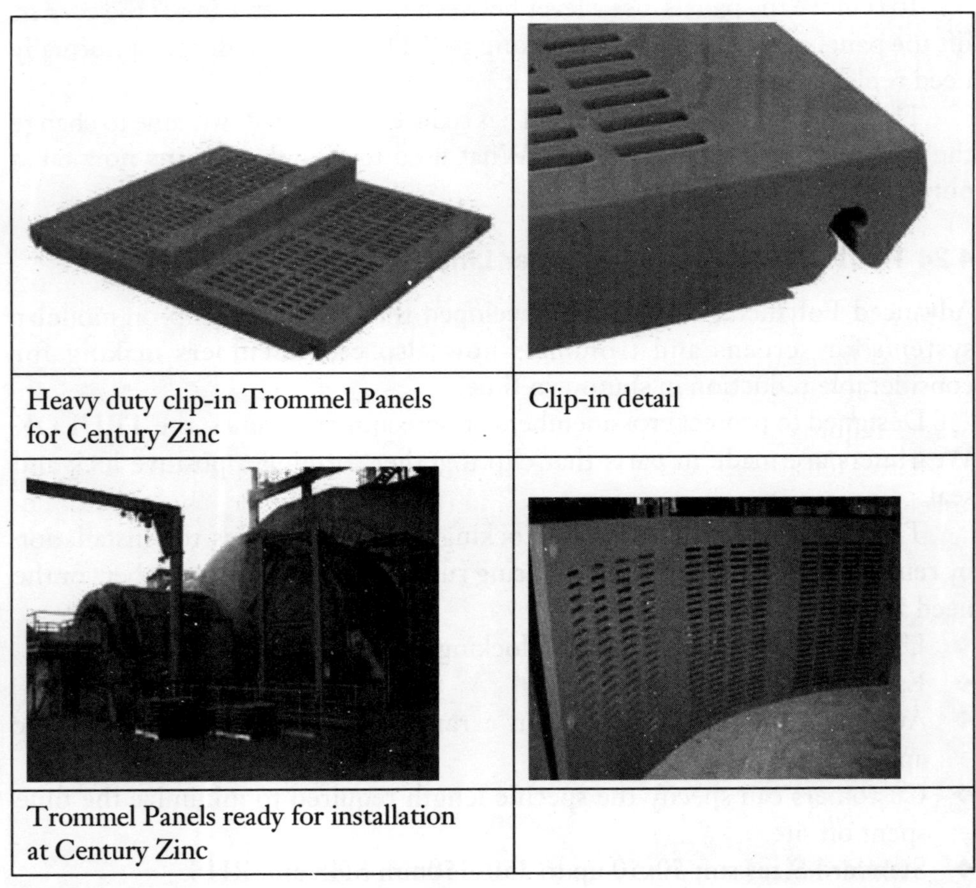

Heavy duty clip-in Trommel Panels for Century Zinc	Clip-in detail
Trommel Panels ready for installation at Century Zinc	

4.23 Screen Cloths

Advanced Polymer Products custom make a wide range of polyurethane screencloths. Each cloth is manufactured specifically for the customer's screen and specific requirements. There are a wide range of apertures available and our engineers are expert in designing a screening solution for any application.

The advantages of polyurethane screens are many ...

Cut and Tear Resistant: Far superior to rubber and steel. It has excellent resistance to sliding and impact abrasion.

Tough but Resilient: It absorbs the impact energy of large falling rocks—its resilience acts like a cushion to protect lined parts from breaking apart or being damaged on impact.

Low Coefficient of Friction: When compared to rubber—extremely slippery when wet.

Temperature Range: Excellent from -40°C to +80°C.

High Load Bearing Capabilities: Far superior to rubber, metals and other plastics in its ability to absorb Rnergy, sound and vibration.

Chemical Resistance: Generally resistant to a wide range of solvents, oils and chemicals.

Hydrolytic Stability: Formulated to resist water absorption without losing its performance properties.

Noise Abatement: Excellent to suppress sound.

Light Weight: When compared to steel.

Screening Performances: Comparable to steel wire meshes—reduced clogging, reduced blinding and excellent aggregate/ore cleaning.

Easy Installation: Using standard cross tension fixing systems.

We are also able to supply punched rubber cross tensioned screen cloths where required.

Cross tensioned screen cloths a generally installed according to the screening machine construction. Wire rope within the cloths keep the cloths tensioned with elongation of 0.5-1 per cent. We provide 5mm mild steel or 3mm stainless steel hooks for fastening with tension plates. Other fastening methods include wedge fastening, screw fastening and plug fastening.

For a quick price, please supply

- Over hook width of existing cloth or width of screen deck (inside side plates)
- Length of existing cloths or overall length of screen deck Required aperture size or required particle pass size and incline of deck *(Thickness of screen is based on aperture size)*

Additional information required to manufacture

- Stringer bar centres
- Centre clamp bar width, bolt size and centres

Rubber cross tensioned screen cloth Rubber cross tensioned screen cloth

4.24 Screen Panels

Similar to trommel panels, screens panels are manufactured with an internal steel frame to your specified size and can be supplied flat or rolled (DSM Screen).

Screen panels can be manufactured using our existing aperture tooling or with injection moulded aperture elements for fine apertures (below 1mm slot or 5mm square).

The fixing method used can be to suit your equipment be it bolt-on, hold down pin, clamp down or clip-on.

Boral Emu screen deck

Fine grinding mill discharge panels

4.25 Wear Liners

Wear liners can be manufactured in just about any configuration to suit your application. Generally we cast the polyurethane to a 4MS backing plate to provide a rigid liner plate that won't allow a build up of product behind it.

We can also provide our own Polycap bolts which incorporate a Ø40 polyurethane head that sits flush with the top face of the liner plate with a M10 cup square bolt to enable ease of fitment. If provided with hole centre details, we can produce the wear liners with corresponding square holes through the steel and Ø40 c/bores through the poly to simplify the installation process.

Alternatively, we can manufacture wear liners with any hole detail you require or blank so that they can be drilled on site.

Below is an example of our Wearblazer style of product that enables the wear of the liner to be easily monitored. Handles can be added to allow ease of handling and are trimmed off after the panel is installed.

Wearblazer liner plate

4.26 Others

Advanced Polymer Products isn't restricted to the manufacture of screens for the mining industry. We are also able to manufacture any product that may be required on site.

The flexibility of the materials we use and the production method combined with the relative inexpensive tooling methodology means that we can often provide a cost competitive solution.

Granulator outlet weir sections **Trough liners**

5

Polymer Processing, Testing and Analysis

5.1 Polymer Processing Operation

Extrusion

- ❖ Single Screw Extrusion
- ❖ Twin Screw Extrusion

Fiber Spinning

- ❖ Dry Spinning
- ❖ Melt Spinning
- ❖ Wet Spinning

Filament Winding

Film Blowing

Injection Molding

Mixing and Compounding

- ❖ Batch Intensive Mixing
- ❖ Twin Screw Extrusion
- ❖ Pultrusion
- ❖ Reaction Injection Molding
- ❖ Spin Coating
- ❖ Transfer Molding

5.2 Single Screw Extrusion

Single screw extrusion is one of the core operations in polymer processing and is also a key component in many other processing operations. The foremost goal of a single screw extrusion process is to build pressure in a polymer melt so that it can be extruded through a die or injected into a mold. Most machines are plasticating: they bring in solids in pellet or powder form and melt them as well as building pressure.

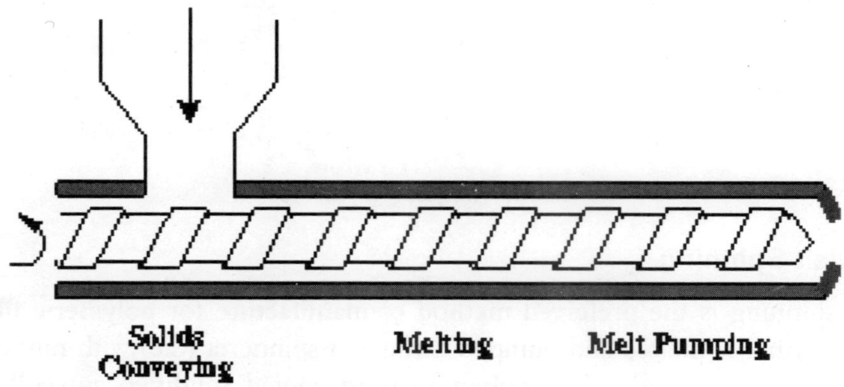

Suppliers of Single Screw Extruders

- ❖ Battenfeld
- ❖ Cincinnati Milacron
- ❖ Davis-Standard
- ❖ Randcastle
- ❖ Wayne Machine and Die Company

5.3 Dry Spinning

Dry spinning is used to form polymeric fibers from solution. The polymer is dissolved in a volatile solvent and the solution is pumped through a spinneret (die) with numerous holes (one to thousands). As the fibers exit the spinneret, air is used to evaporate the solvent so that the fibers solidify and can be collected on a take-up wheel. Stretching of the fibers provides for orientation of the polymer chains along the fiber axis. Cellulose acetate (acetone solvent) is an example of a polymer which is dry spun commercially in large volumes. Due to safety and environmental concerns associated with solvent handling this technique is used only for polymers which cannot be melt spun.

Process Schematic

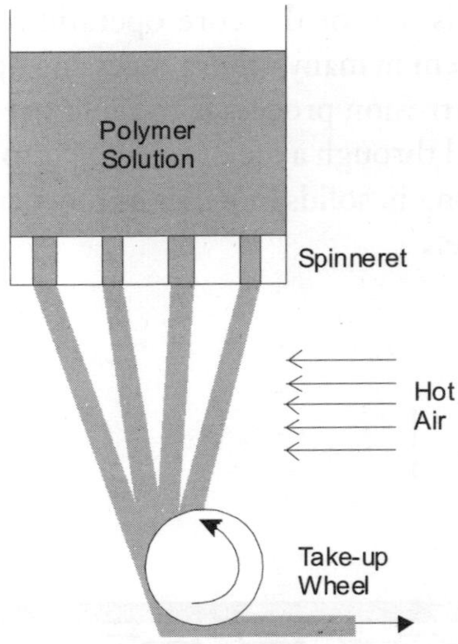

5.4 Melt Spinning

Melt spinning is the preferred method of manufacture for polymeric fibers. The polymer is melted and pumped through a spinneret (die) with numerous holes (one to thousands). The molten fibers are cooled, solidified, and collected on a take-up wheel. Stretching of the fibers in both the molten and solid states provides for orientation of the polymer chains along the fiber axis. Polymers such as poly (ethylene terephthalate) and nylon 6,6 are melt spun in high volumes.

Process Schematic

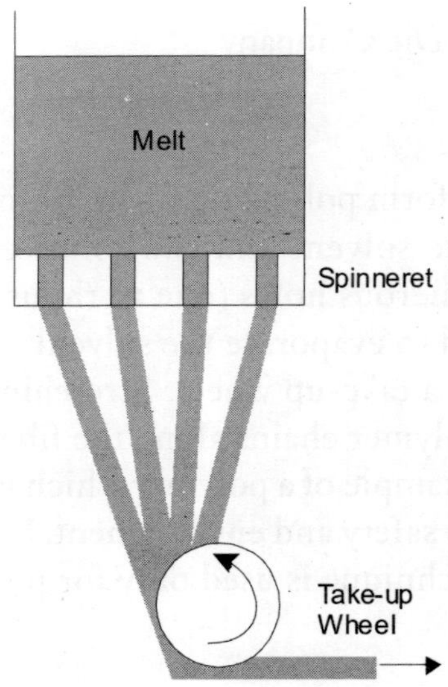

5.5 Filament Winding

Filament winding is used for the manufacture of parts with high fiber volume fractions and controlled fiber orientation. Fiber tows are immersed in a resin bath where they are coated with low or medium molecular weight reactants. The impregnated tows are then literally wound around a mandrel (mold core) in a controlled pattern to form the shape of the part. After winding, the resin is then cured, typically using heat. The mold core may be removed or may be left as an integral component of the part.

Process Schematic

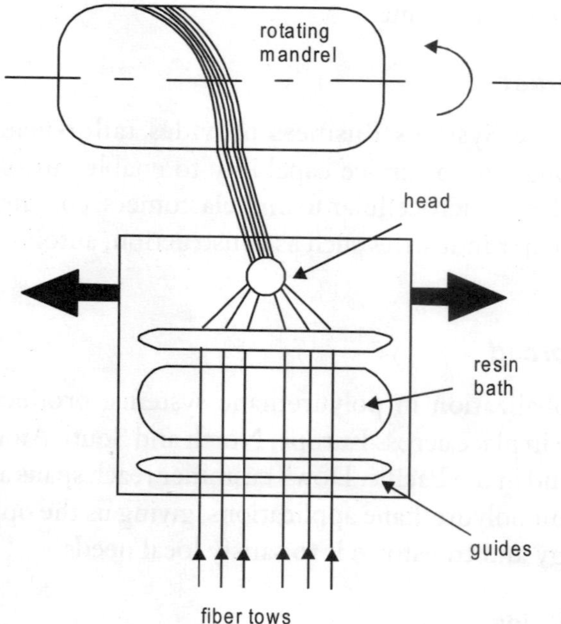

5.6 Dow's Polyurethane Systems

Dow's Polyurethane Systems business, with worldwide headquarters in Correggio, Italy and North American headquarters in Marietta, Georgia, manufactures and markets custom formulated rigid and semi-rigid, flexible, integral skin, and microcellular polyurethane foams and systems as well as elastomers, coatings, adhesives and binders. These products are used in applications ranging from appliances, furniture and shoes to decorative molding, boats and athletic equipment. Dow also offers state-of-the-art, patented foam dispensing equipment and related software.

Through Polyurethane Systems, customers can purchase foams in a variety of sizes, from a 16 oz. (150 ml) can all the way up to a truckload.

Committed to the principles of Sustainable Development, Dow and its 50,000 employees seek to balance economic, environmental and social responsibilities.

For further information, please select a topic from the list below. Also, please visit Dow Polyurethanes.

- Channel to Market
- Geographic Spread
- Global Capabilities
- Quality System
- Delivery System
- Environmental Commitment

Channel to Market

Dow Polyurethane Systems Business provides tailor-made polyurethane formulations along with a service capability to enable customers to produce the flexible, rigid and microcellular foams, elastomers, coatings or binders that become part of major industries such as construction, automotive, footwear or furniture.

Geographic Spread

With Dow's globalization in polyurethane systems, production and service facilities now are in place across Europe, North and South America, the Middle East and Africa and in the Pacific. Dow's customer reach spans all the geographic areas and different polyurethane applications, giving us the opportunity to take global technology and transform it to satisfy local needs.

Global Capabilities

Our capability to provide local service is enhanced by development centers throughout the world that are interconnected through a state-of-the-art information technology network. Opportunities can be rapidly developed in any part of the world and benefit from the deep experience and know-how accomplished in our global network.

Our objective is to provide a fast response time to customers through a rapid flow of information to service the market needs through local development laboratories and versatile production plants.

All systems are developed, produced and supplied to meet present or future market requirements and are tailored according to the specific needs of each single customer.

Quality System

All functions, from production to research, from technical-commercial functions to logistics, work together in a synchronized fashion to achieve complete customer satisfaction.

All different phases of the operating process, including systems design, production, quality control, shipment and customer service, are based on a total quality system, certified according to international quality standards, assuring consistent high level quality products and services.

Delivery System

Through order management and material supply, we provide on time delivery to our customers. The formulated polyurethane systems are delivered in different packaging according to every specific need: from traditional drums to returnable intermediate bulk containers that eliminate disposal problems. Additionally bulk supply with tanks is chosen for those customers equipped with the adequate storage facilities.

Environmental Commitment

Dow engineers cooperate with customers in a continuous improvement process to identify and accomplish new opportunities: thus our job continues after the delivery of the product.

Dow's accountability and commitment are not only addressed to the market, but also to environment, health and safety. Therefore Dow has joined the Responsible Care® program and is strongly engaged in the development of products and processes that during their life cycle have reduced environmental impact and can consequently be successful in the future.

More than 35 years of experience in the polyurethanes industry have made Dow a dynamic business organization capable of developing, in conjunction with our customers, tailor-made solutions providing a high quality service for success today and into the future.

5.7 Film Blowing

The majority of polymer films are manufactured by film blowing (blown film extrusion). A single screw extruder is used to melt the polymer and pump it into a tubular die, as shown in cross-section at right. Air is blown into the center of the extruded tube and causes it to expand in the radial direction. Extension of the melt in both the radial and down-stream direction stops at the freeze line (frost line) due to crystallization of the melt. The nip rolls collect the film, as well as sealing the top of the bubble to maintain the air pressure inside. This process is used extensively with polyethylene and polypropylene.

Process Schematic

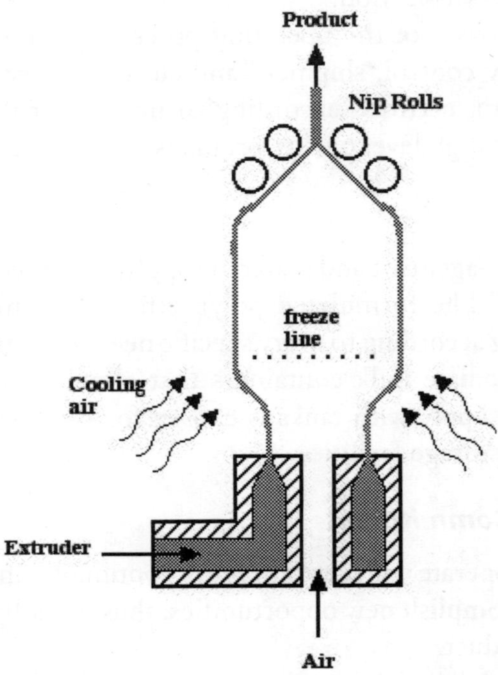

5.8 Injection Molding

Injection molding is used extensively for the manufacture of polymeric items. A reciprocating/rotating screw both melts polymer pellets and provides the pressure required to quickly inject the melt into a cold mold. The polymer cools in the mold and the part is ejected. Injection molding machines are sized primarily by the force available in the mold clamping unit; ranging from 20 tons in laboratory machines to over 5000 tons in large commercial machines.

Process Schematic

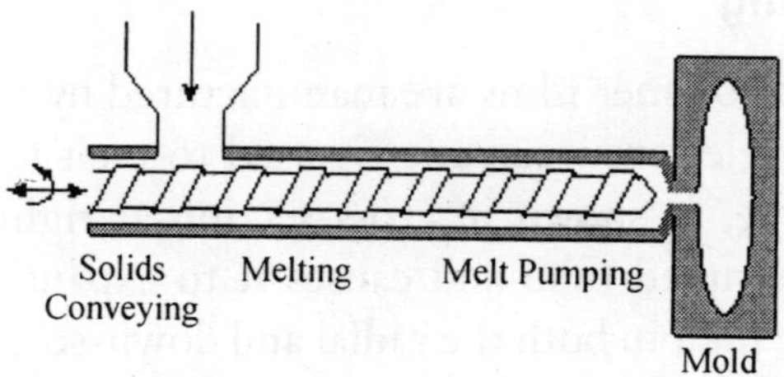

5.9 Batch Mixing

Batch mixing is generally carried out using two specially designed blades inside a temperature controlled chamber. It is most commonly used for compounding and mixing of rubber formulations. Its most important commercial use is the incorporation of carbon black and other additives into rubber for the manufacture of automobile tires. Laboratory size batch intensive mixers are extensively used for characterization of materials and processes.

5.10 Reaction Injection Molding (RIM)

Reaction injection molding (RIM) is a processing technique for the formation of polymer parts by direct polymerization in the mold through a mixing activated reaction. A simplified process schematic is shown at right. Two reactive monomeric liquids, designated in the figure as A and B, are mixed together by impingement and injected into the mold. In the mold, polymerization and usually phase separation occur, the part solidifies, and is then ejected. Primary uses for RIM products include automotive parts, business machine housings, and furniture.

Process Schematic

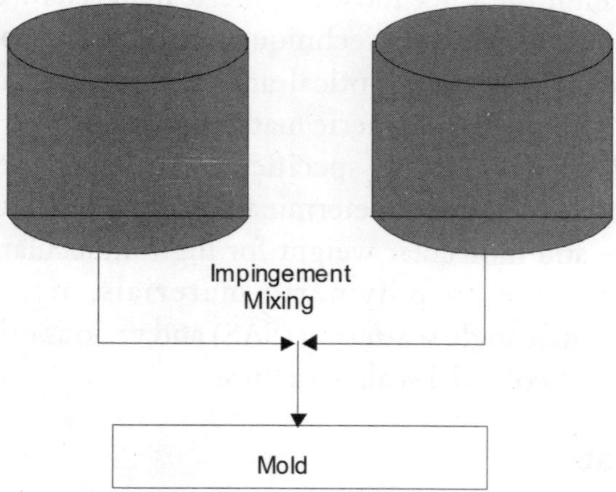

5.11 Spin Coating

Spin coating is the preferred method for application of thin, uniform films to flat substrates. An excess amount of polymer solution is placed on the substrate. The substrate is then rotated at high speed in order to spread the fluid by centrifugal force. Rotation is continued for some time, with fluid being spun off the edges of the substrate, until the desired film thickness is achieved. The solvent is usually volatile, providing for its simultaneous evaporation.

Process Schematic

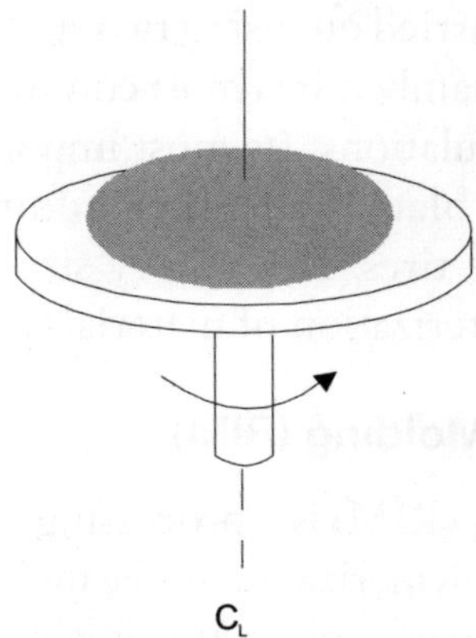

5.12 Polymer Analysis

Analysis of polymeric systems is essentially a subtopic of the field of chemical analysis of organic materials. Because of this, spectroscopic techniques commonly used by organic chemists are at the heart of Polymer Analysis, e.g. infra-red (IR) spectroscopy, Raman spectroscopy, nuclear magnetic resonance (NMR) spectroscopy and to some extent ultra-violet/visible (UV/Vis) spectroscopy. In addition, since most polymeric materials are used in the solid state, traditional characterization techniques aimed at the solid state are often encountered, x-ray diffraction, optical and electron microscopy as well as thermal analysis. Unique to polymeric materials are analytic techniques which focus on viscoelastic properties, specifically, dynamic mechanical testing. Additionally, techniques aimed at determination of colloidal scale structure such as chain structure and molecular weight for high molecular weight materials are somewhat unique to polymeric materials, i.e. gel permeation chromatography, small angle scattering (SAS) and various other techniques for the determination of colloidal scale structure.

5.13 Polymer Lab

The Polymer Lab offers thermal, chemical and rheological testing of polymer raw materials through finished plastic parts. Our laboratory has plastic testing capability for:

- ❖ Material Identification
- ❖ Contamination Identification
- ❖ Degradation Analysis
- ❖ Crystallinity Verification
- ❖ Reverse Engineering

- ❖ Product Failure Analysis
- ❖ Process Validation
- ❖ Ongoing Quality Assurance

The Polymer Lab's material analysis procedures include:

- ❖ Gel Permeation Chromatography—GPC
- ❖ Thermogravimetric Analysis—TGA
- ❖ Microstructural Analysis
- ❖ Differential Scanning Calorimetry—DSC
- ❖ Fourier Transform Infrared Spectrometry—FTIR
- ❖ Ash Content—Burn-off Testing
- ❖ Melt Flow Rate
- ❖ Relative Viscosity

5.15 Chemical Analysis of Polymers

ipolytech have diverse analytical capabilities for the characterisation of polymeric materials and their additive systems. The main techniques employed are described below. Please contact us to **discuss specific polymer analysis requirements**:

- ❖ **FT-IR**—Fourier Transform Infrared Spectroscopy (FT-IR) is a popular tool for identifying and characterising polymer materials and their additives. Independent Polymer Technology uses a Nicolet Impact 400 Fourier transform infrared bench driven by "Omnic" software. Spectra from unknown samples can be readily compared with extensive reference spectra using the library search facility, to aid identification.

Infrared Microscope (IR-Plan)

The infrared bench and software is also used to drive an infrared microscope. Infrared micro-spectroscopy is a powerful technique in problem solving—particularly for the identification of inclusions or defects in moulded parts or plastic films. The microscope can also be a valuable tool for forensic investigations when the available sample may be limited. See the case study on **regrind contaminant identification**

- **DSC**—Differential Scanning Calorimetry measures heat flow to a polymer. By monitoring the heat flow as a function of temperature phase transitions such a glass transition temperatures and crystalline melt temperatures can be characterised.

Thermal Analysis Suite including DSC, TGA and TMA

From a knowledge of both the glass transition temperature and crystalline melt properties information regarding the degree of crystallinity miscibility of blends and alloys and processing conditions for component manufacture can be gained. The technique can also be used to determine the thermal stability of polymers through determination of the OIT (oxidation induction time/temperature)

- **TGA—**Thermogravimetric Analysis (TGA) measures changes in the weight of a sample as a function of temperature and/or time. TGA is used to determine polymer degradation temperatures, residual solvent levels, absorbed moisture content, and the amount of inorganic (non-combustible) filler in polymer or composite material compositions. It can also assist in de-formulation of complex polymer products.
- **TMA**—This technique measures changes in the length or volume of a sample as a function of temperature and/or time. TMA is commonly used to determine thermal expansion coefficients and the glass transition temperature of polymer or composite materials. A weighted probe is placed on the surface of the specimen. The vertical movement of the probe is continuously monitored while the sample is heated at a controlled rate.
- **Atomic Force Microscopy**—AFM is a powerful surface imagining technique which uses a fine probe to scan the sample surface. Deflections of the probe are detected by a laser reflecting off the back surface of the probe tip and are used to characterise the surface topography of the sample.

Glass Fibre on PTFE surface

The is technique allows high resolution imaging of samples to the region of nano-meters with significant simplification of sample preparation in comparison to alternative techniques such as SEM.

5.16 Polymers Analysis Capabilities

Polymer Testing Expertise and Resources

Intertek provides clients with advanced analysis of polymers, plastics and composites. Global laboratory capabilities include identification and quantitation of plastics, resins, fillers, additives and more.

Intertek laboratories are staffed by experienced chemists and technicians with years of industry experience, providing a significant range of polymer testing capabilities. Expertise includes polymer characterization, polymer process chemistry, composites, additives, impact modifiers, alloys and blends. Intertek polymer testing laboratory locations are located on a global basis.

Polymer analytical and physical test techniques include:

- ❖ Polymer Testing Services A to Z
- ❖ PTLI Testlopedia TM
- ❖ Chromatography
- ❖ Elemental Analysis
- ❖ Electrical Properties

- Mass Spectrometry
- Mechanical Tests
- Microscopy
- Microstructure Analysis
- Nuclear Magnetic Resonance
- Consultancy Expertise
- Physical Tests
- Rheology
- FTIR
- Spectroscopy
- Structural Analysis
- Surface Analysis
- Thermal Analysis
- Global Analytical Capabilities

Additional Polymer Testing Capabilities:

- Polymer Capability Overview
- Polymer Analysis Review
- Polymers and Plastics
- Polymer Analysis Techniques Review
- Packaging Materials Testing
- Polymer Characterisation
- Polymer Testing
- Plastics Accelerated Weathering Laboratory
- Polymer Analysis Case Studies
- Polymer Laboratory

5.17 Polymers and Plastics Testing Services

Testing Services for Polymers and Plastics

Intertek provides polymers and plastics testing services through a global network of laboratories, staffed with experienced polymer chemists using state-of-the-art instrumentaton. The polymer labs run chemical composition, routine testing, non-routine analysis, failure analysis, formulation, physical properties plus many other polymer and monomer tests. Plastics contamination issues are identified and characterised, including identification of physical and chemical defects, inorganic fillers, lubricants, gels, nibs, fish eyes, deposits, stains, particles and residue.

A-Z Service Listing

- Adhesives analysis
- Ageing Laboratory
- ASTM - Plastics and Polymers

- Bio-Polymers Analysis
- Bisphenol-A Testing
- Bisphenol-A Migration Testing
- Birefringence Properties in Polymers

- Chemical analysis
- Chemical compostion
- Chemical resistance

- Data interpretation
- Dielectric Testing
- Dispersions analysis
- DSC Testing of Plastics

- Electrical Properties Tests

- Films testing for Packaging
- Films and Sheets Testing
- Flammability Testing of Plastics
- Flexural Properties Testing
- Food Contact Testing
- Forensic Science Expertise
- Formulation, Deformulation
- FTIR Polymer Analysis

- Gel permeation chromatography

- Identification of polymers
- Injection Moulding Facility

- Laboratory capabilities (global)
- Linear Thermal Expansion Kunststoffe polymere (deutsch)

- Measurement Science Group UK
- Mechanical testing of plastics

- Mechanical properties of polymers
- Methods development
- Migration testing, chemicals
- Molecular weight distribution
- Monomer ratios in mixed copolymers
- Monomor and oligomer characterization GLP

- Non-routine testing and analysis of polymers
- NMR analysis of polymers

- Optical and surface properties
- On-line polymer analyzers expertise
- Organoleptic analyses and compliance tests based on FDA and European regulations

- Packaging Materials Testing
- Particle analysis
- Physical characteristics and strength
- Physical testing of plastics
- Pilot Plant processing
- Plastics Testing
- Polychemlab
- Polymer characterisation
- Polymer failure analysis
- Polymer chemical composition
- Polymer Services
- Process chemistry, polymer catalyst testing
- Process analysis with rheology
- PTLI Laboratory USA
- PTLI Testlopedia TM
- Polymer mixtures identification
- Polymer impurities and dimensional variations in powder, granulate and articles
- Plastics and polymer processing expertise
- Plastic materials processing
- Plastic processing services
- Polymer extruders
- Polymer analysis in a GLP laboratory
- Polymers Tested,

- Resistively Testing
- Rheology
- Rheology Tests
- Rheology for process problems
- Routine testing, quality control of polymers
- Research and development
- Reaction kinetic studies

- Seal Compatibility Tests
- Sheets and Films Testing
- Surface Analysis
- Surface Free Energy

- Tensile Testing
- Testing Services
- Testlopedia TM
- Thermal Analysis for Polymers
- Troubleshooting, characterization

- VOC analysis
- Volatile organic compounds auto
- VOC (deutsch)

- Weathering Laboratory
- X-Ray Diffraction for Polymers and Composites

6

Polymer Blends and Composites

6.1 Polymer and Composite Materials

There are many types of polymers and composites of commercial interest. These materials are reviewed below. We help clients with product and process challenges involving all of these types of materials, for a vast range of end use applications. In working with us, you will receive the quality of services that you require and deserve from a team of professionals who care about your specific needs and who have the ability to address these needs successfully.

Amorphous and Semicrystalline Thermoplastics, and Their Melts

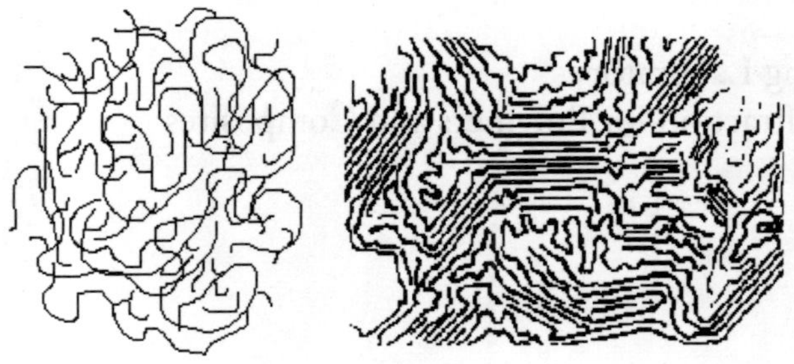

Schematic illustration. Typical **amorphous random coil** (left) and **semicrystalline** (right) morphologies of **thermoplastic polymers**. Atactic polystyrene and bisphenol-A polycarbonate are examples of amorphous thermoplastics. High-density polyethylene and poly (ethylene terephthalate) are examples of semicrystalline thermoplastics. The amorphous and crystalline domains are connected to each other by covalent chemical bonds in a semicrystalline polymer so that a chain may traverse both amorphous and crystalline regions. Chain segments linking two crystalline domains are often called "tie chains". The boundaries between amorphous and crystalline domains are usually diffuse. They are sometimes called "interphase" regions.

Conventional (Thermoset) Rubbers/Elastomers

Covalent crosslinks provide a three-dimensional **elastic network** with "junctions". When the specimen is heated, these junctions survive until their thermal or thermooxidative degradation temperatures are reached. The material is hence a "thermoset", in the sense that once its network structure is formed it can no longer be processed by using melt processing techniques. However, it is also an elastomer (a "rubbery" material) because its glass transition temperature is low and so it has a low stiffness (low elastic moduli, and a lot of "bounce") at typical end use temperatures.

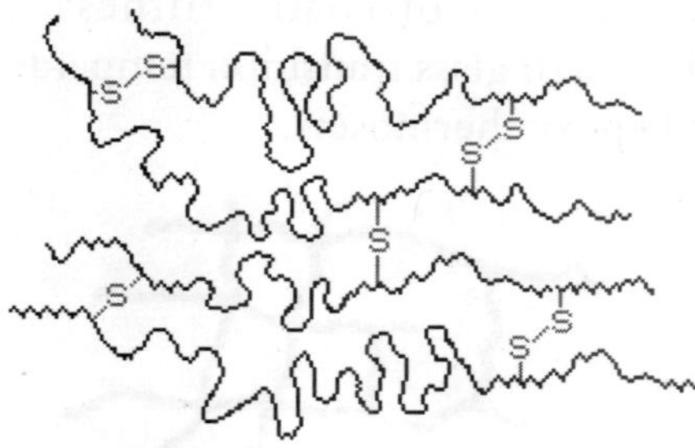

Schematic illustration. The chains of natural rubber (*cis*-polyisoprene) are crosslinked slightly with sulfur (-S-) linkages by using the "vulcanization" process. There are many other types of elastomers as well, differing from this example by the use of different polymer structures, crosslink types, and preparation processes.

Thermoplastic elastomers

A three-dimensional **elastic network** is provided by **"physical crosslinks"**, such as crystalline and/or rigid glassy amorphous domains in a soft amorphous matrix. Unlike conventional (thermoset) elastomers, thermoplastic elastomers can be processed by using melt processing techniques, since their crystalline and/or rigid glassy amorphous domains can be melted during processing and can then reestablish themselves upon cooling.

Schematic Illustration. Evolution of thermodynamic **equilibrium morphology** of a typical AB-diblock copolymer [such as poly (styrene-*block*-butadiene)] with immiscible A and B components as the volume fraction of the

B component (black) is changed. From left to right: spheres of B domains in a matrix of A, cylinders of B domains in a matrix of A, ordered bicontinuous gyroid, lamellar bicontinuous, ordered bicontinuous gyroid, cylinders of A domains in a matrix of B, and spheres of A domains in a matrix of B.

Rigid thermosets

These materials are, in general, **amorphous** polymers where **covalent crosslinks** provide a three-dimensional **network**. However, unlike conventional (thermoset) elastomers, rigid thermosets are stiff (have high elastic moduli) at their use temperatures (and often up to much higher temperatures) because their particular combinations of chain stiffness and crosslink density characteristics result in a high glass transition temperature. The most familiar examples are the cured epoxy thermosets.

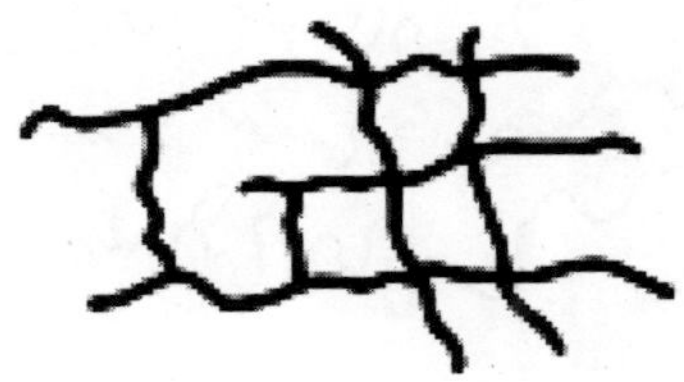

Schematic illustration. Note the similarity to the example of a conventional (thermoset) elastomeric network that was shown earlier. The main practical difference between these two classes of polymers is that rigid thermosets have much higher glass transition temperatures than conventional (thermoset) elastomers.

Gels

A gel is a semi-solid jelly-like state of a material, similar to gelatin in its consistency. There are many different types of **polymeric gels**. For example, during the synthesis of a crosslinked thermoset from liquid oligomer and curing agent reactants, it increases in stiffness (modulus) rapidly when the "gel point" is reached so that a three-dimensionally percolating network of crosslinks is established. On the other hand, a covalently or physically crosslinked "solid" polymer can form a gel by absorbing a sufficient amount of an appropriately chosen solvent. A **hydrogel** is a gel where water is the solvent. A **smart gel** or **stimuli-responsive gel** (such as certain synthetic polyacrylamide gels) can change its volume (expand or contract), its elastic properties (soften or stiffen), and/or its optical properties (colour) rapidly and drastically as functions of small changes in solvent composition, pH, temperature, lighting, electric field, magnetic field, etc.

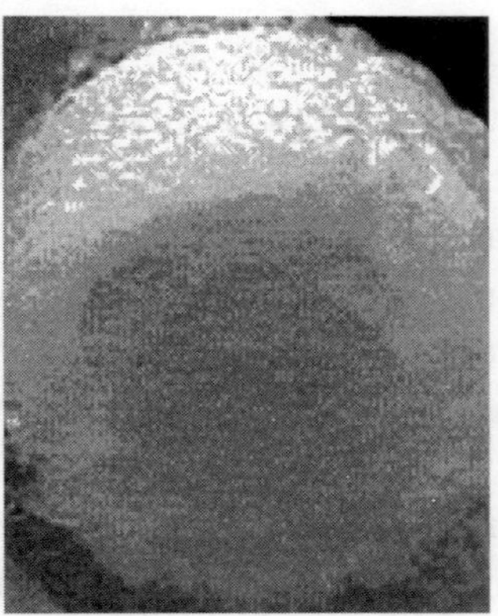

A gel.

Solutions

Depending on the chemical structure of the solvent and the chemical structure and chain architecture of the polymer, one can find a wide range of behaviours and morphologies, including homogeneous solution, precipitation of the polymer, and a wealth of ordered phases similar to those shown above for block copolymers. An example is shown below.

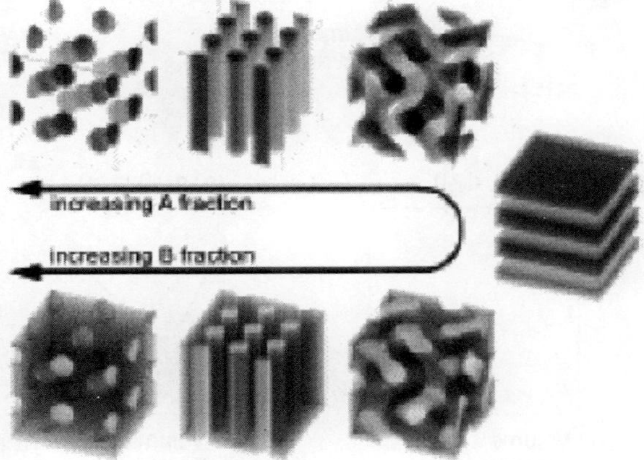

Schematic illustration. Morphologies in mixtures of polyol-based triblock copolymer surfactants with water. Going clockwise from the top left, micellar cubic, hexagonal, bicontinuous cubic, lamellar, reverse bicontinuous cubic, reverse hexagonal, and reverse micellar cubic. By definition, the hydrophobic ("oil") phase domains are enclosed in the continuous water phase in the micellar cubic and hexagonal morphologies, while the water domains are enclosed in the continuous hydrophobic ("oil") phase in the reverse micellar cubic and reverse hexagonal morphologies.

Blends

Blends span the entire range from fully **miscible** to completely **immiscible.** The thermodynamic drive towards **phase separation** increases with increasing inherent incompatibility and as with increasing average molecular weights of polymer chains. Unlike block copolymers where highly ordered morphologies (as shown above) are found, one does not normally find ordered arrangements of regularly-shaped domains in a blend since the polymer chains of different blend components are not bonded to each other. The blend morphology can be affected significantly by many factors. These factors include the incorporation of compatibilizers, the kinetic "freezing in" of nonequilibrium morphologies by the application of shear during processing, and annealing at an elevated temperature in order to release the kinetically frozen-in morphological features and thus approach thermodynamic equilibrium.

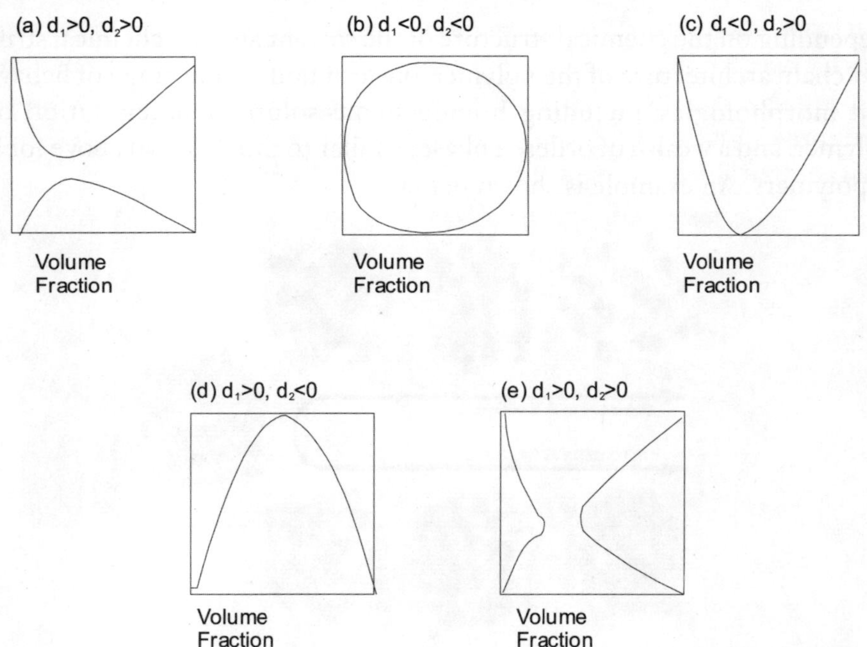

Schematic illustration. Types of possible polymer blend **phase diagrams,** for binary blends where additional complications that can be introduced by competing processes (such as crystallization of a component) are absent. The coefficients d_1 and d_2 refer to a general functional form (as a function of temperature and component volume fractions) of the binary interaction parameter that quantifies deviations from ideal mixing.

Polymers derived from fossil fuel based raw materials

Polymers spanning the entire range of performance/price balance, from the cheapest **commodity polymers** with low profit margins intended for large volume markets to the most expensive **specialty polymers** intended for niche markets. Polymers synthesized by using different polymerization methods.

Biobased polymers

Such polymers are obtained from **renewable resources**. They include **biopolymers**, such as the polyhydroxyalkanoates and cellulose, which are obtained directly from living organisms such as **plants** and/or **bacteria**. They also include polymers prepared in the laboratory by modifying biopolymers (such as cellulose triacetate) or by polymerizing **biobased monomers** [such as poly (lactic acid)]. The industrial importance of biobased polymers is expected to increase gradually in the future. The main driving forces are the continuing high prices for petroleum-based raw materials and the potential environmental advantages (possibly resulting in increasing governmental regulations and incentives) of using renewable resources whenever it is cost-effective to do so without sacrificing critical performance attributes.

Examples of biobased sources of polymers. (Top) From left to right, a paper birch tree, cotton, soybeans. (Bottom) From left to right, corn, and a culture of *Wautersia eutropha* bacteria.

Foams

Foams can be **flexible** or **rigid**, **thermoplastic** or **crosslinked**, and **physically foamed** or **chemically foamed**. The most familiar examples of both rigid and flexible crosslinked and chemically foamed materials (where the polymeric structure develops via chemical reactions along with expansion into a foam) are polyurethane foams manufactured from different formulations by using different process conditions. The most familiar example of a physically foamed system (where a previously synthesized molten polymer is expanded into a foam) is an expanded atactic polystyrene rigid foam.

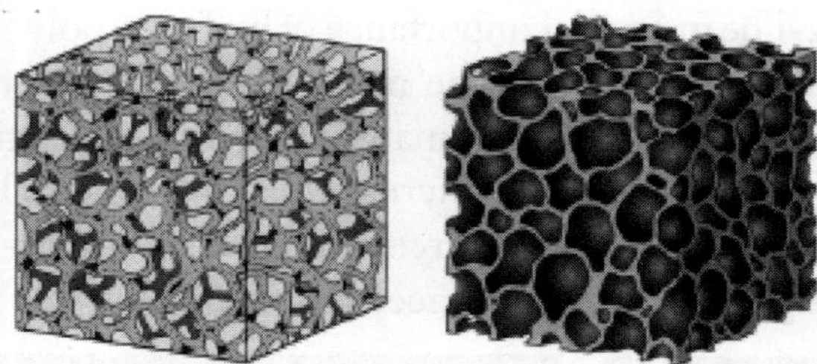

Illustration. Models of slabs of typical **open-cell foams** (left) and **closed-cell foams** (right). The foam cells adopt polyhedral shapes which result from the combination of the physical factors governing the foaming process and the geometrical constraints of packing.

Composites

The term **composite** is often used to describe **matrix** polymers containing conventional **fillers** (such as glass spheres, short fibers, continuous fibers, or talc particles). Composites where (most often highly anisotropic) **nanofillers** (such as carbon nanotubes or "exfoliated" clay platelets) have been dispersed in a polymer are commonly called **nanocomposites**.

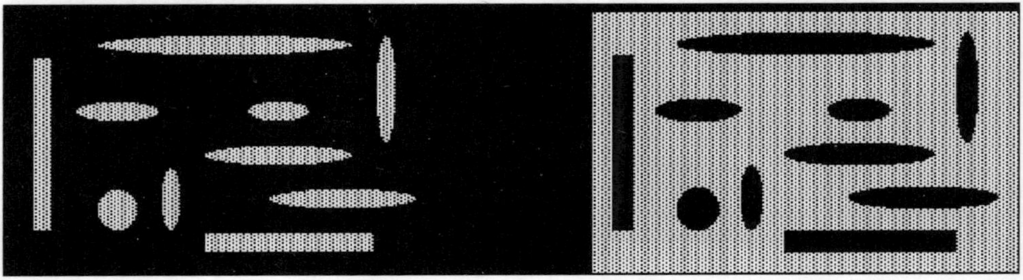

Schematic illustration. A composite consisting of stiff (high-modulus, darkly-shaded) matrix polymer containing flexible (low-modulus, lightly-shaded) discontinuous fillers (left), and a flexible matrix polymer containing stiff discontinuous fillers (right). The properties of a composite depend on the material properties and volume fractions of each component; the shape, orientation and size distributions of each type of filler present in the composite; the strengths of the matrix-filler interfaces, and the sample preparation conditions.

Suspensions

Often also referred to as **dispersions**, the systems of interest most often consist of various types of **particles in ordinary or polymeric fluids**. The particles can be spherical, fibrous or discoidal in shape. Particles of more than one type of material and/or different shapes may be present in the same suspension. The morphology and the rheological properties of a suspension depend on the same factors as those listed in the caption of the figure shown above as affecting the properties of composites.

Liquid crystalline polymers

From left to right, the key differences between **crystalline order**, **liquid crystalline order**, and liquidlike disorder are shown below. The molecular arrangement manifests both translational order and orientational order in a crystalline arrangement. Orientational order along a preferred direction is also observed in liquid crystalline arrangements, while the full three-dimensional translational periodicity is lost. There are several types of liquid crystalline order (**nematic**, **smectic**, **cholesteric** and **columnar** phases), differing in the extent and the details of the order manifested by them.

Schematic illustration. Polymers manifesting liquid crystalline order include various "rigid rodlike" aromatic polyamides and polyaromatic heterocyclics. These polymers have very stiff chains. These chains do not readily adopt the random coil conformation found in typical amorphous polymers. However, they also do not pack well into a fully three-dimensionally periodic crystalline lattice because of their molecular shapes. The result is a preference for liquid crystalline order.

6.2 Designing Polymer Blends

Finding the appropriate ternary system using Edisonian methods is time consuming and not always productive. This experimentally proven modeling method, which can be run on fast PCs, will help.

by Timothy J. Cavanaugh, A. Peter Russo, and E. Bruce Nauman

Multiphase polymer blends are of major economic importance in the polymer industry. The most widespread examples involve the impact modification of a thermoplastic by the microdispersion of a rubber into a brittle polymer matrix. Most commercial blends consist of two polymers combined with small amounts of a third, compatibilizing polymer—typically a block or graft copolymer. Multiphase blends can be made using a variety of methods; however, research has focused on compositional quenching (1), melt compounding, and reactive technologies such as precipitation polymerization and reactive extrusion (2, 3).

Multiphase polymer blends can be easier to process than a single polymer with similar properties. The possible blends from a given set of polymers offer many more physical properties than do the individual polymers. Selections for blend production are currently made by Edisonian experimentation. This approach has shown some success but becomes cumbersome when more than a few components are involved.

Here we describe an emerging technique for computer-aided design of multiphase polymer blends. Polymer compositions are chosen to produce interesting and desired morphologies. We also provide a classification scheme for specific morphologies and methods of experimental confirmation with material properties.

Modeling blending processes

Compositional quenching. In this relatively new blending process, two or more incompatible polymers are dissolved in a common solvent to form a homogeneous mixture. The solvent is then rapidly removed by flash devolatilization. The initial, single-phase mixture is plunged quickly and deeply into a multiphase region in which phase separation occurs by a process known as spinodal decomposition. Our discussion begins with this process because it can be accurately modeled. The model consists of three equations: a modified Flory-Huggins equation for the free energy of mixing, the Landau-Ginzburg functional equation, and a modified Cahn-Hilliard diffusion equation.

Many polymer pairs are incompatible and form two phases when mixed. When a common solvent is added, a limited region of two-phase miscibility occurs. The formation of a single-phase solution—two or more polymers in a single solvent—is the starting point for compositional quenching.

The first quantitative theory of polymer-solvent equilibria was presented by Flory (4) and Huggins (5). They modeled the Gibbs free energy of mixing (g_{mix}) in a binary polymer-solvent system. According to this theory, the Gibbs free energy of mixing per unit volume is written as

$$\frac{g_{mix}}{RT} = \frac{(a)\ln(a)}{n_a} + \frac{(1-a)\ln(1-a)}{n_B} + \chi a(1-a)$$

where a is the volume fraction of polymer A; n_A and n_B are the number of repeating units for the two polymers; and χ is an interaction parameter.

The Flory-Huggins theory is similar to regular solution theory except that volume fractions replace mole fractions as the fundamental composition variable. The drawback of this theory is that it cannot predict the low critical solution temperatures that have been observed in many polymer systems. Eliminating this drawback, however, makes the equation more complex. The Flory-Huggins theory is adequate for our purposes.

In compositional quenching, the solvent is removed by flash devolatilization. In a deep adiabatic flash, most of the solvent is vaporized, generating an enormous ratio of vapor volume to liquid volume. The vapor volume that is generated completely disrupts the foam structure, so the partially devolatilized polymer is dispersed in the vapor as fine droplets or as small-diameter fibers (Figure 6.1) (6).

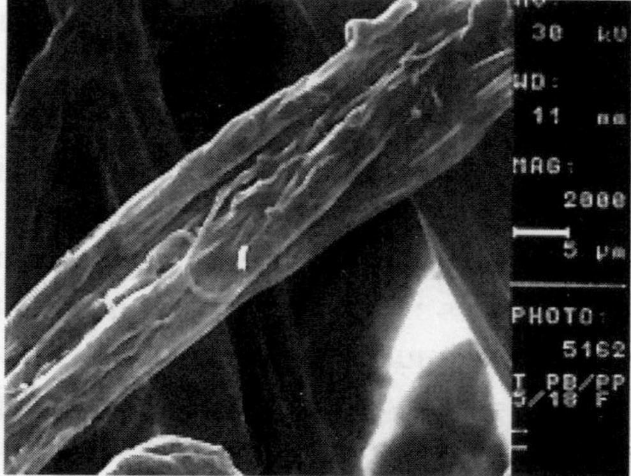

Fig. 6.1: Flash devolatilization processing forms fibers in polymer blends

To form a microdispersion of one polymer in another by compositional quenching, a common solvent must be found (e.g., xylene for polystyrene and polybutadiene). A single-phase solution is formed; it typically contains 5-10 per cent total polymer and 90-95 per cent solvent. This mixture is heated under pressure and then flashed.

The rapid removal of solvent by flash devolatilization causes a large displacement from equilibrium. Phase separation occurs by spinodal decomposition (7-9). The free energy of the system depends on the component concentrations and the gradients in concentration. Spinodal decomposition is governed by a fourth-order partial differential equation known as the modified Cahn-Hilliard equation:

$$\frac{\delta a}{\delta t} = \nabla D a(1-a)\nabla\left(\frac{\delta g}{\delta a} - \kappa\nabla^2 a\right)$$

where a is the volume fraction of polymer A, D is the diffusivity, g is the free energy of mixing, and κ is the gradient energy parameter.

This equation applies to a binary blend. Another partial differential equation is added for each additional component. Details have been given in the literature (10). Although these equations are mathematically complex, it is important to remember that they contain few adjustable parameters: the volume fraction of the polymers, the interaction parameter, and the chain lengths of the polymers. Diffusivity is a parameter that disappears on scaling. These equations can, in fact, be solved easily. A two-dimensional simulation of a ternary polymer blend requires a few hours on a fast personal computer.

The volume fractions of the components are most important in determining the morphology of a polymer system. For a ternary polymer system, a component with a volume fraction of 0.6 or higher always forms a continuous phase, whereas no continuous phase is observed for volume fractions <0.33. The semicontinuous phase is limited to the range of 0.3-0.5, and dispersed morphologies are confined to volume fractions <0.45.

The dependence of the equations on the interaction parameter is more difficult to describe. We know that the polymers dislike each other more with an increasing χ values. In determining the overall morphology, the ratio of the interaction parameters is believed to be more important than the absolute values themselves. Simulations performed with high χ values (e.g., $\chi_{ab} = 6$, $\chi_{ac} = 12$, and $\chi_{bc} = 6$) yield the same morphology but with faster ripening (particle growth) than those performed with low χ values (e.g., $\chi_{ab} = 3$, $\chi_{ac} = 6$, and $\chi_{bc} = 3$).

Simulations run on the effect of chain length for a binary system have shown interesting results (11). For off-critical quenches, the effects of the chain length of the matrix material on the domain size are negligible. However, a smaller chain length of the minor component greatly increases the ripening rate and therefore the domain size.

Conventional compounding. The most common method for generating multicomponent polymer blends is melt compounding. Even blends produced by compositional quenching must be processed by conventional means to form useful products. This type of processing often exposes the blend to forces that may affect its morphology. He and Nauman (12) addressed this concern by examining spinodal decomposition under shear flows. The geometry used in their simulations is a starting approximation of the region of a single-screw extruder near the barrel wall.

The effect of hydrodynamics on spinodal decomposition was first studied by Farrell and Valls (13). Rousar and Nauman (14, 15) and Vasishtha and Nauman (16) expanded this study by including the continuity equation and flows induced by pressure gradients. The work of He and Nauman (12) was based on all of these results.

The modeling equations included a convective diffusion form of the modified Cahn-Hilliard equation, in which an additional term accounts for flows induced by concentration gradients. This equation was coupled with the two-dimensional equations of motion and the continuity equation. A special volumetric body force term was included in the equations of motion to account for the heterogeneity of the fluid mixture (15, 16).

The geometry of interest consists of two concentric cylinders representing the walls of an extruder. The inner cylinder could be rotated to apply shear in the circumferential direction. Numerical simulations were performed on this geometry for conditions that are initially homogeneous (i.e., extrusion of a compositionally quenched blend) or phase separated (i.e., blend of two homopolymers).

Simulation results showed that including body force hastened phase ripening and that the resultant morphologies were similar to diffusion-only models. Applying shear had a tremendous effect on morphology. The appearance of phase boundaries was delayed, and there was a strong tangential orientation. This preferential orientation occurs because shear effects dominate at short and long times, whereas diffusion effects dominate the intermediate stages. The final morphology appears as alternating concentric regions of each phase (Figure 6.2). The starting condition of the system seems to have little effect on the final morphology—thus the importance of spinodal decomposition in traditional melt blending and compositional quenching.

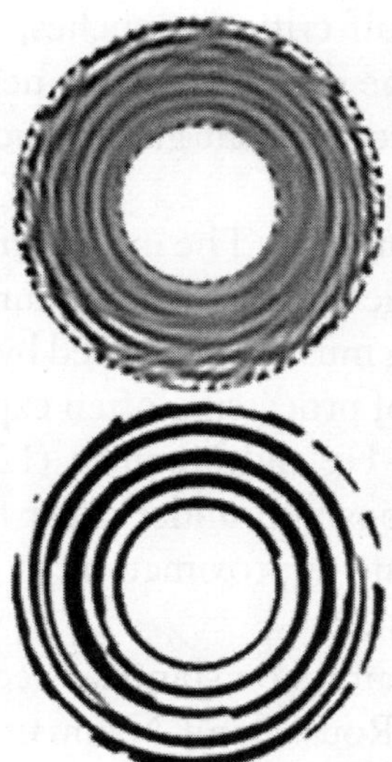

Fig. 6.2: Simulation of a two-phase extrusion showing alternating concentric regions of each phase. Shown is the morphology after 1 s (top) and 20 s (bottom).

After the final morphology was established, rotation was stopped. When the internal body force was present, the circumferential orientation began to break up and evolve toward compact domains. Such structures are expected in actual extrusion.

The work done so far provides a starting point for the modeling of melt blending based on the principles of spinodal decomposition. Actual extruders have a three-dimensional (3-D) geometry with more complex flow fields than do the models; the scale of the structures formed in actual extruders and processing times are similar to those used in these studies. Therefore, these results are a reasonable first approximation of the morphology to be expected from polymer blends processed by conventional means.

Reactive techniques. Techniques involving chemical reactions provide another method for producing polymer blends. The classic example is the production of high-impact polystyrene (HIPS) by precipitation polymerization. HIPS is manufactured by polymerizing styrene in the presence of dissolved polybutadiene (PB). Early in the polymerization, the PB phase separates because of the immiscibility of the rubber with the polystyrene (PS). As the polymerization continues, the styrene grafts to the PB, and as the volume ratio of PS to PB in styrene approaches 1, a phase inversion occurs. The rubber-phase morphology is strongly influenced by the shearing agitation during and shortly after phase inversion. During removal of the residual monomer, the PB in the rubber phase is cross-linked.

The size and morphology of the rubber phase can be controlled to obtain the desired properties. Besides the shearing action, initiators, chain-transfer agents, and diluents play an important role in controlling the morphology of the rubber phase (17). This third technique has to be modeled to obtain an overall understanding of the production of polymer blends.

Morphologies of interest

Of the morphologies in the work of Nauman and He, three types have significant potential for industrial polymer blends: core-shell, multiple discrete particle, and dual semicontinuous. The core-shell structure (see Sidebar "Classifying Morphology") has the ternary classification cp-dc-si. This morphology can simultaneously increase the volume of the dispersed phase and the matrix-elastomer interface for a given amount of elastomer. If the matrix-elastomer interface adhesion is strong, the material should have a higher impact strength than the same blend without adhesion (18). For component C, commercial blenders use a block copolymer of components A and B to increase matrix-elastomer adhesion, but it is clear from the simulations that many polymer compositions can give this structure.

TO SIDEBAR: Classifying morphology

The phase classification for the multiple-discrete-particle structure is cp-dc-dc; phase A contacts both B and C, and phases B and C do not contact each other (Figure 6.3). On a commercial scale, it would be possible to produce a bimodal particle size distribution of two different rubbers. It is believed that, for polymers that fail by crazing and shear yielding (such as ABS), the presence of the two particle sizes—one below the critical size and one above—would enhance the toughness of the matrix compared with the same matrix with only one particle size (19).

Fig. 6.3: Ternary blend simulation yield a multiple discrete particle structure. The phase classification for this structure is cp-dc-dc; phase A contacts phases B and C, whereas phases B and C do not contact each other. For this simulation, a = 0.6, b = 0.2, c = 0.2, $\chi\alpha_b = 3$, $\chi_{ac} = 3$, and $\chi_{bc} = 6$.

Currently, multiple-discrete-particle blends are produced by compositional quenching of HIPS with a mixture of PS, PB, and PS-PB diblock copolymer. One set of rubber particles is produced by precipitation polymerization and the other by compositional quenching. Excellent impact strength can be obtained by adding small amounts of HIPS to the blend of PS, PB, and diblock (Figure 6.4). The volume fraction of the rubber phase was maintained at 23 per cent.

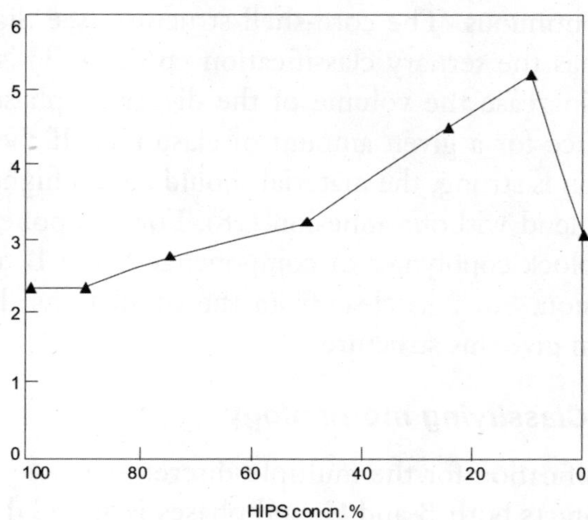

Fig. 6.4: Excellent impact strength can be obtained with compression molding by adding small amounts of high-impact polystyrene (HIPS) to the blend of polystyrene (PS), polybutadiene (PB), and PS-PB diblock polymers. The volume fraction of the rubber phase was maintained at a constant 23 per cent. Compression molding was done for 2 h at 200°C.

The particulate morphology of ternary mixtures has been studied; dual semicontinuous morphology prepared by spinodal decomposition, however, has rarely been reported. A dual semicontinuous morphology with component C at the interface is shown in Figure 6.5. If stabilized, these blends represent a much better trade-off between Izod strength and tensile modulus than the particulate morphology. Besides mechanical properties, these blends can possess other desired features, such as a selective permeability.

Kilogram-scale Preparations

Although modeling can give an excellent basis for the design of a polymer blend, experimental verification is still necessary. Also, even though excellent predictions of the morphologies are available, the relationship between the morphology and the physical properties is not always clear.

Fig. 6.5: A dual semicontinuous morphology with component C at the interface. If stabilized, these blends represent a much better trade-off between izod strength and tensile modulus than the particulate morphology. Besides mechanical properties, these blends can have other desired features, such as a selective permeability. For this simulation, a = 0.45, b = 0.45, c = 0.10, $\chi_{ab} = 6$, $\chi_{ac} = 3$, and $\chi_{bc} = 3$.

To produce blends by conventional melt compounding, devices such as twin-screw extruders and Banbury mixers are often used. The Banbury mixer, which consists of two spiral shafts that counter-rotate about a fixed axis, is capable of producing batch blends from a few to several thousand kilograms. The ingredients are pushed into the top of the mixer by an air cylinder ram. The time of mixing and the pressure exerted by the ram can be varied.

Twin-screw extrusion is a continuous process in which the premixed feed material enters the extruder, usually in the form of a solid. The material is subjected to high shear by the co-rotating screws. This high shear aids in melting the material and provides excellent mixing.

In compositional quenching, the blends of interest are produced by dissolving the polymers in a solvent at room temperature or, in some cases, slightly elevated temperatures. Typically, solutions of 3-10 wt per cent are made. The solution is pumped through a heat exchanger and then flashed across a needle valve into a low-pressure chamber. The needle valve provides back pressure so that the solution does not boil in the heat exchanger. The inlet temperature and flash pressure are monitored to control the domain size of the morphology. Commercially, the effluent polymer blend would be a pumpable liquid. The apparatus shown in Figure 6.6 can flash to a nearly dry solid and produces 5- to 10-kg samples.

Fig. 6.6: This compositional quenching apparatus can flash to a nearly dry solid and produces 5- to 10-kg samples

Analysis and Images

After a blend is produced, we analyze the material to determine physical properties and confirm the predicted morphology. Mechanical properties such as impact strength and tensile strength must be measured. In most cases, we want increased impact strength over a wide temperature range without a significant loss in tensile modulus. Thermal properties such as melt flow rate are tested to determine the required processing parameters of the new blend. The combination of these properties determines the blend's marketability.

Because the microstructure of these blends can be on the order of 0.1-10 µm, it is often necessary to combine several microscopic techniques, including optical, scanning, and transmission electron microscopy. An optical phase-contrast microscope can distinguish the refractive indices of the three phases (20); Figure 6.7 was created using this technique. It offers ease of sample preparation, but there are two main drawbacks. First, if the refractive indices of two of the phases are close, it is difficult to distinguish the difference. Second, the optical resolution is poor for objects smaller than ~1 µm. Therefore, optical phase-contrast microscopy is unsuitable for the morphological study of materials with small particle sizes (21).

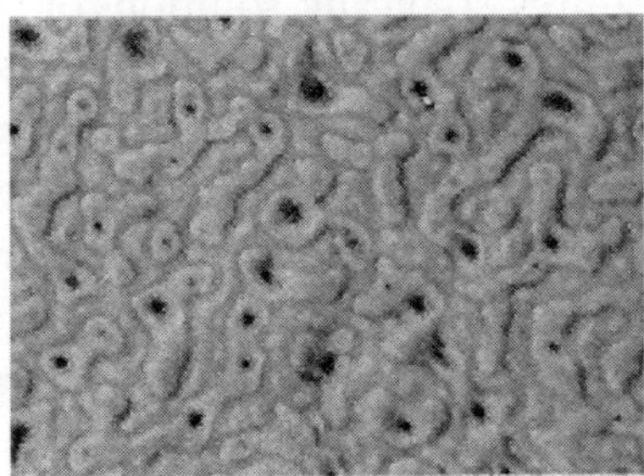

Fig. 6.7: An optical-phase contrast microscope can distinguish among the refractive indices of three phases. The dark, continuous phase is polymethyl methacrylate (PMMA). The lighter, irregular regions are PS. The dark particles inside the PS are PB

Scanning electron microscopy (SEM) offers much better resolution (~10 nm) than optical microscopy, and large samples may be examined with only minimal preparation. However, SEM is limited to the examination of sample surfaces, making it difficult—if not impossible—to distinguish the morphology of a ternary blend. SEM is used extensively to examine fracture surfaces. The fracture surface of a blend of PS, PB, and polypropylene (PP) is shown in Figure 6.8.

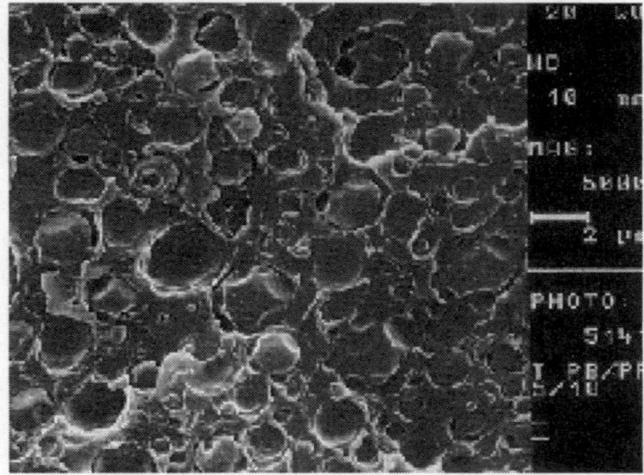

Fig. 6.8: Scanning electron microscopy shows the fracture surface of a ternary blend of PS, PB, and polypropylene (PP). The large particles consist of PP surrounded by smaller PB spheres

Transmission electron microscopy (TEM) is analogous to optical microscopy except that a beam of electrons, rather than light, is used as the illuminating source. Typical resolutions are on the order of 1 nm. TEM requires tedious sample preparation and phase-contrast enhancement by staining methods (22). Typical specimens for conventional TEM are thin films <100 nm thick.

High-voltage electron microscopy (HVEM) can penetrate thicker specimens than a conventional TEM can because its beam is stronger. Figure 6.9 was obtained using HVEM on a 0.75-µm-thick sample of a PS-PB-PP blend. With HVEM, stereoscopic images of the sample can be made and then reconstructed into a 3-D image. Several systems are available to reconstruct from serial sections; however, STERECON (STEReoscopic RECONstruction) has the unique ability to reconstruct a 3-D image by stereoscopic contour imaging (23). The photographs are scanned into a computer, and the digitized images are displayed on two monitors, arranged at 45° from a half-silvered mirror. The mirror combines the two images into a stereo pair. The contours are drawn with polarizing glasses in stereo overlay bit planes using a digitized pad. The reconstruction can be displayed many ways, the simplest of which is a contour drawing.

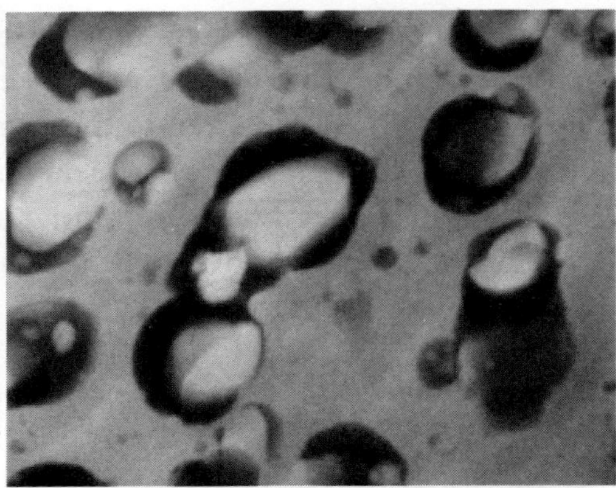

Fig. 6.9: High-voltage electron microscopy (HVEM) of a core-shell structure. HVEM can penetrate much thicker sections than transmission electron microscopy because of its stronger beam. We obtained this image of a PS, PB, and PP blend on a sample 0.75 μm thick. The PB forms the dark rings around the white spheres, which are PP

A Valuable Tool

Polymer blends designed using spinodal decomposition theory offer several advantages over more traditional methods. First, minimal experimentation is necessary to produce a desired morphology. Second, a wider range of morphologies and properties may be achieved via spinodal decomposition than via reactive techniques, for example.

It is important to confirm predicted morphologies with experimental results. A correlation must also be made between blend properties and blend morphology. When this is done, we have a powerful means with which to design marketable blends in the polymer community.

6.3 Types of Polymer Clay

There are many types (or brands) of polymer clay available on the market today. I will cover the properties of only a few most common ones here. Please refer to Polymer Clay: Resources page for a list of places where you can buy these and other types of polymer clay.

Sculpey III is soft and easy to work with. It is the easiest one to push through clay guns. The colours blend easily, which is good for colour mixing, but is not so good for cane-work (millifiore). Being soft right out of the package, it is good clay for kid's projects. However, this clay is more brittle than others after baking.

Premo! Sculpey is softer than FIMO, but stiffer than Sculpey. It retains flexibility in thin areas, making small details less vulnerable to breakage. It has a rich colour palette.

Fimo Classic is stiffer than any of Sculpey clays, so it keeps the shapes and colours you want, making it a very good choice for cane-work.

Fimo Soft is very similar to Sculpey III in its properties.

Kato is the stiffest of all of polymer clays, but it is also the strongest when baked. This is my clay of choice for all my jewelry projects.

Tips and Tricks: Here is one thing some new polymer clay enthusiasts may not realize right away:

6.4 Polymer Drying: The Basics

Many plastics, especially so-called engineering materials, are "hygroscopic." As soon as the polymer is manufactured and exposed to the atmosphere, it begins to absorb moisture. Water vapor continues to migrate from the surrounding air until equilibrium is reached between the moisture content of the polymer and the surrounding air. This moisture absorption takes place on a molecular level. The natural attraction between polymer chains and water molecules is what causes hygroscopic resins, exposed to a humid atmosphere, to take up and retain water. It is also what makes drying them difficult.

Fortunately, the process can be reversed. A few fundamental factors govern this process:

- ❖ Polymer temperature;
- ❖ Relative humidity/dew point of the air; and
- ❖ Drying time.

In addition, there must be some mechanism that extracts the moisture released by the polymer during the drying process and, in most dryer designs, this is accomplished by a stream of dry air. Let's look at each of these closely.

Polymer Temperature

The temperature of a hygroscopic polymer determines the rate at which water molecules move through it. Therefore, temperature is probably the most important consideration in typical drying applications. As the temperature of a hygroscopic polymer is increased, the molecules move about more vigorously and the attraction between the polymer chains and the water molecules is decreased. Thus, heating used in the drying process allows the water molecules to escape from the polymer chains. Generally, the higher the drying temperature, the faster the polymer will dry. There is, however, a practical temperature limit. If the polymer is exposed to a drying temperature that is too high, the pellets may stick together and bridge in the drying hopper. Some polymers, such as nylon, may oxidize and discolour at drying temperatures above those recommended by the material supplier.

Conversely, there is also a practical limit to how low drying temperatures can be and still be effective. The lower the drying temperature is, the longer it will take for a polymer to give up its moisture.

Dewpoint/Relative Humidity

Once heat has freed water molecules from their bond with the polymer, it is necessary to induce them to migrate out of the plastic pellet. This is accomplished by surrounding them by air that has a lower "vapor pressure" than the pellet. In most dryers, a stream of low-dewpoint (dry) air provides this low-vapor-pressure environment. The drier the air surrounding the pellet, the higher the vapor pressure inducing the water molecules out of the pellet.

Most conventional dryer designs use this dry air to convey water away from the polymer and, because hot air can hold much more moisture than cool air (it has a lower "relative humidity"), it is advantageous to heat the air. In fact, most dryers use hot dry air as the heat input to the polymer.

Drying Time

Plastic pellets do not dry instantaneously. It takes time to raise the temperature of the pellets and, once the pellets are subjected to lower vapor pressure conditions, it takes more time for moisture to diffuse and migrate to the surface. Resin manufacturers have defined how long this process takes for their particular resin types and grades. Remember, however, that effective drying time is the time the pellets are exposed to optimum drying conditions. The time the polymer is in the drying hopper at anything less than the recommended temperature and dew point cannot be considered drying time, and you run the risk on insufficiently drying the material. This, in turn, can be comparable to not drying the resin at all.

6.5 Composite

A composite is any material made of more than one component. There are a lot of composites around you. Concrete is a composite. It's made of cement, gravel, and sand, and often has steel rods inside to reinforce it. Those shiny balloons you get in the hospital when you're sick are made of a composite, which consists of a **polyester** sheet and an aluminum foil sheet, made into a sandwich.

Modern composites are usually made of two components, a *fiber* and *matrix*. The **fiber** is most often **glass**, but sometimes **Kevlar**, **carbon fiber**, or **polyethylene**. The matrix is usually a **thermoset** like an **epoxy resin**,

polydicyclopentadiene, or a **polyimide**. The fiber is embedded in the matrix in order to make the matrix stronger. Fiber-reinforced composites have two things going for them. They are strong and light. They're often stronger than steel, but weigh much less. This means that composites can be used to make automobiles lighter, and thus much more fuel efficient. This means they pollute less, too.

What Fibers Do

A common fiber-reinforced composite is Fiberglas™. Its matrix is made by reacting a **polyester** with carbon-carbon double bonds in its backbone, and **styrene**. We pour a mix of the styrene and polyester over a mass of glass fibers.

O O O O n + → crosslinked resin

The styrene and the double bonds in the polyester react by **free radical vinyl polymerization** to form a crosslinked resin. The glass fibers are trapped inside, where they act as a reinforcement.

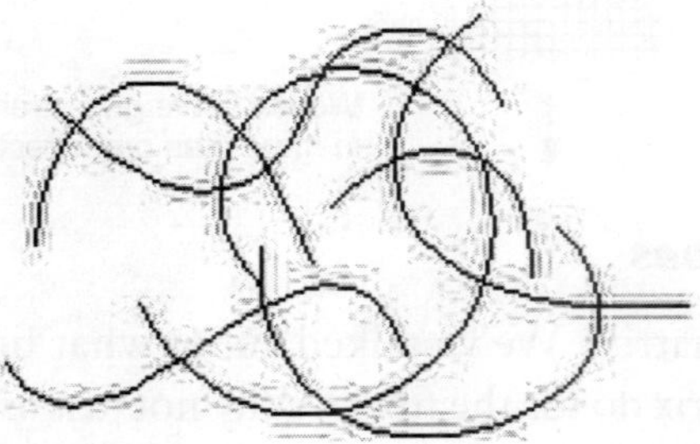

Glass fibers in Fierglas

In Fiberglas™ the fibers aren't lined up in any particular direction. They're just a tangled mass, like you see on the right. But we can make the composite stronger by lining up all the fibers in the same direction. Oriented fibers do some weird things to the composite. When you pull on the composite in the direction of the fibers, the composite is very strong. But if you pull on it at right angles to the fiber direction, it isn't very strong at all.

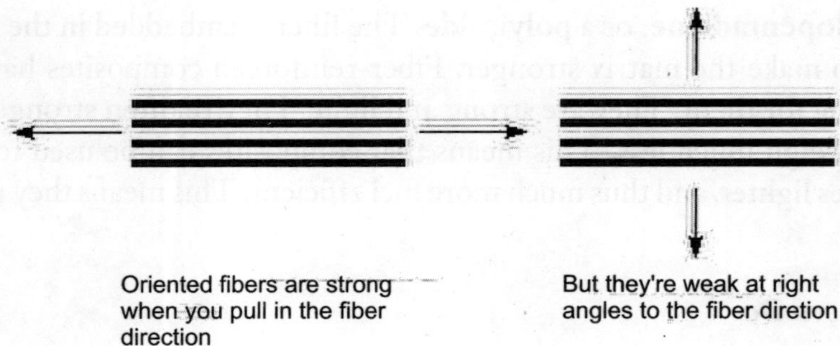

This isn't always bad, because sometimes we only need the composite to be strong in one direction. Sometimes the item you're making will only be under stress in one direction.

But sometimes we need strength in more than one direction. So we simply point the fibers in more than one direction. We often do this by using a woven fabric of the fibers to reinforce the composite. The woven fibers give a composite good strength in many directions.

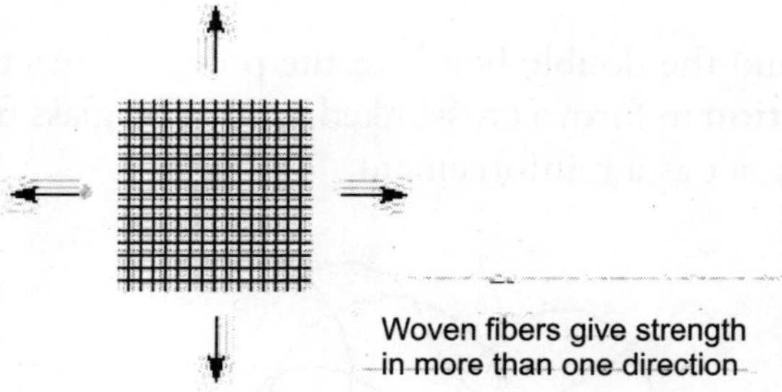

What the Matrix Does

But what about the matrix? We've talked about what fibers do for the matrix, but what does the matrix do for the fiber? Why not just use fibers by themselves? First of all, the matrix holds the fibers together. A loose bundle of fibers wouldn't be of much use. Also, though fibers are strong, they can be brittle. The matrix can absorb energy by deforming under stress. This is to say, the matrix adds *toughness* to the composite. And finally, while fibers have good *tensile* strength (that is, they're strong when you pull on them), they usually have awful *compressional* strength. That is, they buckle when you squash them. The matrix gives compressional strength to the composite.

If you'd like to know more about just what it means to be strong and tough, and what the difference between strength and toughness is, visit our page dealing with the **Mechanical Properties of Polymers.**

Measuring up the Fibers

Not all fibers are the same. They all have good points and bad points. Glass is the most popular fiber. Why? Because it's really cheap. Now it may seem strange that glass is used as a reinforcement, as glass is really easy to break. I know this from years of destructive experience in the laboratory. But for some reason, when glass is spun into really tiny fibers, it acts very different. Glass fibers are strong, and flexible.

As a matter of fact...You know those glass slippers that Cinderella wore to the ball? They weren't made of plain old glass. They were actually made from a glass fiber-reinforced composite material. I mean, think about it. Glass shoes? They'd break the first time Cindy stepped on a piece of gravel in the driveway of Prince Charming's palace! And even if they didn't they'd definitely shatter when the clumsy prince stepped on Cinderella's foot dancing. But slippers made from glass-reinforced composites would be strong enough to handle a stomp by the biggest royal foot!

Still, there are stronger fibers out there. This is a good thing, because sometimes glass just isn't strong and tough enough. For some things, like airplane parts, that undergo a lot of stress, you need to break out the fancy fibers. When cost is no object, you can use stronger, but more expensive fibers, like Kevlar™, carbon fiber, or Spectra™. Carbon fiber is usually stronger than Kevlar®, that is, it can withstand more force without breaking. But Kevlar™ tends to be *tougher*. This means it can absorb more energy without breaking. It can stretch a little to keep from breaking, more so than carbon fiber can. But Spectra™, which is a kind of **polyethylene**, is stronger *and* tougher than both carbon fiber and Kevlar™.

Measuring up the Matrices

Different jobs call for different matrices. When we want to save money, there are some cheap matrices that have decent properties. The unsaturated polyester/styrene systems we've already looked at are one example. They're fine for everyday applications. Chevrolet Corvette bodies are made from composites using unsaturated polyester matrices and glass fibers. But they have some drawbacks. They shrink a good deal when they're cured, they can absorb water very easily, and their impact strength is low. They're also not very chemically resistant.

Another low-cost system is the so-called vinyl ester resin. The first step to making a vinyl ester resin is reacting a di-epoxide with acrylic acid, or methacrylic acid:

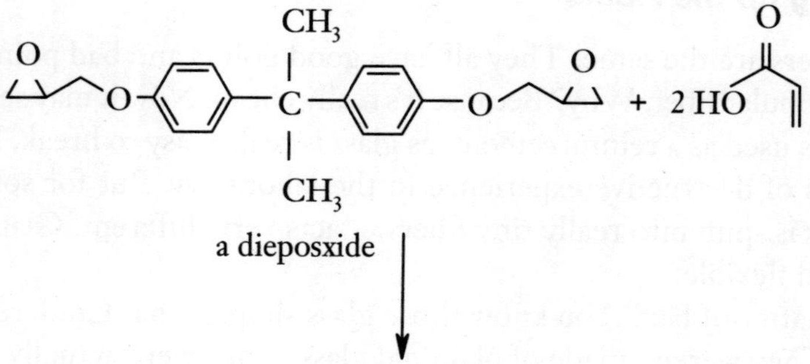

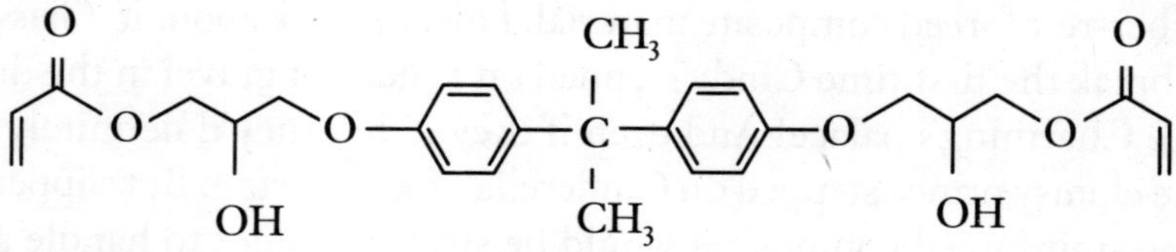

a dieposxide

Then you polymerize the vinyl groups, and you've got a crosslinked resin. Sometimes we use bigger oligomers that look like this:

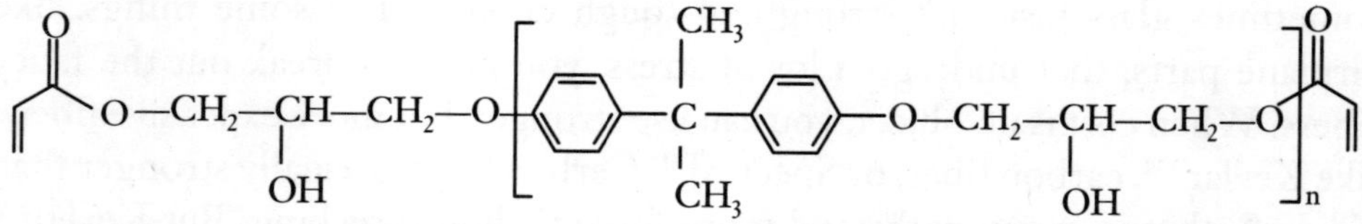

And they're cured the same way, by polymerizing the vinyl groups. Vinyl ester resins have some advantages over unsaturated polyesters. They don't absorb as much water, and they don't shrink nearly as much when cured. Plus, they have very good chemical resistance. Also, because of the hydroxyl groups, it bonds well to glass. This comes in handy if you're using glass as your fiber.

But neither vinyl ester resins nor unsaturated polyesters are very good for high temperature applications. For high temperatures, we need to use matrices like **epoxy resins**. To make these, we start with a di-epoxide, just like we did when we were making vinyl ester resins. Only this time, we don't react it with acrylic acid. Instead, we cure it with a diamine. The epoxide groups will react with the diamine, and the whole system becomes crosslinked:

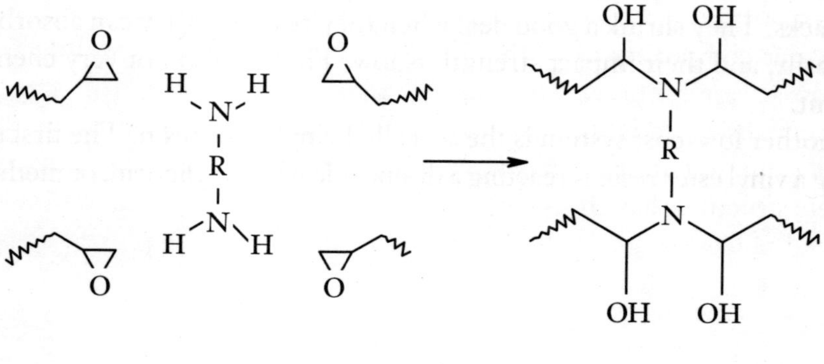

Because of all those hydroxyl groups, epoxy resins can bond well to glass fibers. But they have some properties you can't get with cheaper matrices. They don't absorb water, and they don't shrink very much when they're cured. And they can be used at higher temperatures, up to 160 °C in some cases.

But for REALLY high temperature applications, we're not sure what to do. There are a lot of options. **Polyimides** are good at resisting temperature, but they absorb a lot of water, which causes them to break down. Polybenzoxazoles are good at resisting temperatures, but they are almost impossible to process. Some people are interested in all-hydrocarbon matrices. Research is still underway in this field.

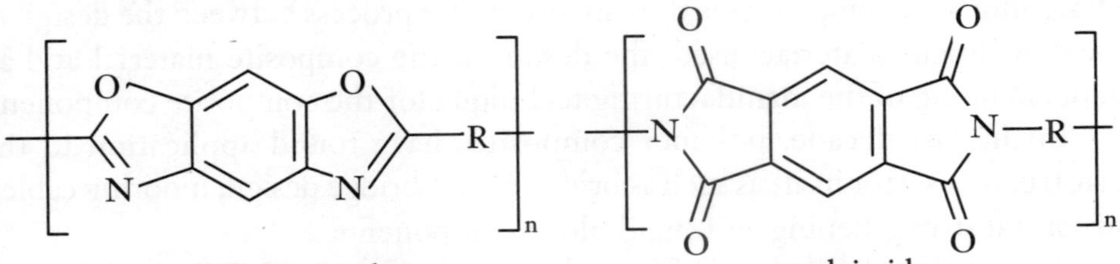

a polybenzoxazole a polyimide

Most of the interest is in aerospace. There's a lot of interest right now in making a space-plane that can fly from Tokyo to Los Angeles in three hours. This plane would enter a low earth orbit during its flight. That means it would have to re-enter the atmosphere, and re-entry will generate a lot of heat on the surface of airplane. For the plane to make it through, it needs advanced composites that can stand up to the torture of all that heat.

6.6 Polymer Composites in Construction

About 30 per cent of all polymers produced each year are used in the civil engineering and building industries. Polymers offer many advantages over conventional materials including lightness, resilience to corrosion and ease of processing. They can be combined with fibres to form composites which have enhanced properties, enabling them to be used as structural members and units. Polymer composites can be used in many different forms ranging from structural composites in the construction industry to the high technology composites of the aerospace and space satellite industries.

Polymer composites were first developed during the 1940's, for military and aerospace applications. Considerable advances have been made since then in the use of this material and applications developed in the construction sector. Load bearing and infill panels have been manufactured using composites. Complete structures have been fabricated where units manufactured from glass-reinforced polyester are connected together to form the complete system in which the shape provides the rigidity. Glass-reinforced plastics have been used in many other applications including pressure pipes, tank liners, and roofs.

In contrast to the design procedures used for the more traditional construction materials, those for polymer composite materials in structural applications require greater development effort and a wider understanding of the material. The material properties of the final component are the result of a design process that considers many factors which are characterised by the anisotropic behaviour of the material and cover the micro-mechanical, elasticity, strength and stability properties. These properties are influenced by manufacturing techniques, environmental exposure and loading histories. Designing with composites is thus an interactive process between the design of the constituent materials used, the design of the composite material and an understanding of the manufacturing technique for the composite component.

In the last decade, polymer composites have found application in the construction sector in areas such as bridge repair, bridge design, mooring cables, structural strengthening and stand-alone components

This project, funded by DTI, was launched in July 2001 with the following key objectives:

- collate, interpret and analyse international research knowledge of FRP materials in construction from 1990 onwards.
- repackage the key information into informative and digestible forms, to establish and highlight synergy's with traditional construction materials for achieving design and cost efficiencies

The principal target audience for this project is individuals involved in application of building materials within the construction industry that have little or no knowledge of composites or are for whatever reason predisposed against them. Dissemination plays a major role within the project and the team will be working closely with NGCC to ensure this is carried out efficiently and effectively. Trend2000 and BRE will lead the project with support from Brookes Stacey Randall, Concrete society, Fibreforce Composites ltd, Tony Gee and Partners, University of Sheffield, and Saint-Gobain Vetrotex. The project will run for 18 months from its start date.

6.7 Polymer Composites Group

The Polymer Composites Group operates within the School of Mechanical, Materials and Manufactuiing Engineering and focuses on the processing and performance of polymer matrix composites. We have interests in both thermoset and thermoplastic matrices.

The group includes 8 members of academic staff, 3technical staff and around 30 full time research workers. Industrial links have always been a strong feature of our work, much of which involves collaboration with Ford Motor Company. The work is by no means exclusively automotive however and we have projects running which are linked with many other industrial sectors including aerospace, power generation and biomedical materials.

We have recently been awarded a prestigious EPSRC Platform Grant for our internationally leading work on processing and performance of textile composites.

Current Projects

We have several ongoing composites research projects, most of which are funded by collaborative research grants, although a few are supported directly by industry. The programme spans design, processing, performance and end-of-life disposal.

PLATFORM GRANT: Processing and Performance of Textile Composites

Funding body: EPSRC (GR/T18578/01)
Research grant: £445k—C D Rudd, A C Long, R Brooks, I A Jones, S J Pickering, N A Warrior, M J Clifford, M S Johnson C A Scotchford, G A Walker

UV Curing of Textile Composites (CURE-TEX)

Funding body: DTI Technology Programme (TP/2/MS/6/I/10062)
Partners: Pera, Uvasol, Composite Integration, VT Halmatic, Ciba, Formax, BI Composites, Litetec
Research grant: £150k (total project value £1,050k)—A C Long, M S Johnson

High Value Composite Materials from Recycled Carbon Fibre (HIRECAR)

Funding body: DTI Technology Programme
Partners: Advanced Composites Group, Ford Motor Company, Technical Fibre Products, Toho Tenax
Research grant: £333k (total project value £927k)—N A Warrior, S J Pickering, E Lester

Development of Novel Anti-slash Technical Textiles

Funding body: DTI Technology Programme (TP/3/SMS/6/I/16992)
Partners: Cunningham Covers, Monarch, JD Wilkie, Montgomery, William Clark & Sons, Soltex, Pera
Research grant: £ 220k (total project value £1,412k)—A C Long

Fatigue, Damping and Impact Properties of Textile Composites

Funding body: EPSRC (EP/C538137/1)
Partners: Dowty Propellers, Rolls Royce, J R Technology Ltd, Cambridge University, Cranfield University, Auburn University, University of Alabama at Birmingham
Research grant: £ 168k - I A Jones, A C Long, N A Warrior

Simulation and Modelling of 3D Woven Fabrics for Structural Composites (3DSimComs)

Funding body: DTI Technology Programme (TP/3/DSM/6/I/16666)
Partners: Rolls Royce, Advanced Composites Group, BAE Systems, Deepsea Engineering, Dowty Propellers, Sigmatex, Bristol University, Ulster University
Research grant: £ 189k (total project value £1,850k)—N A Warrior, A C Long, I A Jones

Fibre Reinforced Healthcare

Funding body: EPSRC via NIMRC (EP/E001904/1)
Partners: University of Leeds, EPFL, Smith and Nephew1, UCL
Research grant: £477k - C D Rudd, I A Jones, G S Walker, C A Scotchford

Multi-scale Integrated Modelling for High Performance Flexible Materials

Funding body: DTI Technology Programme (TP/5/MAT/6/I/H0558C)
Partners: Unilever, OCF, Croda, Scotweave, Remploy, Carrington Career & Workwear, Hield Brothers, Airbag International, Technitex Faraday, University of Manchester, Heriot-Watt University
Research grant: £318k (total project value £1,703k) - M J Clifford, A C Long, H Lin

Analysis and Modelling of Advanced Preform Processing (AMAPP)

Funding body: EPSRC via NIMRC (EP/E001904/1)
Partners: Airbus UK, BAE Systems, Bentley Motors, Carr Reinforcements, Dowty Propellers, ESI Software, Rolls Royce
Research grant: £495k—A C Long

Warpage compensation in composite laminate manufacture (WILMA)

Funding body: DTI Technology Programme (TP/5/MAT/6/I/H0473F)
Partners: FEA Ltd, Advanced Composites Group, BAE Systems, Bombardier Aerospace UK
Research grant: £242k (total project value £2,256k)—I A Jones, A C Long

Carbon Fibre Recycling—Materials and Product Development

Funding body: Boeing Corporation.
Research Grant: \$260k—S J Pickering, N A Warrior

Damage Modelling to Develop Novel Hybrid Structures/Materials Comprising Textile Composites and Shock Absorbing Layers

Funding body: EPSRC via NIMRC (EP/E001904/1)
Partners: AIGIS Ltd, Carr Reinforcements, St-Gobain Vetrotex, Security Composites
Research grant: £378k—R Brooks

Affordable Innovative Rapid Production of Off-shore Wind Energy Rotor-blades (AIRPOWER)

Funding body: DTI Technology Programme
Partners: Insensys Limited, Gamesa Innovation & Technology SLU, BAE Systems, GE Aviation, NaREC, Solent Composite Systems, Hexcel Composites, Magnum Venus, Plastech, Wind Force GmbH, WindSupply
Research grant: £302k (total project value £1,405k) - P J Schubel, N A Warrior

7

Polymer Fibre Composite Materials

7.1 Fibre-reinforced Plastic

A **fibre-reinforced plastic** (**FRP**) (also *fibre-reinforced polymer*) is a composite material comprising a polymer matrix reinforced with fibres. The fibers are usually fiberglass, carbon, or aramid, while the polymer is usually an epoxy, vinylester or polyester thermosetting plastic. FRPs are commonly used in the aerospace, automotive, marine, and construction industries.

7.2 Glass-reinforced Plastic

Glass-reinforced plastic (**GRP**), is a composite material or fiber-reinforced plastic made of a plastic reinforced by fine fibers made of glass. Like graphite-reinforced plastic, the composite material is commonly referred to by the name of its reinforcing fibers (fiberglass). The plastic is chemosetting, most often polyester or vinylester, but other plastics, like epoxy (GRE), are also used. The glass is mostly in the form of chopped strand mat (CSM), but woven fabrics are also used.

An individual structural glass fiber is both stiff and strong in tension and compression—that is, along its axis. (Although one might intuitively imagine the fiber to be weak in compression, it is actually only the long aspect ratio of the fiber which makes it seem so; i.e., because a typical fiber is long and narrow, it buckles easily.) On the other hand, the glass fiber is relatively unstiff and unstrong in shear—that is, across its axis. In other words, the fiber is stiff and strong in a preferred direction, namely, along its length. Therefore if a collection of fibers can be arranged permanently in a preferred direction within a material, and if the fibers can be prevented from buckling in compression, then that material will become preferentially strong in that direction. Furthermore, by laying multiple layers of fiber on top of one another, with each layer oriented in various preferred directions, the stiffness and strength properties of the overall material can be controlled in an efficient manner. In the case of glass-reinforced plastic, it is the plastic matrix which permanently constrains the structural glass fibers to directions chosen by the designer. With chopped strand mat, this directionality is essentially an entire two dimensional plane; with woven fabrics or unidirectional layers, directionality of stiffness and strength can be more precisely controlled within the plane. A glass-reinforced plastic component is typically of a thin "shell" construction, sometimes filled on the inside with structural foam, as in the case of surfboards. The component may be of nearly arbitrary shape, limited only by the complexity and tolerances of the mold used for manufacturing the shell.

Applications

GRP was developed in the UK during the Second World War as a replacement for the molded plywood used in aircraft radomes (GRP being transparent to microwaves). Its first main civilian application was for building of boats, where it gained acceptance in the 1950s. Its use has broadened to the automotive and sport equipment sectors, although its use there is being taken over by carbon fiber which weighs less per given volume and is stronger both by volume and by weight. GRP uses also include hot tubs, pipes for drinking water and sewers, office plant display containers and flat roof systems.

Advanced manufacturing techniques such as pre-pregs and fiber rovings extend the applications and the tensile strength possible with fiber-reinforced plastics.

GRP is also used in the telecommunications industry for shrouding the visual appearance of antennas, due to its RF permeability and low signal attenuation properties. It may also be used to shroud the visual appearance of other equipment where no signal permeability is required, such as equipment cabinets and steel support structures, due to the ease with which it can be molded, manufactured and painted to custom designs, to blend in with existing structures or brickwork. Other uses include sheet form made electrical insulators and other structural components commonly found in the power industries.

Storage Tanks

A wide variety of storage tanks are made of GRP with capacities up to about 300 tonnes. The smaller tanks can be made with chopped strand mat cast over a thermoplastic inner tank which acts as a preform during construction. Much more reliable tanks are made using woven mat or filament wound fibre with the fibre orientation at right angles to the hoop stress imposed in the side wall by the contents. They tend to be used for chemical storage because the plastic liner (often polypropylene) is resistant to a wide range of strong chemicals. GRP tanks are also used for septic tanks.

House Building

Glass reinforced plastics are also used in the house building market for the production of roofing laminate, door surrounds, over-door canopies, window canopies & dormers, chimneys, coping systems, heads with keystones and cills. The use of GRP for these applications provides for a much faster installation and due to the reduced weight manual handling issues are reduced. With the advent of high volume manufacturing processes it is possible to construct GRP brick effect panels which can be used in the construction of composite housing. These panels can be constructed with the appropriate insulation which reduces heat loss.

Piping

GRP and GRE pipe systems can be used for a variety of applications, underground as well as above.

- Firewater systems
- Cooling water systems
- Drinking water systems
- Waste water systems/Sewage systems
- Gas systems

Laminating Operations

Filament Winding Operation

Fiberglass Sheet Laminating Operation

First, you have to mix resin with catalyst (e.g. butanox LA) or hardener if working with epoxy, otherwise it won't cure (become hard) for days/weeks. Next, you need to wet out the mold with the resulting mixture, placing the sheets of fiberglass over it. The sheets are rolled down into to mold using more resin, the operator must make sure it is securely attached to the mold, air must not be trapped in between the fiberglass layers. Steel rollers are useful to make sure the resin is between all the layers, and the glass is wet with epoxy throughout the entire thickness of the laminate. One must work quickly depending on the curing time of the epoxy used (various cure times are available depending on the amount of catalyst employed), otherwise the epoxy will cure prematurely and the process will need to be started anew.

Fiberglass Spray Lay-Up Operation

The fiberglass spray lay-up process is similar to the hand lay-up process but the difference comes from the application of the fiber and resin material to the mold. Spray-up is an open-molding composites fabrication process where resin and reinforcements are sprayed onto a mold. The resin and glass may be applied separately or simultaneously "chopped" in a combined stream from a chopper gun. Workers roll out the spray-up to compact the laminate. Wood, foam or other core material may then be added, and a secondary spray-up layer imbeds the core between the laminates. The part is then cured, cooled and removed from the reusable mold.

Fiberglass Hand Lay-Up Operation

Pultrusion Operation: Pultrusion is a manufacturing method used to make strong light weight composite materials, in this case fiberglass. Fibers (the glass material) are pulled from spools through a device that coats them with a resin. They are then typically heat treated and cut to length. Pultrusions can be made in a variety of shapes or cross-sections such as a W or S cross-section. The word pultrusion describes the method of moving the fibers through the machinery. It is pulled through using either a hand over hand method or a continuous roller method. This is opposed to an extrusion which would push the material through dies.

Chopped strand mat

Chopped strand mat or **CSM** is a form of reinforcement used in glass-reinforced plastic. It consists of glass-fibers laid randomly across each other and held together by a binder.

It is typically processed using the hand lay-up technique, where sheets of material are placed in a mold and brushed with resin. Because the binder dissolves in resin, the material easily conforms to different shapes when wetted out. After the resin cures, the hardened product can be taken from the mold and finished.

Using chopped strand mat gives a glass-reinforced plastic with isotropic in-plane material properties.

Examples of GRP Use

- Sailplanes, kit cars, Sports cars, microcars, karts, bodyshells, boats, kayaks, flat roofs, lorries, wind turbine blades.
- Pods, domes and architectural features where a light weight is necessary.
- Bodies for automobiles, such as the Chevrolet Corvette and Pontiac Fiero.
- A320 Radome.

Kayaks made of GRP

7.3 Carbon Fiber Reinforced Plastic

Carbon fiber reinforced plastic (**CFRP** or **CRP**), is a very strong, light and expensive composite material or fiber reinforced plastic. Similar to glass-reinforced plastic, sometimes known by the genericised trademark fiberglass, the composite material is commonly referred to by the name of its reinforcing fibers (carbon fiber). The plastic is most often epoxy, but other plastics, such as polyester, vinyl ester or nylon, are also sometimes used. Some composites contain both carbon fiber and other fibres such as kevlar, aluminium and fiberglass reinforcement. The terms **graphite-reinforced plastic** or **graphite fiber reinforced plastic** (**GFRP**) are also used but less commonly, since glass-(fibre)-reinforced plastic can also be called GFRP.

It has many applications in aerospace and automotive fields, as well as in sailboats, and notably in modern bicycles and motorcycles, where its high strength to weight ratio is of importance. Improved manufacturing techniques are reducing the costs and time to manufacture making it increasingly common in small consumer goods as well, such as laptop computers, tripods, fishing rods, paintball equipment, racquet sports frames, stringed instrument bodies, classical guitar strings, and drum shells.

Composite

Materials produced with the above-mentioned methodology are often generically referred to as *composites*. The choice of matrix can have a profound effect on the properties of the finished composite. One method of producing graphite-epoxy parts is by layering sheets of carbon fiber cloth into a mold in the shape of the final product. The alignment and weave of the cloth fibers is chosen to optimize the strength and stiffness properties of the resulting material. The mold is then filled with epoxy and is heated or air cured. The resulting part is very corrosion-resistant, stiff, and strong for its weight. Parts used in less critical areas are manufactured by draping cloth over a mold, with epoxy either preimpregnated into the fibers (also known as *prepreg*), or "painted" over it. High performance parts using single molds are often vacuum bagged and/or autoclave cured, because even small air bubbles in the material will reduce strength.

Process

The process in which most carbon fiber reinforced plastic is made varies, depending on the piece being created, the finish (outside gloss) required, and how many of this particular piece are going to be produced.

For simple pieces of which relatively few copies are needed, (1-2 per day) a vacuum bag can be used. A fiberglass or aluminum mold is polished, waxed, and has a release agent applied before the fabric and resin are applied and the vacuum is pulled and set aside to allow the piece to cure (harden). There are two ways to apply the resin to the fabric in a vacuum mold. One is called a wet layup, where the two-part resin is mixed and applied before being laid in the mold and placed in the bag. The other is a resin induction system, where the dry fabric and mold are placed inside the bag while the vacuum pulls the resin through a small tube into the bag, then through a tube with holes or something similar to evenly spread the resin throughout the fabric. Wire loom works perfectly for a tube that requires holes inside the bag. Both of these methods of applying resin require hand work to spread the resin evenly for a glossy finish with very small pin-holes. A third method of constructing composite materials is known as a dry layup. Here, the carbon fiber material is already impregnated with resin (prepreg) and is applied to the mold in a similar fashion to adhesive film. The assembly is then placed in a vacuum to cure. The dry layup method has the least amount of resin waste and can achieve lighter constructions than wet layup. Also, because larger amounts of resin are more difficult to bleed out with wet layup methods, prepreg parts generally have fewer pinholes. Pinhole elimination with minimal resin amounts generally require the use of autoclave pressures to purge the residual gases out.

A quicker method uses a compression mold. This is a two-piece (male and female) mold usually made out of fiberglass or aluminum that is bolted together with the fabric and resin between the two. The benefit is that, once it is bolted together, it is relatively clean and can be moved around or stored without a vacuum until after curing. However, the molds require a lot of material to hold together through many uses under that pressure.

Many carbon fiber reinforced plastic parts are created with a single layer of carbon fabric, and filled with fiberglass. A chopper gun can be used to quickly create these types of parts. Once a thin shell is created out of carbon fiber, the chopper gun is a pneumatic tool that cuts fiberglass from a roll and sprays resin at the same time, so that the fiberglass and resin are mixed on the spot. The resin is either external mix, where the hardener and resin are sprayed separately, or internal, where they are mixed internally, which requires cleaning after every use.

For difficult or convoluted shapes, a filament winder can be used to make pieces.

Automotive Uses

Carbon fiber reinforced plastic is used extensively in high end automobile racing. The high cost of carbon fiber is mitigated by the material's unsurpassed strength-to-weight ratio, and low weight is essential for high-performance automobile racing. Racecar manufacturers have also developed methods to give carbon fiber pieces strength in a certain direction, making it strong in a load-bearing direction, but weak in directions where little or no load would be placed on the member. Conversely, manufacturers developed omnidirectional carbon fiber weaves that apply strength in all directions. This type of carbon fiber assembly is most widely used in the "safety cell" monocoque chassis assembly of high-performance racecars.

Several supercars over the past few decades have incorporated CFRP extensively in their manufacture, using it for their monocoque chassis as well as other components.

Until recently, the material has had limited use in mass-produced cars because of the expense involved in terms of materials, equipment, and the relatively limited pool of individuals with expertise in working with it. Recently, several mainstream vehicle manufacturers have started to use CFRP in everyday road cars.

Use of the material has been more readily adopted by low-volume manufacturers who used it primarily for creating body-panels for some of their high-end cars due to its increased strength and decreased weight compared with the glass-reinforced plastic they used for the majority of their products.

Often street racers or hobbyist tuners will purchase a carbon fiber reinforced plastic hood, spoiler or body panel as an aftermarket part for their vehicle. However, these parts are rarely made of full carbon fiber. They are often just a single layer of carbon fiber laminated onto fiberglass for the "look" of carbon fiber. It is common for these parts to remain unpainted to accentuate the look of the carbon fiber weave.

Civil Engineering Applications

Carbon fiber reinforced plastic has over the past two decades become an increasingly notable material used in structural engineering applications. Studied in an academic context as to its potential benefits in construction, it has also proved itself cost-effective in a number of field applications strengthening concrete, masonry, steel and timber structures. Its use in industry can be either for retrofitting to strengthen an existing structure, or as an alternative reinforcing (or prestressing material) to steel from the outset of a project.

Retrofitting has become the increasingly dominant use of the material in civil engineering, and applications include increasing the load capacity of old structures (such as bridges) that were designed to tolerate far lower service loads than they are experiencing today, seismic retrofitting, and repair of damaged structures. Retrofitting is popular in many instances as the cost of replacing the deficient structure can greatly exceed its strengthening using CFRP. Due to the incredible stiffness of CFRP, it can be used underneath bridge spans to help prevent excessive deflections, or wrapped around beams to limit shear stresses.

When used as a replacement for steel, CFRP bars are used to reinforce concrete structures. More commonly they are used as prestressing materials due to their high stiffness and strength. The advantages of CFRP over steel as a prestressing material, namely its light weight and corrosion resistance, enable the material to be used for niche applications such as in offshore environments.

CFRP is a more costly material than its counterparts in the construction industry, glass fibre reinforced polymer (GFRP) and aramid fibre reinforced polymer (AFRP), though CFRP is generally regarded as having superior properties.

Much research continues to be done on using CFRP both for retrofitting and as an alternative to steel as a reinforcing or prestressing material. Cost remains an issue and long term durability questions still remain. Some are concerned about the brittle nature of CFRP, in contrast to the ductility of steel. Though design codes have been drawn up by institutions such as the American Concrete Institute, there remains some hesitation among the engineering community about implementing these alternative materials. In part this is due to a lack of standardisation and the proprietary nature of the fibre and resin combinations on the market, though this in itself is advantageous in that the material properties can be tailored to the desired application requirements.

Other Applications

Carbon fiber reinforced plastic has found a lot of use in high-end sports equipment such as racing bicycles. For the same strength, a carbon-fiber frame weighs less than a bicycle tubing of aluminum or steel. The choice of weave can be carefully selected to maximize stiffness. The variety of shapes it can be built into has further increased stiffness and also allowed aerodynamic considerations into tube profiles. Carbon fiber reinforced plastic frames, forks, handlebars, seatposts and crank arms are becoming more common on medium- and higher-priced bicycles. Carbon fiber reinforced plastic forks are used on most new racing bicycles. Despite carbon fiber reinforced plastic's advantages, it has been known to fail suddenly in bicycles, causing devastating crashes. Other sporting goods applications include rackets, fishing rods and rowing shells.

Much of the fuselage of the new Boeing 787 Dreamliner and Airbus A350 XWB will be composed of CFRP, making the aircraft lighter than a comparable aluminum fuselage, with the added benefit of less maintenance thanks to CFRP's superior fatigue resistance.

Due to its high ratio of strength to weight, CFRP is widely used in micro air vehicles (MAVs). In MAVSTAR Project, the CFRP structures reduce the weight of the MAV significantly. In addition, the high stiffness of the CFRP blades overcome the problem of collision between blades under strong wind.

CFRP has also found application in the construction of high-end audio components such as turntables and loudspeakers, again due to its stiffness.

It is used for parts in a variety of musical instruments, including violin bows, guitar pickguards, and a durable ebony replacement for bagpipe chanters. It is also used to create entire musical instruments such as Blackbird Guitars carbon fiber rider models, Luis and Clark carbon fiber cellos, and Mix carbon fiber mandolins.

In firearms it can substitute for metal, wood, and fiberglass in many areas of a firearm in order to reduce overall weight. However, while it is possible to make the receiver out of synthetic material such as carbon fiber, many of the internal parts are still limited to metal alloys as current synthetic materials are unable to function as replacements.

Nike often uses carbon fiber as a shank plate in their basketball sneakers to keep the foot stable. It usually runs the length of the sneaker just above the sole and is left exposed in some areas, usually in the arch of the foot.

CFRP is used, either as standard equipment or aftermarket parts, in high performance radio controlled vehicles and aircraft, i.a. for the main rotor blades of radio controlled helicopters—which should be light and stiff to perform 3D manoeuvres.

End of Useful Life/Recycling

Carbon fiber reinforced plastics (CFRPs) have an almost infinite service lifetime when protected from the sun. When it is time to decommission CFRPs they cannot be melted down in air like many metals. When free of vinyl (PVC or polyvinyl chloride) and other halogenated polymers, CFRPs can be thermally decomposed via thermal depolymerization in an oxygen free environment. This can be accomplished in a refinery in a one-step process. Capture and reuse of the carbon and monomers is then possible. CFRPs can also be milled or shredded at low temperature to reclaim the carbon fiber, however this process shortens the fibers dramatically. Just as with downcycled paper, the shortened fibers cause the recycled material to be weaker than the original material. There are still many industrial applications that do not need the strength of full-length carbon fiber reinforcement. For example, chopped reclaimed carbon fiber can be used in consumer electronics, such as laptops. It provides excellent reinforcement of the plastics used even if it lacks the strength-to-weight ratio of an aerospace component.

7.4 Polymer and Fiber Engineering

Many engineering achievements develop in conjunction with advances in engineered materials. Strong yet lightweight structures for aircraft, automobile components, and high performance racing bikes are a few examples. Increasingly, these engineered materials utilize polymers and fibers. The relationship between the structure, properties, and performance of these materials is critical to advances in technology.

With major input from our industry board and alumni, our faculty has restructured the undergraduate engineering curriculum to meet the changing requirements of the industrial, government, and academic employers of our graduates. The new curriculum, **Polymer and Fiber Engineering**, reflects the importance of polymers, composite materials, and fibrous materials in such diverse fields as plastics, elastomers (rubber), adhesives, surface coatings (paints), paper, packaging, insulation, filtration, biomedical, automotive, aerospace, marine, construction, environmental, industrial, nonwoven, recreational, and safety materials.

Research and instruction in polymer and fiber engineering includes:

- polymer synthesis and processing
- characterization and evaluation of structure and properties of polymeric materials using advanced techniques and state-of-the-art instrumentation
- modeling of structure-property-performance relationships emphasizing correlation of properties with the structure across nano-, micro-, and macro-length scales
- design, analysis, engineering, and assembly of polymeric fibrous materials into advanced engineered materials with novel compositions and tailored microstructures
- product, mold, and die design

Graduate offerings in the department include the masters degree in polymer and fiber engineering and the doctoral degree in integrated textile and apparel science, which is jointly administered with the Department of Consumer Affairs.

7.5 Bi-component Polymer Fibers Made by Rotary Process

TECHNICAL FIELD This invention relates in general to the manufacture of polymer fibers, and specifically to a method for manufacturing bicomponent polymer fibers by a modified rotary process.

BACKGROUND Bicomponent mineral fibers, such as glass, have previously been made by a modified rotary process. Two different types of molten glass are supplied to a rotating spinner having an orificed peripheral wall. The two types of molten glass are centrifuged through the orifices to form bicomponent glass fibers. The fibers are particularly useful in insulation products.

The manufacture of glass fibers is a different field from the manufacture of polymer fibers. The two materials have different physical properties such as different viscosities and melting points. The technologies for making the fibers are also different. Bicomponent polymer fibers have previously been made by a textile process. In this process, two molten polymers are supplied to a stationary spinneret having holes from which fibers are pulled or drawn. The polymers are usually combined to form fibers having a core of one polymer and a surrounding sheath of the other polymer. The fibers are useful in products such as fabrics and hosiery. For example, in a typical process two different types of nylon are formed into bicomponent fibers for making hosiery. The textile process usually makes bicomponent fibers having a relatively large diameter.

For some applications it is desirable to make bicomponent fibers from polymers that are difficult to fiberize together, or difficult to fiberize at all. The polymers may be difficult to fiberize at all because they easily break apart during fiberizing. They may be difficult to fiberize together because they require different fiberizing conditions in view of their different physical properties. It would be advantageous to provide a method which, more easily than a textile process, can make bicomponent fibers from difficult to fiberize polymers.

For other applications, there are advantages to using bicomponent polymer fibers having a relatively small diameter. Therefore, it would also be advantageous to provide a method which can make small diameter bicomponent fibers more easily than a textile process.

DISCLOSURE OF INVENTION This invention relates to a method for making multicomponent polymer fibers, and particularly bicomponent polymer fibers. In the method, first and second molten polymers are supplied to a rotating spinner having an orificed peripheral wall. The molten polymers are centrifuged through the orifices as molten bicomponent polymer streams. Then the streams are cooled to make bicomponent polymer fibers.

The bicomponent polymer fibers of this invention can be formed from polymers that are difficult to fiberize together, or difficult to fiberize at all. For example, the fibers can be formed from two polymers that have different coefficients of thermal expansion, to make curvilinear fibers for high loft wool packs or webs having excellent insulating properties. As another example, the fibers can be formed from two polymers that have different melting points to make heat fusible fibers. The method of this invention can easily form fibers having a small diameter.

Best Mode for Carrying Out the Invention

It is to be understood, however, that various fabrication processes can be used with the bicomponent polymer fibers to make textiles, filtration products, and other products. Such processes include stitching, needling, hydro-entanglement, and encapsulation. It is also understood that multicomponent fibers other than bicomponent fibers are included in the invention, and that the fibers can be formed from other thermoplastic materials such as asphalt in addition to polymers.

In the illustrated process, two distinct molten polymer compositions (polymer A and polymer B) are supplied to polymer spinners 10. The molten polymer compositions are supplied from any suitable source. For example, hoppers 12 containing polymer granules can be connected to extruders 14 where the polymers are melted and then supplied to the spinners. As will be described below, the spinners produce veils 16 of bicomponent polymer fibers. The fibers are directed downwardly by any means, such as by annular blower 18. As the fibers are blown downwardly, they are attenuated and cooled. The fibers are collected as a wool pack 20 on any suitable surface, such as conveyor 22. A partial vacuum, not shown, can be positioned beneath the conveyor to facilitate fiber collection.

The wool pack of bicomponent polymer fibers may then optionally be passed through a station for further processing, such as oven 24. While passing through the oven, the wool pack is preferably shaped by top conveyor 26 and bottom conveyor 28, and by edge guides (not shown). The wool pack exits the oven as insulation product 30.

The spinner is rotated on any suitable means, such as spindle 36, as is known in the art. The rotation of the spinner centrifuges molten polymer through orifices in the peripheral wall to form bicomponent polymer fibers 38, in a manner described in greater detail below. The spinner preferably rotates at a speed from about 1200 rpm to about 3000 rpm. Spinners of various diameters can be used, and the rotation rates adjusted to give the desired radial acceleration at the inner surface of the peripheral wall. The spinner diameter is preferably from about 20 centimeters to about 100 centimeters. The radial acceleration (velocity/radius) of the inner surface of the peripheral wall is preferably from about 4,500 meters/second^2 to about 14,000 meters/second^2, and more preferably from about 6,000 meters/second to about 9,000 meters/second".

Preferably the interior of the spinner is heated by any heating means (not shown) such as by blowing in hot air or other gas. The temperature of the spinner is preferably from about 150°C to about 300°C but can vary depending on the type of polymers.

A heating means such as annular hot air supply 42 can optionally be positioned outside the spinner to heat either the spinner or the fibers, to facilitate the fiber attenuation and maintain the temperature of the spinner at the level for optimum centrifugation of the polymers. The interior of the spinner is supplied with two separate streams of molten polymer, a first stream containing polymer A and a second stream containing polymer B. Preferably the streams of molten polymer are supplied by injection under pressure. The polymer A in the first stream drops from a first delivery tube 44 directly onto the bottom wall and flows outwardly due to the centrifugal force toward peripheral wall to form a head of polymer A as shown. Polymer B. delivered via a second delivery tube 46, is positioned closer to the peripheral wall than the first stream, and polymer B is intercepted by annular horizontal flange 48 before it can reach the bottom wall. Thus, a build-up or head of polymer B is formed above the horizontal flange as shown. It is understood that

the polymers could also be supplied so that polymer A is intercepted by the annular horizontal flange and polymer B drops to the bottom wall.

A series of vertical baffles 52, positioned between the peripheral wall and vertical interior wall, divide that space into a series of generally vertically-aligned compartments 54 which run substantially the entire height of the peripheral wall. It can be seen that the horizontal flange, vertical interior wall, and vertical baffles together comprise a divider for directing polymers A and B into alternate adjacent compartments so that every other compartment contains polymer A while the remaining compartments contain polymer B.

The peripheral wall is adapted with orifices 56 which are positioned adjacent the radially outward end of the vertical baffle 52. Each orifice has a width greater than the width of the vertical baffle, thereby enabling a flow of both polymer A and polymer B to emerge from the orifice as a single bicomponent polymer fiber. Preferably, the peripheral wall has from about 200 to about 5.000 orifices, depending on the spinner diameter and other process parameters. Where polymers A and B have different viscosities at the temperature of the spinner peripheral wall, an orifice perfectly centered about the vertical baffle 52 would be expected to emit a higher throughput of the lower viscosity polymer than the throughput of the higher viscosity polymer. One method to counteract this tendency and to balance the throughputs of the molten polymers, is to increase the height of the head of the higher viscosity polymer relative to the height of the head of the lower viscosity polymer in the spinner. Another method to balance the throughputs of the molten polymers is to position the slot orifice so that it is offset from the centerline of the vertical baffle. Another method to balance the throughputs of the molten polymers is to restrict the flow of polymer into the alternate compartments containing the low viscosity polymer, thereby partially starving the holes so that the throughputs of polymers A and B are roughly equivalent. The orifice can also be centered about the vertical baffle when the polymers have similar viscosities or when different throughputs are desirable.

The peripheral wall is adapted with rows of orifices 70 which are positioned adjacent the radial outward end of the vertical baffle. The orifices are in the shape of a "V", with one end or leg leading into a compartment containing polymer A and one leg leading into a compartment containing polymer B. The flows of both polymer A and polymer B join and emerge from the orifice as a single bicomponent polymer fiber.

The bottom wall slants upwardly as it approaches the peripheral wall. The interior of the spinner is supplied with two separate streams of molten polymer, a first stream containing polymer A and a second stream containing polymer B. The polymer in the first stream drops from a first delivery tube 78 directly onto the bottom wall and flows outwardly and upwardly due to the centrifugal force toward the peripheral wall to form a head of polymer A as shown. Polymer B, delivered via a second delivery tube 80, is positioned closer to the peripheral wall than the first stream, and polymer B is intercepted by annular horizontal flange 82 before it can reach the bottom wall. Thus, a build-up or head of polymer B is formed above the horizontal flange as shown.

The peripheral wall is adapted with a row of orifices 84 around its circumference, the orifices being positioned adjacent the radially outward end of the horizontal flange. Each orifice is in the shape of a "Y", with one arm leading to polymer A. the other arm leading to polymer B, and the base leading to the exterior of the peripheral wall. The flows of both polymer A and polymer B join and emerge from the orifice as a single bicomponent polymer fiber 86.

Other spinner configurations can also be used to supply dual streams of polymers to the spinner orifices.

The thermoplastic materials can be any heat softenable thermoplastic materials such as polymers or asphalt, including amorphous thermoplastic materials. In many applications it is desirable to use thermoplastic materials that have similar physical

properties and are relatively easy to fiberize. However, the bicomponent fibers of this invention can also be formed from thermoplastic materials that are difficult to fiberize together, or difficult to fiberize at all. Advantageously, the present rotary process can form bicomponent fibers from difficult to fiberize thermoplastic materials much more easily than a textile process. The thermoplastic materials may be difficult to fiberize at all because they easily break apart during fiberizing. They may be difficult to fiberize together because they require different fiberizing conditions in view of their different physical properties.

For example, bicomponent fibers can be formed from two polymers that have different coefficients of thermal expansion. As each fiber cools, the polymer with the greater coefficient of thermal expansion contracts at a faster rate than the other polymer. The result is stress upon the fiber, and to relieve the stress, the fiber must bend into a curve. As a result, the bicomponent polymer fibers have an irregular, curvilinear nature. Such a curvilinear nature is particularly advantageous for giving the fibers excellent insulating properties when they are used in insulating materials or textiles. Preferably the coefficient of thermal expansion of one polymer is different from that of the other polymer by an amount greater than about 5.0 ppm/°C, and more preferably greater than about 10.0 ppm/°C. Examples of two polymers having significantly different coefficients of thermal expansion are polypropylene (68 ppm/°C) and poly (ethylene terephthalate) (17 ppm/°C).

As another example, bicomponent fibers can be formed from two polymers that have different melting properties. For purposes of this invention, melting points of thermoplastic materials such as polymers are determined using DSC (Differential Scanning Calorimetry). It is understood that use of the term "melting point" does not strictly apply to some classes of thermoplastic materials, specifically amorphous materials. In such cases, the term "melting point" means the temperature at which the material softens and is easily flowable so that it can be fiberized, as known to persons skilled in the art.

One application requiring polymers having different melting points is heat fusible bicomponent polymer fibers. A wool pack or web of the fibers can be fused together by heating to a temperature sufficient to melt the lower melting polymer but not the higher melting polymer. Such heat fusible bicomponent polymer fibers are useful in many nonwoven applications.

Preferably the melting point of the first thermoplastic material is at least about 10°C greater than the melting point of the second thermoplastic material, and more preferably at least about 25°C greater. Examples of relatively high melting or softening thermoplastic materials include, but are not limited to, poly (phenylene sulfide) ("PPS"), polyethylene terephthalate) ("PET"), poly (butylene terephthalate) ("PBT"), polycarbonate, polyamide, and mixtures thereof. Examples of relatively low melting or softening thermoplastic materials include, but are not limited to, polyethylene, polypropylene, polystyrene, asphalt, and mixtures thereof.

The rotary process of this invention can also form bicomponent fibers from two thermoplastic materials having significantly different viscosities. The viscosity of the first thermoplastic material can be different from that of the second thermoplastic material by a factor within the range of from about 5 to about 1000, and usually from about 50 to about 500. For purposes of this invention, the viscosity is measured at the temperature of the peripheral wall of the spinner. Bicomponent polymer fibers having a small diameter can be formed more easily by the rotary process of this invention than by a textile process. This advantage is provided because the rotary process uses centrifugal force to attenuate the fibers instead of the mechanical attenuation of the textile process. Preferably the bicomponent polymer fibers have an average outside diameter of from about 5 microns to about 50 microns, and more preferably from about 5 microns to about 35 microns.

The rotary process of this invention can also produce a high loft nonwoven product similar to products made by a melt blowing process, without requiring the secondary processing steps typical of textile processes.

Each of the bicomponent polymer fibers of the present invention is composed of two different polymer compositions, polymer A and polymer B. If one were to make a cross-section of an ideal bicomponent polymer fiber, one half of the fiber would be polymer A, with the other half polymer B. In reality, a wide range of proportions of the amounts of polymer A and polymer B may exist in the fibers, or perhaps even over the length of an individual fiber. The percentage of polymer A may vary within the range of from about 5 per cent to about 95 per cent by weight of the total fiber, with the remainder being polymer B. In general, a group of fibers such as a wool pack will have many different combinations of percentages of polymer A and polymer B, including a small fraction of fibers that are single component. The preferred composition of the bicomponent fibers

will differ depending on the application. For some applications, preferably the bicomponent fibers comprise, by weight, from about 40 per cent to about 60 per cent polymer A and from about 40 per cent to about 60 per cent polymer B.

Cross-section photographs of fibers can be obtained by mounting a bundle of fibers in epoxy with the fibers oriented in parallel as much as possible. The epoxy plug is then cross-sectioned and polished. The polished sample surface is then coated with a thin carbon layer to provide a conductive sample for analysis by scanning electron microscopy (SEM). The sample is then examined on the SEM using a backscattered- electron detector, which displays variations in average atomic number as a variation in the gray scale. This analysis may reveal the presence of two polymers by a darker and lighter region on the cross-section of the fiber, and shows the interface of the two polymers.

If the ratio of polymer 90 to polymer 92 is 50:50, the interface 88 between polymer 90 and polymer 92 passes through the center 94 of the fiber cross-section. This requires that the bicomponent polymer fiber stream emanating from the spinner be maintained at a temperature sufficient to enable the low viscosity polymer 92 to flow around the higher viscosity polymer 90. Adjustments in the spinner operating parameters, such as hot air flow rate, blower pressure, and polymer temperature, may be necessary to achieve the desired wrap of the low viscosity polymer.

In some cases the lower viscosity polymer flows around the higher viscosity polymer to form an angle alpha of at least 270 degrees, i.e., the lower viscosity polymer flows around the higher viscosity polymer to an extent that at least 270 degrees of the circumferential surface 96 of the bicomponent polymer fiber is made up of the second polymer.

Under certain conditions the polymer 92 can flow all the way around the polymer 90 so that the polymer 92 encloses the polymer 90 to form a cladding. In that case, the entire circumferential surface 96 (360 degrees) of the bicomponent polymer fiber is the polymer 92 or the lower viscosity polymer.

The method of the invention is not limited to bicomponent fibers, but rather includes other multicomponent fibers such as the tricomponent fiber. To form this tricomponent fiber, separate streams of first, second and third molten polymers 97. 98 and 99 are supplied to a rotating spinner having an orificed peripheral wall. The polymers are maintained separate until combined in the orifices. One method is to use a spinner having a single row of orifices but where the area above the annular horizontal flange 82 is separated into alternate compartments. Thus, two streams could be fed into each orifice from above the flange while a third stream is fed into each orifice from below the flange. Other spinner structures can also be used. The first, second and third molten polymers are centrifuged through the orifices as a molten tricomponent stream, and the tricomponent stream is maintained at a temperature sufficient to enable one of the lower viscosity polymers to flow around at least one of the other polymers. Upon cooling of the tricomponent stream, a tricomponent fiber is formed. Another method to form a tricomponent fiber is to form a molten bicomponent stream of a first polymer and a blend of second and third polymers, where the second and third polymers have different physical properties so that they separate from one another upon cooling to form fibers. The multicomponent fibers can also include more than three components. The above descriptions and comparisons of the physical properties of the thermoplastic materials apply to each of the materials of a multicomponent fiber.

Bicomponent fibers in accordance with this invention include fibers in which the thermoplastic materials are disposed in side by side relation with one another. The rotary apparatus described above usually forms such side by side bicomponent fibers. The bicomponent fibers of this invention also include fibers in which one of the thermoplastic materials forms a core, while the other forms a sheath surrounding the core. The rotary apparatus can be specially constructed by methods known in the art to form sheath and core bicomponent fibers. In general, such apparatus feeds one molten component through orifices which form a sheath, and feeds the other molten component into the interior of the sheath to form a core. Combinations of different kinds of fibers can also be formed. The multicomponent fibers of the invention can also be shaped fibers, produced by shaping the orifice so that fibers are formed having a non-circular cross section. Methods of manufacturing shaped fibers are disclosed in U.S. Patent Nos. 4,636,234 and 4,666,485 to Huey *et al.*

Example Bicomponent polymer fibers of this invention were formed from poly (phenylene sulfide) ("PPS") and polyethylene terephthalate) ("PET"). The PPS had a melting point of about 285°C, and the PET had a melting point of about 270°C. Separate streams of molten PPS and PET were supplied to the spinner having a temperature of about 205°C at the peripheral wall. At the temperature the polymers were delivered to the spinner, the PPS had a viscosity of about 4,000 poise and the PET had a viscosity of about 300 poise. The spinner had a diameter of about 20.3 centimeters and was rotated to provide a radial acceleration of about 7,600 meters/second^2. The spinner peripheral wall was adapted with 350 orifices. Bicomponent streams of molten PPS and PET were centrifuged through the orifices. The streams were cooled to make bicomponent polymer fibers which were collected as a wool pack. The average outside diameter of the fibers was about 25 microns.

The principle and mode of operation of this invention have been explained and illustrated in its preferred embodiment. However, it must be understood that this invention may be practiced otherwise than as specifically explained and illustrated without departing from its spirit or scope. INDUSTRIAL APPLICABILITY

The multicomponent fibers of this invention are useful in many applications including apparel products, thermal and acoustical insulation products, filtration products, and as binders in composite materials.

7.6 Unsheathed Polymer Fibre

Core Material: Polymethyl Methacrylate
Cladding Material: Fluorinated polymer
Refractive Index: 1.49
Numerical Aperture: 0.5
Temperature Range: –55°C to +70°C
Attenuation: typically max. 200dB/km for 1mm fibre

Product Code	*Outer Diameter*	*Core Diameter*	*Reel Length*	*Approximate Weight (exc. packaging)*
CK10	0.25mm	240μm	6000m	0.06g/m
CK20	0.50mm	486μm	6000m	0.4g/m
CK30	0.75mm	735μm	2700m	0.9g/m
CK40	1.00mm	980μm	1500m	1.5g/m
CK60	1.50mm	1470μm	700m	3.2g/m
CK80	2.00mm	1960μm	250m	4.0g/m

7.7 Fiber Reinforced Concrete

Fiber reinforced concrete (FRC) is concrete containing fibrous material which increases its structural integrity. It contains short discrete fibers that are uniformly distributed and randomly oriented. Fibers include steel fibers, glass fibers, synthetic fibers and natural fibers. Within these different fibers that character of fiber reinforced concrete changes with varying concretes, fiber materials, geometries, distribution, orientation and densities.

Historical Perspective

The concept of using fibers as reinforcement is not new. Fibers have been used as reinforcement since ancient times. Historically, horsehair was used in mortar and straw in mud bricks. In the early 1900s, asbestos fibers were used in concrete, and in the 1950s the concept of composite materials came into being and fiber reinforced concrete was one of the topics of interest. There was a need to find a replacement for the asbestos used in concrete and other building materials once the health risks associated with the substance were discovered. By the 1960s, steel, glass (GFRC), and synthetic fibers such as polypropylene fibers were used in concrete, and research into new fiber reinforced concretes continues today.

Effect of Fibers in Concrete

Fibers are usually used in concrete to control plastic shrinkage cracking and drying shrinkage cracking. They also lower the permeability of concrete and thus reduce bleeding of water. Some types of fibers produce greater impact, abrasion and shatter resistance in concrete. Generally fibers do not increase the flexural strength of concrete, so it can not replace moment resisting or structural steel reinforcement. Some fibers reduce the strength of concrete.

The amount of fibres added to a concrete mix is measured as a percentage of the total volume of the composite (concrete and fibres) termed volume fraction (V_f). V_f typically ranges from 0.1 to 3 per cent. Aspect ratio (l/d) is calculated by dividing fibre length (l) by its diameter (d). Fibres with a non-circular cross section use an equivalent diameter for the calculation of aspect ratio. If the modulus of elasticity of the fibre is higher than the matrix (concrete or mortar binder), they help to carry the load by increasing the tensile strength of the material. Increase in the aspect ratio of the fibre usually segments the flexural strength and toughness of the matrix. However, fibres which are too long tend to "ball" in the mix and create workability problems.

Some recent research indicated that using fibers in concrete has limited effect on the impact resistance of concrete materials. This finding is very important since traditionally people think the ductility increases when concrete reinforced with fibers. The results also pointed out that the micro fibers is better in impact resistance compared with the longer fibers.

The High Speed 1 tunnel linings incorporated concrete containing 1 kg/m^3 of polypropylene fibres, of diameter 18 & 32 ìm, giving the benefits noted below:

Benefits

Polypropylene fibres can:

- Improve mix cohesion, improving pumpability over long distances
- Improve freeze-thaw resistance
- Improve resistance to explosive spalling in case of a severe fire
- Improve impact resistance
- Increase resistance to plastic

Some developments in fiber reinforced concrete

The newly developed FRC named Engineered Cementitious Composite (ECC) is 500 times more resistant to cracking and 40 percent lighter than traditional concrete. ECC can sustain strain-hardening up to several percent strain, resulting in a material ductility of at least two orders of magnitude higher when compared to normal concrete or standard fiber reinforced concrete. ECC also has unique cracking behaviour. When loaded to beyond the elastic range, ECC maintains crack width to below 100 μm, even when deformed to several percent tensile strains.

Recent studies performed on a high-performance fiber-reinforced concrete in a bridge deck found that adding fibers provided residual strength and controlled cracking. There were fewer and narrower cracks in the FRC even though the FRC had more shrinkage than the control. Residual strength is directly proportional to the fiber content.

A new kind of natural fiber reinforced concrete (NFRC) made of cellulose fibers processed from genetically modified slash pine trees is giving good results. The cellulose fibers are longer and greater in diameter than other timber sources. Some studies were performed using waste carpet fibers in concrete as an environmentally friendly use of recycled carpet waste. A carpet typically consists of two layers of backing (usually fabric from polypropylene tape yarns), joined by $CaCO_3$ filled styrene-butadiene latex rubber (SBR), and face fibers (majority being nylon 6 and nylon 66 textured yarns). Such nylon and polypropylene fibers can be used for concrete reinforcement.

7.8 BS 4994

BS4994 (formally: **British Standard 4994:1987**) is the "specification for the design and construction of vessels and storage tanks in reinforced plastics". It specifies a code of practice for use by manufacturers of such containers.

An earlier version of the specification was BS 4994:1973.

8

Design and Manufacture of Composites

8.1 Polymer Composites and Composite Design

When should polymer composites be used?

Polymeric composites are one class of engineering material. As with all other engineering materials, they have particular strengths and particular weaknesses. The matrix protects the strong stiff fibres and together the composite material improves on the properties of either the matrix material or the fibres alone. A major driving force behind the development of composites has been to produce materials with improved specific mechanical properties over existing materials. Specific stiffness can be defined as the stiffness of a material divided by the density of material and specific strength can be defined as the strength of a material divided by the density of the material. It is these good specific properties of composites that allow the design of high performance structural components. Polymer composite material structures can also be engineered so that the directionality of the reinforcement material is arranged so as to match the loading on a given component or structure. In addition, polymer composites are useful in applications where the environment would be detrimental to other materials. A wide selection of resins and coatings are available to match appropriate environmental conditions. Cost is ever present in the engineering equation and it is the balance of cost and performance that determine whether or not to use polymer composites over an alternative structural material option.

Composite Design

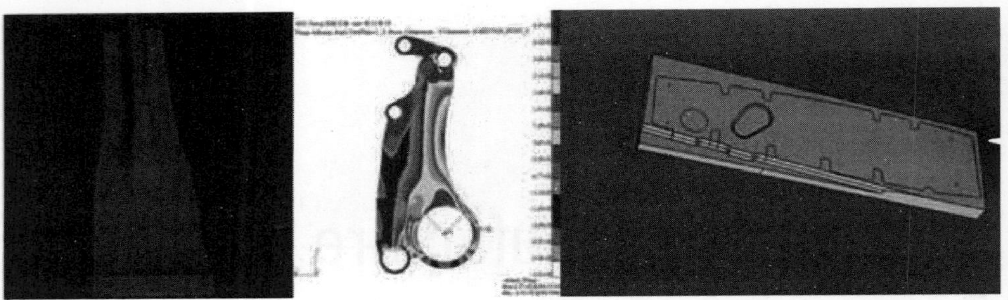

EireComposites currently provides Stress and Design services to leading aerospace and space companies under industrial product contracts and on a Research & Development basis.

EireComposites has a pool of experienced staff engineers with a strong background in the aerospace industry, many of whom have direct previous employment history with leading OEMs & material suppliers.

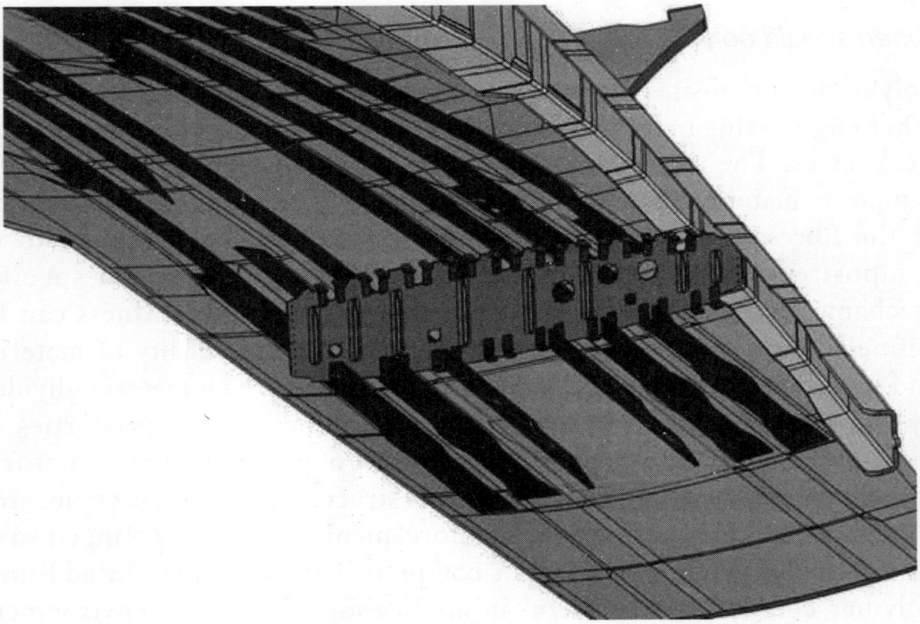

The company can provide Stress and Design Office support with industry recognised tools:

MSC/PATRAN and NASTRAN finite element analysis, CATIA V5 and EireComposites can also integrate with customer's specific proprietary requirements.

EireComposites design engineers work in close cooperation with CTL engineers, thus allowing the company to offer state of the art structural testing capabilities in conjunction with active design projects.

All EireComposites engineers have 'hands-on' experience working with advanced thermoplastic and thermoset composite materials and manufacturing processes. This capability is exploited in a team that can deliver innovative and cost effective design solutions in an applied Research & Development environment.

8.2 Thermoplastic Composites Explained

Thermoplastic composites are composites that use a thermoplastic polymer as a matrix. These composites can be reinforced with glass, carbon, aramid or metal fibres. A thermoplastic polymer is a long chain polymer that can be either amorphous in structure or semi-crystalline. These polymers are long chain, medium to high molecular weight materials, whose general properties are those of toughness, resistance to chemical attack and recyclability.

Thermoplastic polymers used in thermoplastic composites can be divided into two classes, high temperature thermoplastics and the engineering thermoplastics. The classification is based on the maximum service temperature of the polymers, which in turn is based on the Glass Transition (Tg) temperature. This is the temperature at which the amorphous portion of the polymer changes from a glassy to a rubbery phase on heating. Thermoset polymers may not usefully carry mechanic loads above Tg, but semi-crystalline thermoplastic polymers may carry load above Tg, as only the amorphous phase of the polymer has become rubbery. The crystalline portion of the polymer remains solid until the melt temperature, Tm.

Table 8.1 shows the most commonly used high temperature thermoplastic polymers for thermoplastic composites.

Table 8.1: High Temperature Thermoplastics

Matrix	*Morphology*	*Tg (°C)*	*Process Temp (°C)*	*Cost (Relative)*
PEEK	SC	143	390	$$$
PEI	A	217	330	$$
PPS	SC	89	325	$
PEKK	SC	156	340	$$

Table 8.2 shows the engineering thermoplastic polymers used in composites.

Table 8.2: Engineering Thermoplastics

Matrix	*Morphology*	*Tg (°C)*	*Process Temp (°C)*	*Cost (Relative)*
PBT	SC	56	190	$$
PA-6	SC	48	220	$
PA-12	SC	52	190	$
PP	SC	–20	190	$

Thermoplastics Vs Thermoset Matrices

Thermoset polymers are the matrix of choice for most structural composite materials. The single biggest advantage of thermoset polymers is that they have a very low viscosity and can thus be introduced into fibres at low pressures. Impregnation of the fibres is followed by chemical curing to give a solid structure, which can usually be carried out isothermally. An advantage of thermoplastics is that the moulding can be carried out non-isothermally, i.e. a hot melt into a cold mould, in order to achieve fast cycle times. However, polymerised thermoplastics tend to have melt viscosities between 500 and 1000 times that of thermosets, which necessitates higher pressures, causes processing difficulties and adds expense.

Thermoplastic composite polymers can, however, be readily recycled, an increasingly important issue in many markets, but especially in the automotive sector. For instance, an advanced thermoplastic composite component can be chopped to pellet-size and injection-moulded to yield long-fibre reinforced mouldings, which can in turn be recycled at the end of their life. Thermoset composite materials, on the other hand, can only be ground and used as filler, a process which decreases the value of the composite enormously.

Another advantage of thermoplastic composites are their superior impact and damage resistance properties. Over 90 per cent of polymers used in composites are thermosets, with thermoplastic composites still a niche market, mainly due to the difficulties in processing.

There are also significant environmental issues associated with thermoset processing, as a chemical reaction is necessary to form the solid structure of the polymer. Approximately 65 per cent of thermoset matrices used in structural composites are unsaturated polyesters. Environmental regulations regarding the styrene emissions of unsaturated polyester are affecting the total cost associated with using them. Because of this, many people are willing to consider a substitute for unsaturated polyester at a higher price.

Thermoplastic composites to date have needed high processing pressures, and hence expensive product tooling, as well as significant energy input in heating and cooling the tooling. These disadvantages have in many areas outweighed the advantages of these materials such as ease of recycling and high toughness, and limited their applications. On the other hand, thermoset composites are easier to process, requiring less energy and pressure, but are inherently brittle and cannot be usefully recycled.

A new composites processing technology, known as liquid monomer processing, has been developed. The advantage of the liquid monomer thermoplastics is that they combine the processing advantages of thermoset materials with the mechanical, durability and environmental advantages of thermoplastic polymers. In particular, one of the liquid monomer materials (Cyclics PBT) can be processed isothermally: injected, polymerised, crystallised and de-moulded at the same temperature, but yielding a thermoplastic polymer. The cycle time is therefore only limited by the injection, polymerisation and crystallisation time of the material itself.

Commingled Thermoplastic Composites

The alternative method of processing thermoplastic composites is to introduce the polymer in solid form to the fibres in such a manner that an intimate mixing of the two parts of the composite is achieved, normally in either fibre or powder form. In the fibre commingling process, the reinforcing fibres and the polymer fibres are blended in as fine a manner as possible. The subsequent "hybrid-yarn" is processed into fabrics or other textile forms, and impregnation is achieved by application of sufficient heat and pressure to cause the polymer to flow the short distances around the fibres. Cooling of the impregnated material results in a solid thermoplastic composite.

The main advantages of the commingling route are that the textile preform is now quite drapeable over complex shapes, and is significantly lower in cost than the pre-impregnated tapes. Disadvantages can include higher pressures and longer times to process because of the extra infiltration/consolidation process. Problems can also be associated with excessive fibre movement as the commingled yarns undergo much debulking during the melting process.

Twintex™ Glass-fibre Reinforced Polypropylene

Twintex™ is a trade name of St Gobain Vetrotex and refers to a commingled fabric of glass and polypropylene fibres. The fibre volume fraction of this material is around 20-25 per cent and the material is processed under vacuum or in a press at around 190°C. The low melt viscosity of the polypropylene at this temperature makes it relatively easy to process under vacuum conditions.

Sheet-forming of Thermoplastic Composites

Sheet-forming of thermoplastic composites is a process much like sheet-forming of metals or plastics, where a solid composite laminate is heated above its melt temperature and rapidly formed over, or into, a complex-shaped mould. A typical process is shown in Figure 8.1, where the composite sheet is heated rapidly in an infra-red oven, and then indexed between two cool tools, which close rapidly to form and cool the sheet. The main advantage of this process is that very fast cycle times can be achieved, but it is limited to components with simple or medium shape complexity.

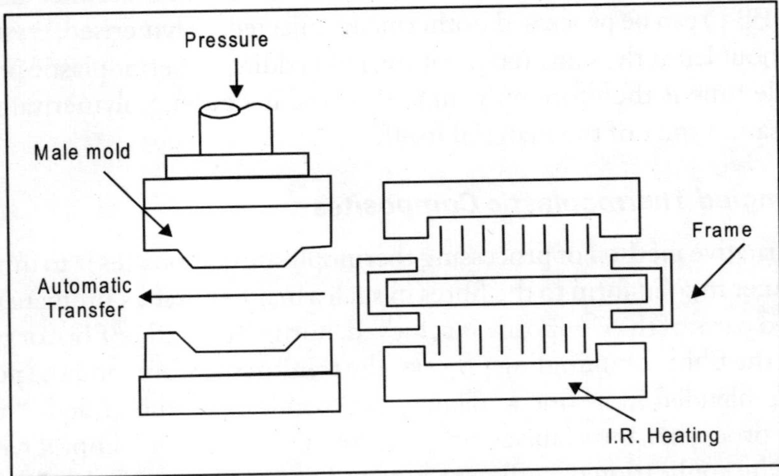

Fig. 8.1: Press-Forming of Thermoplastic Composite Sheets

Tape-laying of Thermoplastic Composites

Thermoplastic composites, because of their chemistry, can be rapidly heated and rapidly cooled without any damaging effects to their microstructure. The tape-laying process uses this principle to locally heat and melt, consolidate, and cool a tape of thermoplastic pre-preg, while placing it in position (see Figure 8.2). Thermoset tape-placement has been developed for many years, with large, seven-axis robotically-controlled fibre-placement systems in operation in many aerospace plants worldwide. The difference between a thermoplastic and a thermoset system is the extra heating, consolidation and cooling equipment, needed with the head. The main advantage of the thermoplastic system is that the product is completely finished once the tape has been laid, whereas the thermoset product must be further bagged and cured in an autoclave.

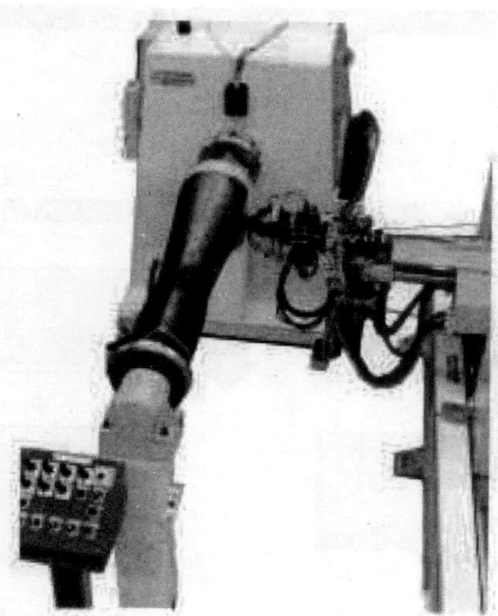

Fig. 8.2: Tape-Laying of Thermoplastic Composites

Open Vs Closed Moulds—Liquid Moulding

Advanced thermoset composites have traditionally been hand-laminated and cured under vacuum or in autoclaves. While suitable for low-volume applications in the aerospace sector, recent trends have sought to use closed-mould processes such as liquid moulding (Figure 8.3), where the dry preform is laid between two rigid tools (or between one rigid and one soft tool or bag material), and the liquid thermoset resin injected or infused through the reinforcement. The main advantage of this process for aerospace is that the dimensional tolerances achieved are much better than with autoclaving. The potential for other sectors is in the automation of the process, leading to higher volumes and faster cycle times. Moulding pressures can be low (1-5 bar) and thus light, inexpensive tooling can be used. The process is currently limited to thermoset materials, and their cycle times are relatively long because of the necessity for chemical curing.

Liquid Moulding Of Thermoplastics

The new generation of thermoplastic materials are processed in a water-like state, and thus need much lower pressures, less expensive tooling, and lower energy input, while retaining all of the attractive properties of thermoplastic materials. Examples of the new liquid-moulded thermoplastic composites are Cyclics PBT from Cyclics Corp. (US) and Grilamid PA-12 materials from Ems-Chemie AG (Switzerland) and. End-of-life recycling is easily achieved by chemical means, or by re-melting and use as injection-moulding compounds.

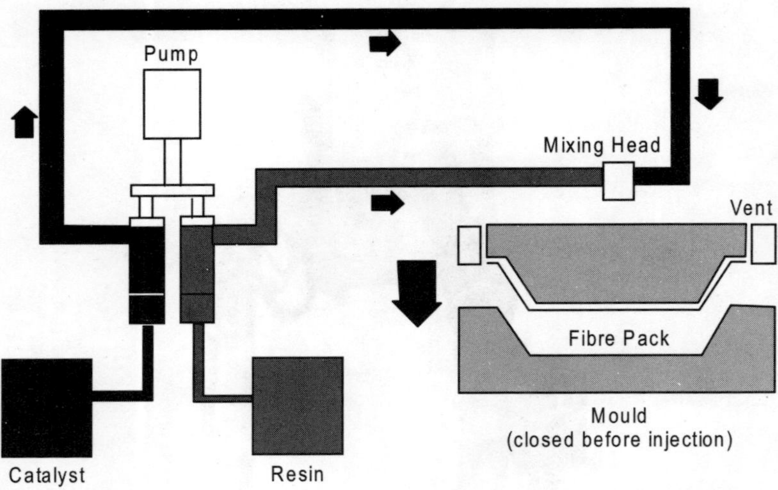

Fig. 8.3: Liquid Moulding of a Thermoset Resin into a Dry Fibre Preform

The traditional disadvantage of thermoplastic resins is that the melts have a high viscosity, typically above 500 Pa.sec, which is much too viscous to infiltrate a high-volume fraction of fibres. Figure 8.4 illustrates the principles of liquid moulding of thermoplastics. The low viscosity resins now available are injected into the composite as an activated monomer, with a resulting low viscosity. Once infiltrated, polymerisation takes place in-situ, yielding a semi-crystalline thermoplastic composite with all the inherent advantages of toughness, solvent resistance, dielectric strength and recyclability.

Manufacturing of high content continuous fibre reinforced composites by direct impregnation of the fibre bed by a liquid matrix can be industrially applied if each of the following conditions is fulfilled (Connor, SAMPE 99):

(i) The matrix viscosity during the impregnation stage is very low (e.g., $<$ 1Pa.s)
(ii) Once the fibre bed is fully impregnated, the matrix can be solidified chemically (e.g. curing, polymerisation) or physically (e.g. cooling, crystallisation) in a sufficiently short time (order of minutes).
(iii) The final matrix has high enough physical properties to transmit good mechanical stability to the composite part.

Two liquid thermoplastic materials in particular are known to be developed which meet these criteria, APLC-12 and Cyclic PBT/PC. In both cases, the injection material is a pre-activated monomer melt with low viscosities, which polymerises in-situ to form tough, solvent-resistant, semi-crystalline polymer matrices.

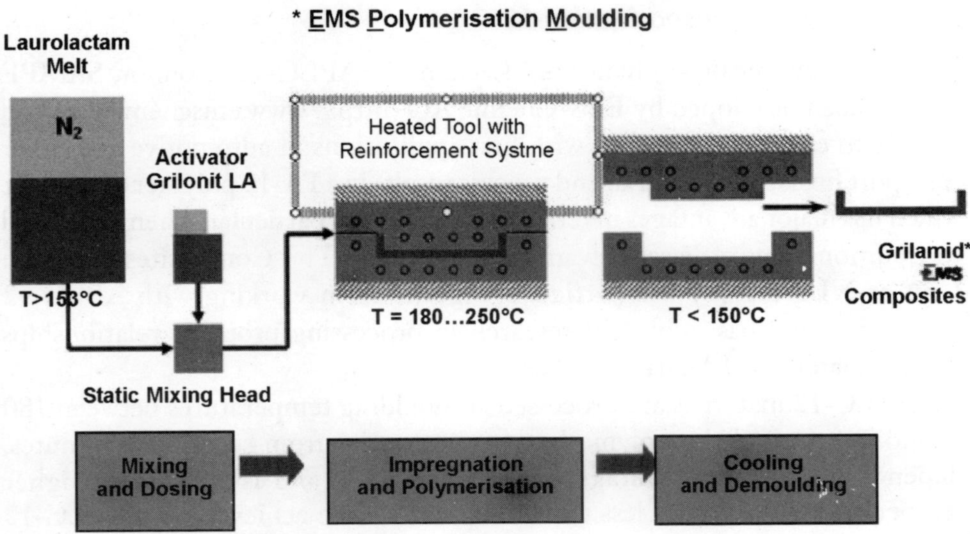

Fig. 8.4: Liquid-Moulding of Thermoplastic Composites

Cyclic PBT (CBT™)

Cyclic PBT and PC resin systems were developed by GE in the late 1980s. The technology was purchased from GE, resulting in Cyclics Corporation (http://www.cyclics.com), which is now marketing the Cyclic PBT and PC systems for applications as diverse as automotive, electrical, sports and transportation.

The Cyclics resins are available as either 1-part or 2-part systems (Eder, SAMPE 2001). In the 2-part system, the process starts with powdered cyclic oligomers at room temperature. These are heated to a low viscosity liquid at about 140 C, a catalyst is added (either Tin or Titanate), the melt is then injected into the mould at temperatures above 150 C. Typical processing viscosities are between 17 and 150 mPa.sec. polymerisation follows in-situ, with polymerisation times varying from seconds to minutes. 1-part Cyclics resin systems are also available that are pre-mixed with catalyst, and are solid at room temperature. These are simply heated and injected as with the 2-part system.

An important advantage of the Cyclics technology is that the processing can be carried out at near-isothermal conditions, i.e. the material solidifies and crystallises at the processing temperature. It is not, therefore, necessary to cool the tool in order to de-mould the component, which is an important economic consideration, leading to energy savings and shorter cycle times. This behaviour is due to the fact that the Cyclics resins polymerise at temperatures below both the melt temperature and the crystallisation temperature of the PBT polymer.

Anionically-Polymerised Lactam-12

The first, Anionically-Polymerised Lactam-12 (APLC-12, Connor, SAMPE 99) has been developed by Ems-Chemie AG (http://www.emschem.com) and is close to commercialisation, with first applications in automotive and other transport fields. The material and process results in a PA-12 polymer composite which has major advantages in terms of toughness in particular, when compared to traditional epoxy-based advanced composites. The Composites Research Unit at NUI Galway has particular experience in working with APLC-12 composites and has published research in processing/property relationships for this material (Ó Máirtín 2001).

APLC-12 materials are processed at moulding temperatures between 180 C and 240 C, and the polymerisation time varies from seconds to minutes, depending on the percentage of activator used, and is shorter at higher temperatures. Viscosities less that 0.1 Pa.sec can be achieved with APLC-12 moulding, and as a result, infiltration of high fibre volume-fractions under gravity forces only has been demonstrated. The polymerised composite must be cooled to below 100 C before de-moulding.

8.3 Mold Design Inclusion in an Innovative Composites and Polymer Materials Education

At Central Connecticut State University, the Composites and Polymer Materials (C&PM) program is an innovative curriculum leading to a Bachelor of Science degree in engineering technology. Engineering technology differs from engineering by the level of mathematics and physics (Calculus I, II and University Physics I, II) and by the emphasis on applied technology. University faculty, together with an Industrial Advisory Board, (IAB) created the program to provide qualified graduates for the plastics and polymer composites manufacturing industry, as well as to provide opportunity for industrial personnel to update skills in a rapidly advancing discipline.

Integration is the basis for the program's design and curriculum flexibility. The C&PM program integrates disciplinary areas such as chemistry, applied mechanics and materials analysis, with specialized courses such as mold design, plastics and composite manufacturing processes and design analysis. Course content covers theory and application, laboratory experiments using industrial machine tools, testing equipment and computer software to provide in-depth knowledge within the plastics and composites discipline.

Introduction

Plastics and composites, mold design and moldmaking are a natural extension of the established engineering technology programs at Central Connecticut State University (CCSU). Since 1849, the institution has historical significance having evolved as one of the first schools for preparing teachers for public school systems, offering courses in the manual arts, and applied technology for teachers. The institution has demonstrated a strong tradition of practical application and in 1982 the engineering technology programs evolved from technology education and industrial technology.

Connecticut has a diverse manufacturing industry—from aerospace to sporting goods to commercial products—with the plastics molding firms playing a major role in the state's economy. Aware of the dynamic aspects of the industry, the School of Technology at CCSU has established an IAB with representatives from major industries (Pratt & Whitney Aircraft, Sikorsky Aircraft, Hamilton Sundstrand, IBM Global Services and Prototype & Plastics Mold Company, Henkel Loctite and others) to advise faculty with the development of quality programs and curriculum.

This IAB contributes as a team to develop an analytical and applied technical curriculum in C&PM. Leading to a Bachelor of Science degree, the program prepares technologists for the plastics processing, moldmaking, manufacturing and aerospace industries.

CCSU offers a number of Bachelor of Science degree programs including those in manufacturing management, and manufacturing and mechanical engineering Technology. In 1998, the C&PM engineering technology program emerged as an interdisciplinary curriculum integrating courses from the existing engineering technology programs to address the needs of clients—the industry and the students.

A major aspect of implementing educational programs is the logistics of offering courses, laboratory equipment and enrollment, which can often be cost-prohibitive. By integrating courses from disciplinary areas, the curriculum is not only flexible, but also an effectual use of available resources.

Program Curriculum

Curriculum within the engineering technology department at CCSU can be divided into three main components: general education, major technical specialties and additional technical supporting coursework. The entirety of each 130 credit-hour program culminates in a capstone engineering technology senior project.

Under general education, the university provides students with the basic educational foundations seeking to realize several objectives: develop an appreciation for the arts and humanities, foster awareness of the social and behavioural sciences, enhance personal communication skills, strengthen quantitative skills and develop scientific understanding of our natural world. The mathematics (quantitative) and sciences requirements of general education have been specifically geared toward the engineering technology needs for certain technical sciences. In accredited baccalaureate programs, a sequence in calculus is the required level of mathematical proficiency. The science requirement is sufficiently addressed by a physics sequence which is advantageous as a prerequisite in subsequent engineering mechanics and thermodynamics courses. C&PM engineering technology students also are required to complete a set of chemistry courses culminating with organic chemistry, the backbone of polymeric material studies.

Many of the major technical specialty courses are shared among several engineering technology programs. Included is an introduction to engineering technology, a grouping of applied mechanics courses (statics, dynamics, strength of materials, thermodynamics and finite element analysis), an array of design courses—technical drafting, geometric dimensioning and tolerancing, CAD, computer-integrated manufacturing (CIM), machine design, solid modeling, design for manufacture—a materials analysis course, and a course in engineering economy. C&PM engineering technology program-specific specialty courses are: manufacturing with plastics and composites, plastics and composites tool design, and composite design and analysis.

Additional supporting technical coursework uses resources from other departments to provide background in information processing, computer programming, electrical circuits, statistics and continuous process improvement.

Fig. 8.5: Standard injection molded plastic tension specimen is loaded in INSTRON® materials testing system.

C&PM Technical Courses

C&PM processes is an analytical study of thermoplastics, thermosets and polymer matrix composite materials. It also looks at the manufacturing processes used within the plastics and composite molding and fabrication industry. This is a laboratory course using standard materials testing equipment (see Figure 8.5), industrial equipment and introducing problem solving to technical applications. Typical laboratory experiences include the setup, operation and trouble-shooting of plastics molding processes on an Arburg Allrounder® 220-75-250 injection molding machine (see Figure 8.6) and a Carver Monarch™ 50-ton compression molding press (see Figure 8.7), as well as blow molding (BM), vacuum thermoforming and other machine tools. It introduces elements of basic molds and dies for the various processes, including mold materials and components, principles of runner systems, gating, venting and mold fill.

Fig. 8.6: The operation of an Arbug injection molding machine is reviewed.

Fig. 8.7: Carver Monarch™ 50-ton compressoin molding press for forming parts made from thermosetting polymer materials and composites

Plastics and composites tooling design introduces the principles for design of molds and tooling for the production of plastics and composite products. The course uses solid modeling CAD software (Pro/ENGINEER® and Unigraphics®) to design molds for injection molding (IM), BM, compression molding and thermoforming processes (see Figures 8.8 and 8.9).

Fig. 8.8: Unigraphics® solid model of mold A (stationary) plate of an injection mold base

Fig. 8.9: Unigraphics® solid model of mold B (moveable) plate of an injection mold base

The principles of mold design and moldmaking for injection molds, blow molds, compression and transfer molds are addressed, as well as for thermoforming, vacuum forming, RIM and RTM molds. The course analyzes specifics relative to the design and operation for the numerous types of molds, runner and manifold systems, gating systems, venting, cooling and temperature control systems, ejection and knock systems. Also discussed are materials for the moldmaking and toolmaking processes, equipment and methods for mold construction.

Students use solid modeling software to design tooling or molds for the laboratory equipment. Standard D-M-E™mold bases and components are integrated into the design process, with analysis of projected area and volumetric press capacity, temperature control and cooling, runner, gating and venting, shrinkage, surface finish, core and ejector requirements.

Composite design and analysis is aimed at providing students with the analysis techniques and tools for dealing with and designing composite structures. To optimally design composite parts, a thorough review of available constituent materials and the methods of their possible consolidation are required. The various manufacturing and molding processes offer certain advantages and limitations and also impact resultant composite properties. In composite design the student's knowledge base of general isotropic materials is expanded through classical composite theory to account for the anisotropic nature of materials typically encountered in composites. Elements of composite micromechanics, macromechanics, ply mechanics, stiffness and strength design and failure criteria are employed together with computational software to design composite sheet, plate, beam and shell structures.

Engineering technology senior project is a capstone course in which students participate on a teamwork project to study, design, and/or research a project as engineering technologists. Final reports are submitted to the department for archiving, and oral presentations are required. Projects may originate from student, instructor and/or industrial partner.

The moldmaking tasks may often be performed within the context of this course. Students apply experiences and knowledge acquired throughout the program to a research project; they work as teams to develop a technical system or mechanism, such as to manufacture a mold. CNC machine tools are used to manufacture cavity inserts and components for standard D-M-E™ mold bases (see Figure 8.10).

Fig. 8.10: CNC machining center used to manufacture cavity inserts and mold base components

Summary

Graduates of the CCSU C&PM engineering technology program generally function as members of engineering teams in the aerospace, composites and plastics molding or manufacturing industries. Responsibilities may involve design, development, testing and analysis of products and tooling, as well as operation of manufacturing systems for the production of aerospace, automotive, sporting and commercial products. This program is enhanced by the inclusion of a mold design and moldmaking emphasis in the technical specialty courses. The curriculum will continuously improve through on-going university assessment processes and IAB guidance.

8.4 Meaning of Composites?

A composite is a material which is made up of two or more distinct (i.e. macroscopic, not microscopic) materials. A familiar composite is concrete, which is basically made up of sand and cement. Many common materials could be classed as composites, but this website is concerned with fibre reinforced polymer composites.

Polymer composites are plastics within which there are embedded fibres or particles. The plastic is known as the matrix, and the fibres or particles, dispersed within it, are known as the reinforcement.

The reinforcement is usually stiffer than the matrix, thus stiffening the composite material. This stiffer reinforcement will usually be laid in a particular direction, within the matrix, so that the resulting material will have different properties in different directions. This characteristic is usually exploited to optimise the design.

Polymer matrix composites (PMCs) are materials that use a polymer based resin as a matrix material with some form of fibres embedded in the matrix, as a reinforcement. Both thermosetting and thermoplastic polymers can be used for the matrix material. Common polymer composite thermosetting matrix materials include polyester, vinyl ester and epoxy. Polymer composite thermoplastic matrix materials include PEEK, PEI and PPS. Reinforcements include glass, carbonand aramid fibres.

Why Use Polymer Composites?

There can be many secondary reasons why polymer composites may be chosen for the manufacture of particular articles or components, but the primary reason is because of weight saving for their relative stiffness and strength. As an example we can compare a carbon fibre reinforced composite with its steel counterpart. The carbon fibre composite can be five times stronger than 1020 grade steel while having only one fifth the weight. Aluminium (6061 grade) is much nearer in weight to carbon fibre composite (though still somewhat heavier), but the composite can have twice the modulus and up to seven times the strength.

8.5 Polymer Composites Directly Replace Metals?

Many metal articles or components can instead be made from composites, but there are important differences which mean that direct substitution should be made with care.

Most engineering materials are essentially isotropic. That is, they have the same properties such as strength and modulus, in any direction. There may be 'grain' in some metals due to the manufacturing process, but it is only in critical applications that this matters. Most machining or casting processes do not have to take directional differences into account.

Most composites will have very different properties in different directions. This is because, although the matrix material is isotropic, the reinforcement is not. Carbon fibres may be up to 100 times stronger under tension than they are in shear, and the stiffness may differ in the two directions by similar ratios. The properties of the composite will reflect the properties of the reinforcement, so that it can have greatly different properties in different directions. This is exploited in design as manufactured articles rarely require to be equally strong in all directions, and composites can achieve this by particular arrangements of the reinforcement. However, a different design procedure is required for composites compared to that required for metals.

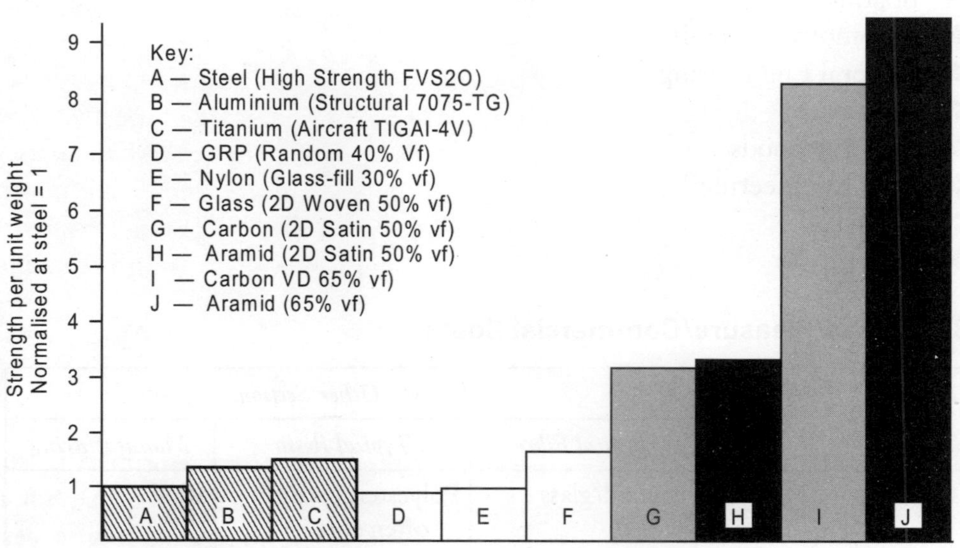

8.6 Type of Polymer Composites Available

Polymer composites can be classified according to the type of MATRIX material, and the type of REINFORCEMENT material.

Because the size of the reinforcement particles, or the type and length of the fibres can be varied, and because the directions in which they can be placed within the matrix can be varied, a very wide variety of properties can be achieved. Where necessary the composite can have different properties in different directions.

When we include the possibility of changing the matrix material, it can be seen that polymer composites form a vast family of engineering materials.

8.7 Main Factors Determining the Costs

Quality control, health and safety considerations, etc. which need to be taken into account but are not dealt with here. Also, there is an increasing emphasis on environmental factors including maintenance, recycling, and disposal, which, over the lifetime of a component, can be important. For many components, the benefits from the use of reinforced polymer composites can become very significant when whole-life costs, rather than just fabrication costs, are included.

8.8 Applications Currently Fabricated from Composites

Many products in a variety of industries are fabricated from composites, from fighter aircraft to bath tubs. This section gives more examples in different industries including:

- Boating
- Automotive and rail
- General Engineering
- Aerospace
- Sporting goods
- Civil Engineering
- Domestic
- Medical

8.9 Sports/Pleasure/Commercial Boats

Item	*Links to Other Sections*		
	Typical Fibre	*Typical Resin*	*Manufacturing*
Deck	Aramid/glass	Polyester, epoxy often core material	RIM
Hulls for racing boats	Carbon/glass	Nomex® honeycomb cores	RIM or Autoclave

(*Contd.*)

Item	*Links to Other Sections*		
	Typical Fibre	*Typical Resin*	*Manufacturing*
Hulls pleasure	Glass	PE/VE/Epoxy	Hand
Hulls commercial	?	?	?
Pressure hull on submersibles	Carbon/glass	PE/Epoxy/PEEK	Filam/Autocl
Propeller shafts	Carbon	Epoxy	Filament
Interior furniture	Glass	PE/VE/Epoxy Nomex® honeycomb cores	VE/Epoxy RIM
Masts	Carbon	Epoxy/PE	FW
Rudders	Carbon	Epoxy/EP	Autoclave
Bulkheads	Glass	VE	Hand
Safety shell cockpit	Carbon/aramid	VE/Epoxy aramid honeycomb core	RIM/Autoc

Glass fibre hull and superstructure of lifeboat

Glass fibre hull and superstructure of pleasure boad

8.10 Automotive and Rail

Item	*Links to Other Sections*		
	Typical Fibre	*Typical Resin*	*Manufacturing*
Battery trays	Glass	PP	Flow/compression moulding
Body panels	SMC		Compression moulding

(*Contd.*)

Item	*Links to Other Sections*		
	Typical Fibre	*Typical Resin*	*Manufacturing*
Bumper fascia	Carbon/glass	Polyester/rubber	Compression moulding
Drive shafts	Carbon/E-glass	VE	FW
Radiator grills	Glass	PBT	RIM
Instrument panes	Glass	PP	Flow/compression moulding
Leaf springs	Glass	VE/Epoxy/PE	FW, Auto
Engine components	Glass	PP/Nylon	IM
Fuel lines	Glass	PP/Nylon	Extruded
Rail car bodies	E-glass fabric	VE/PW	FW

FI sports Car - Carbon fibre

Volvo Tailgate fabricated from SMC

8.11 General Industry and Engineering

Item	*Links to Other Sections*		
	Typical Fibre	*Typical Resin*	*Manufacturing*
Drive shafts for power transmission	carbon	??	FW
Ductwork for air handling	Glass	VE/PE	FW, hand

(Contd.)

Item	*Links to Other Sections*		
	Typical Fibre	*Typical Resin*	*Manufacturing*
Non-conductive ladders	Glass/carbon		pultruded
Pipes	Glass/carbon	VE/PE/Epoxy/Phenolic	FW
Pressure vessels	Glass	VE/PE/Epoxy/Phenolic	FW
Underground storage tanks	Glass	VE/PE	Hand

8.12 Aerospace

Helicopter Rotor hubs and blades

Aero engine, blades and thrust reversets

Eurofighter Jet

Item	*Links to Other Sections*		
	Typical Fibre	*Typical Resin*	*Manufacturing*
Fuselage—military	Carbon (tape and fabric)	Epoxy	Auto/RTM
Fuselage—General aviation	Glass fabric	Epoxy	Auto/RTM/Hand
Wings & control surfaces	Carbon (tape and fabric)	Epoxy	Auto/RTM
Bulkheads & Floors	Glass	??	Sandwich
Cargo liners	Glass	??	Sandwich
Engine cowls - carbon fibre reinforced	Carbon	Epoxy bismaleimide	Auto
Fairings (fuselage/fin/flap track/landing gear doors, etc)	Glass/Carbon	Epoxy	Auto/RTM
Radomes	Aramid/glass	Polyester	Sandwich/Auto
Rotor blades	Carbon/glass	Epoxy	Pultruded spars, sandwich
Rotor hubs	Glass/carbon	Epoxy	Autocl.
Satellite bodies	Glass	Epoxy	Sandwich

8.13 Sport and Recreation

Carbon fibre racing bike

Item	*Links to Other Sections*		
	Typical Fibre	*Typical Resin*	*Manufacturing*
Archery bows	Glass	??	Pultruded
Baseball bats	Carbon	Polycarbonate	??
Bicycle frames	Carbon	Epoxy	??
Canoes, kayaks	Glass (CSM)	PE	Hand
Fishing rods	Glass/carbon	Epoxy/VE	Rolling or FW
Golf club shafts	Carbon	Epoxy	Rolling or FW
Golf club heads	Carbon	Epoxy	Moulded
Fishing rods	Glass/carbon	Epoxy	Rolling or FW
Pole vaulting poles, Ski poles	Glass/carbon	Epoxy/VE	Rolling or FW
Sleds. Surf boards, snow boards, skis	Glass/aramid	Epoxy/VE/PE	Sandwich core
Racquets	Carbon/glass/aramid	?	Injection moulded

8.14 Civil Engineering

Chemical processing plant, Courtesy of Dow Chemicals

Puttruded composites used as soil nails

Item	*Links to Other Sections*		
	Typical Fibre	*Typical Resin*	*Manufacturing*
Bridges	Glass/carbon	Polyester	Pultrusion
Concrete repair	Glass/carbon	?	?
Cladding	Glass	??	??
Column wraps	Carbon	?	Hand lay-up

8.16 Medical

Item	*Links to other sections*		
	Typical Fibre	*Typical Resin*	*Manufacturing*
Orthotics	Carbon/glass	??	
Implants		??	
Wheel chairs/crutches	Carbon	??	
Radiolucent table tops/cradles/positioning	Carbon	??	

9

Biopolymers and Biomaterials

9.1 Biopolymers

Biopolymers are a class of polymers produced by living organisms. Starch, proteins and peptides, DNA, and RNA are all examples of biopolymers, in which the monomer units, respectively, are sugars, amino acids, and nucleic acids.

Biopolymers versus polymers

A major but defining difference between polymers and biopolymers can be found in their structures. Polymers, including biopolymers, are made of repetitive units called monomers. Biopolymers inherently have a well defined structure: The exact chemical composition and the sequence in which these units are arranged is called the primary structure. Many biopolymers spontaneously fold into characteristic compact shapes (see also "protein folding" as well as secondary structure and tertiary structure), which determine their biological functions and depend in a complicated way on their primary structures. Structural biology is the study of the structural properties of the biopolymers. In contrast most synthetic polymers have much simpler and more random (or stochastic) structures. This fact leads to a molecular mass distribution that is missing in biopolymers. In fact, as their synthesis is controlled by a template directed process in most in vivo systems all biopolymers of a type (say one specific protein) are all alike: they all contain the same sequence and number of monomers and thus all have the same mass. This phenomenon is called monodispersity in contrast to the polydispersity encountered in polymers. As a result biopolymers have a polydispersity index of 1.

Conventions and nomenclature

Polypeptides

The convention for a polypeptide is to list its constituent amino acid residues as they occur from the amino terminus to the carboxylic acid terminus. The amino acid residues are always joined by peptide bonds. Protein, though used colloquially to refer to any polypeptide, refers to larger or fully functional forms and can consist of several polypeptide chains as well as single chains. Proteins can also be modified to include non-peptide components, such as saccharide chains and lipids.

Nucleic Acids

The convention for a nucleic acid sequence is to list the nucleotides as they occur from the 5' end to the 3' end of the polymer chain, where 5' and 3' refer to the numbering of carbons around the ribose ring which participate in forming the phosphate diester linkages of the chain. Such a sequence is called the primary structure of the biopolymer.

Sugars

Sugar-based biopolymers are often difficult with regards to convention. Sugar polymers can be linear or branched are typically joined with glycosidic bonds. However, the exact placement of the linkage can vary and the orientation of the linking functional groups is also important, resulting in á- and â-glycosidic bonds with numbering definitive of the linking carbons' location in the ring. In addition, many saccharide units can undergo various chemical modification, such as amination, and can even form parts of other molecules, such as glycoproteins.

Structural Characterization

There are a number of biophysical techniques for determining sequence information. Protein sequence can be determined by Edman degradation, in which the N-terminal residues are hydrolyzed from the chain one at a time, derivatized, and then identified. Mass spectrometer techniques can also be used. Nucleic acid sequence can be determined using gel electrophoresis and capillary electrophoresis. Lastly, mechanical properties of these biopolymers can often be measured using optical tweezers or atomic force microscopy.

Biopolymers as Materials

Some biopolymers—such as polylactic acid, naturally occurring zein, and poly-

3-hydroxybutyrate can be used as plastics, replacing the need for polystyrene or polyethylene based plastics.

Some plastics are now referred to as being 'degradable', 'oxy-degradable' or 'UV-degradable'. This means that they break down when exposed to light or air, but these plastics are still primarily (as much as 98 per cent) oil-based and are not currently certified as 'biodegradable' under the European Union directive on Packaging and Packaging Waste (94/62/EC). Biopolymers, however, will break down and some are suitable for domestic composting.

Biopolymers as Packaging

Biopolymers (also called renewable polymers) are produced from biomass for use in the packaging industry. Biomass comes from crops such as sugar beet, potatoes or wheat: when used to produce biopolymers, these are classified as non food crops. These can be converted in the following pathways:

Sugar beet > Glyconic acid > Polyglonic acid
Starch > (fermentation) > Lactic acid > Polylactic acid (PLA)
Biomass > (fermentation) > Bioethanol > Ethene > Polyethylene

Many types of packaging can be made from biopolymers: food trays, blown starch pellets for shipping fragile goods, thin films for wrapping.

Biopolymers are Renewable, Sustainable, and can be Carbon Neutral

Biopolymers are renewable, because they are made from plant materials which can be grown year on year indefinitely. These plant materials come from agricultural non food crops. Therefore, the use of biopolymers would create a sustainable industry. In contrast, the feedstocks for polymers derived from petrochemicals will eventually run out. In addition, biopolymers have the potential to cut carbon emissions and reduce CO_2 quantities in the atmosphere: this is because the CO_2 released when they degrade can be reabsorbed by crops grown to replace them: this makes them close to carbon neutral.

Biopolymers are Biodegradable, and Some are also Compostable

Some biopolymers are biodegradable: they are broken down into CO_2 and water by microorganisms. In addition, some of these biodegradable biopolymers are compostable: they can be put into an industrial composting process and will break down by 90 per cent within 6 months. Biopolymers that do this can be marked with a 'compostable' symbol, under European Standard EN 13432 (2000). Packaging marked with this symbol can be put into industrial composting processes and will break down within 6 months (or less). An example of a compostable polymer is PLA film under 20ìm thick: films which are thicker than that do not qualify as compostable, even though they are biodegradable. A home composting logo may soon be established: this will enable consumers to dispose of packaging directly onto their own compost heap. The standards for such a home composting logo have not yet been developed.

9.2 Biomaterials

Biomaterials are materials used in close or direct contact with the body to augment or replace faulty materials. Biomaterials must be compatible with the body so that the body does not reject them. However, in some cases, biomaterials like organ transplants do cause rejection, which can be addressed through anti-rejection medications.

Biomaterials do not have to be living or once living materials however. They can be of synthetic origin as well. For example, shunts and pacemakers are both considered biomaterials. Gore-tex® shunts are an excellent example of biomaterials used to either bypass clogged arteries or provide new pathways for the circulatory system. They tend to have the advantage of remaining sound and not disintegrating. However, since they are not living, such shunts placed in children may be outgrown and require replacement.

Some biomaterials are of an organic nature. These include such materials as collagen or fat, often used in plastic surgery applications. Other biomaterials may include arteries or vessels taken from either cadavers, or from porcine, meaning pig, or bovine, meaning cow, sources. If one has a major heart valve replaced, one frequently chooses between a homograph, taken from a cadaver source, or from an allograph, taken from a pig or cow source. An additional choice is an artificial valve, such as one made of Gore-tex®.

Many surgeons prefer biomaterials that are organic over those that are inorganic. However, none of these biomaterials will grow with the body. Homographs and allographs also have a slightly lower rate of blood clotting than do artificial valves, and thus may be preferred by surgeons or patients.

Other biomaterials include certain metals, which might be used in reconstructing bones or joints. For example, metal ball joint sockets can be used in knees or hips, and tend to offer great support for those requiring joint replacement.

Some biomaterials are actually living. This is the case with organ transplants in particular. Organs are expected to grow and develop with the body, and are better replacements than non-living sources. In some cases, a non-living source like an artificial heart or a left ventricular assist device (LVAD) is used while people wait for a heart transplant. These artificial replacements tend not to work for long periods of time, though they can provide someone with the extra days or even a few months they need while waiting to receive a transplant.

Other common biomaterials are used in plastic surgery applications. Calf, breast, cheek, chin, and buttocks implants are all considered to be biomaterials. Occasionally, plastic surgeons will harvest either fat or skin from a patients body to be used in another part of a body. Skin graphs are frequently used to cover scarring, and are most helpful in covering large areas of burned skin, which tends not to regenerate new skin tissue.

One of the most interesting skin biomaterials used recently was the first facial transplant, performed by surgeons in France. The woman receiving the transplant received a partial facial transplant, including new lips and a new nose in 2005. So far, her body has not rejected this transplant. This first successful transplant may prove especially useful for those whose faces have undergone severe and irreparable trauma.

The development of *biomaterials* is not a new area of science, having existed for around half a century. The study of biomaterials is called biomaterial science. It is an exciting field of science, having experienced steady and strong growth over its history with companies such as Smith and Nephew investing large amounts of money in new products. Biomaterials science encompasses elements of medicine, biology, chemistry and materials science.

Definition of a Biomaterial

While a definition for biomaterials has been difficult to formulate, a widely accepted definition for biomaterials is that:

> "A biomaterial is any material, natural or man-made, that comprises whole or part of a living structure or biomedical device which performs, auguments, or replaces a natural function".
>
> " A Biomaterial is a nonviable material used in medical device, intended to interact with a biological systems (William 1987)"

A biomaterial is essentially a material that is used and adapted for a medical application. Biomaterials can have a benign function, such as being used for a heart valve, or may be bioactive and used for a more interactive purpose such as hydroxy-apatite coated hip implants (the Furlong Hip, by Joint Replacement Instrumentation Ltd, Sheffield is one such example—such implants are lasting upwards of twenty years). Biomaterials are also used every day in dental applications, surgery, and drug delivery (a construct with impregnated pharmaceutical products can be placed into the body, which permits the prolonged release of a drug over an extended period of time).

The definition of a biomaterial does not just include man-made materials which are constructed of metals or ceramics. A biomaterial may also be an autograft, allograft or xenograft used as a transplant material.

Biomaterial Applications

Biomaterials are used in:

- Joint replacements
- Bone plates
- Bone cement
- Artificial ligaments and tendons
- Dental implants for tooth fixation
- Blood vessel prostheses
- Heart valves
- Skin repair devices
- Cochlear replacements
- Contact lenses

Biomaterials must be compatible with the body, and there are often issues of biocompatibility which must be resolved before a product can be placed on the market and used in a clinical setting. Because of this, biomaterials are usually subjected to the same requirements of those suffered by new drug therapies. All manufacturing companies are also required to ensure traceability of all of their products so that if a defective product is discovered, others in the same batch may be traced.

Subjects Integral to Biomaterials Science

Toxicology

A material should not be toxic, unless specifically engineered to be so (for example "smart" drug delivery systems that target cancer cells and destroy them).

Biocompatibility

Biocompatibility is difficult to measure, it is defined in terms of success at a specific task.

Functional Tissue Structure and Pathobiology

Understanding of the anatomy and physiology of the action site is essential for a biomaterial to be effective.

Healing

Healing is an essential consideration when using biomaterials. The body may experience what is known as a foreign-body reaction after implementation so immuno-suppression may be required.

Dependence on Specific anatomical sites of implantation

It is important, during design, to ensure that the implement will fit complementarily and have a beneficial effect with the specific anatomical area of action.

Mechanical and Performance requirements

Biomaterials that have a mechanical operation must perform to certain standards and be able to cope with pressures. It is therefore essential that all biomaterials are well designed and are tested. Biomaterials that are used with a mechanical application, such as hip implants, are usually designed using CAD (Computer Aided Design) which allows all of the directional stresses to be calculated, ensuring maximum product life.

Industrial involvement

Companies and researchers push the boundaries and development of science in general, and biomaterials are no exception.

Ethics

Ethical considerations are paramount—as are legal considerations and compliance with the law.

Regulation

As mentioned above, regulation and records are required to be kept by the product manufacturer for much longer than the product life.

A biomaterial is any material, natural or man-made, that comprises whole or part of a living structure or biomedical device which performs, augments, or replaces a natural function. The Society For Biomaterials is a professional society which promotes advances in all phases of materials research and development by encouragement of cooperative educational programs, clinical applications, and professional standards in the biomaterials field.

Biomechanics is a related area of engineering. Biomechanics involves the structure and function of biological systems using the methods of mechanics.

The human total hip system shown below offers a titanium, dual tapered stem design which provides a physiologic proximal load transfer, thus greatly reducing the chance of calcar resorption and distal hypertrophy. No fooling! The system provides an anatomic fit coupled with a straight stem design. The role of cartilage is played by polyethylene in this application. The reference source URL is *Biomet Corporation*. For more about hip replacement and conditions under which this is done, please link to the *Medline plus* website and learn more about hip joint replacement (many great illustrations).

For an example of a B.S. undergraduate study of orthopedic implants, please link to the following *San Jose State University* senior project report.

What other parts of the human anatomy can be replaced? The image on the left, from *ASM-International*, demonstrates other replacement parts. In this way, biomaterials are contributing to longer, higher quality lives.

Biomechanics deals with body movement and medicine. Biomechanical research is performed at many national and international universities. The *Centre of Biomechanics* at the Technical University of Hamburg, Germany, has high importance especially for local health care, and with speciality in the following areas: the quality of implants and medical equipment as well as the development of innovative methods to aid producers of surgical implants and equipment. Surgery methods and materials for implants are improved by the cooperation between biomechanics, materials science, mechanics, informatics, metrology and design. The image below demonstrates one example: hip joint replacement. Determination of the position of the shaft of a hip joint prosthesis in the femur at a planning station of the operating robot is shown. Corresponding to this planning, the milling of the cavitation for the shaft is done during the operation (so that one will have a *tailored* fit).

Both biomaterials and biomechanical expertise are needed to perform *in vitro testing* of spinal implants. The source of the image on the right is the *Institute of Orthopaedic Research and Biomechanics* at the University of Ulm in Germany. Select and open the "Research" folder on the web page.

Materials Scientists continue to develop novel surface modifications to enhance textile performance. Wettability is one important characteristic of a surface. Wettability of textile fibers can be enhanced to assist the dyeing operation during processing; and the fibers can be made to be non-wettable for application in, for example, ski clothing. PhotoLink uses wettability technology, as illustrated below, to, for instance, improve the flow characteristics of a polymer membrane used in a medical, diagnostic test kit. The reference source URL is *SurModics, Inc.*

Endo-vascular stents provide structural support vessels following angioplasty and other major medical procedures. After an angioplasty procedure, vessels can experience re-stenosis and eventually return to their original pre-operative diameter. In as many as 10 per cent of the procedures, the vessels may even collapse immediately. To prevent the vessels from shrinking, endo-vascular prosthesis or stents are used. Stents are tubular structures consisting of a spring, wire mesh or slotted tubes that are deployed inside the vessel. Depending on the design and intended use (coronary/peripheral), they can range in diameter from several millimeters to many times that size. Here, you are inside a vessel, seeing the stent in service.

The **actual stent in service** image is a 1995 copyright of Vital Images, Inc. and the Stanford University *Department of Radiology*. The reference source URL for more information on vascular and orthopaedic, mechanical integrity and durability test devices is *EnduraTEC Systems*. EnduraTEC Systems is a manufacturer of mechanical fatigue testing equipment for the biomedical and general materials industries.

9.3 NovaMatrix—Ultra-Pure Biopolymers and Biomaterials

NovaMatrix is one of the world's leading producers and supplier of ultra-pure, well-characterized and documented biopolymers and biomaterials for use in pharmaceutical, biotechnology and biomedical applications.

The company's current products are special ultra-pure grades of PRONOVA sodium alginate, PROTASAN chitosan base and water-soluble salts, and sodium hyaluronates that fulfil functional and regulatory specifications.

Sodium alginate, chitosan salts and bases, and hyaluronates are manufactured in accordance with cGMP guidelines (Code of Federal Regulations, Title 21, Part 210 and 211), NS-EN ISO 9001:2000 (quality management system) and ISO 13485:2003 (medical device directive) standards.

NovaMatrix is a newly created business unit of FMC BioPolymer, one of the leading suppliers of ingredients to the food, pharmaceutical and specialty products industries.

Ultrapure Biopolymers

NovaMatrix's technology and products are used in the development of novel drug-delivery therapies and its polymers are utilized in the formation of biostructures for tissue engineering, wound healing, implants and advanced biotechnology applications. Alginate, chitosan and hyaluronan have proven properties suitable for parenteral administration of challenging drugs and implants.

PRONOVA UP - Ultrapure Alginates

PRONOVA UP UltraPure (UP) sodium alginates are manufactured in the specially designed NovaMatrix production facility in Oslo, Norway.

This NovaMatrix product group belongs to a family of highly purified and well-characterized alginates developed for use in biomedical and pharmaceutical applications.

The products are characterized by specified parameters reflecting the key properties and purity profile of the material. Process and final product control are based on validated methods of analysis. Safety and toxicology data are reported in a drug master file submitted to the US FDA.

PRONOVA—Sterile Alginates

Sterilizing alginate solutions may often be an obstacle in pharmaceutical and biomedical applications requiring sterility of the applied material. In particular, this relates to problems with polymer degradation during autoclaving or irradiation procedures. Sterile filtration of highly viscous solutions may also be difficult.

To meet customer demands, NovaMatrix offers PRONOVA UP alginate as a sterilized and lyophilized product. Sterile aqueous solutions of sodium alginate may now be made simply by adding sterile water or buffers to the lyophilized alginate. The volume of added solution will thus determine the concentration of the alginate solution. Placing the vial on a shaker facilitates dissolution.

PROTASAN—Ultrapure Chitosans

The PROTASAN UP product series from NovaMatrix consists of ultra-purified and well-characterized water-soluble chitosan chloride and glutamate salts, and chitosan bases, for use in biomedical and pharmaceutical applications and also includes the newly launched series of six different PROTASAN UP chitosan base products. The new series consists of ultra-purified chitosans in the free amine or base form, with different viscosities.

Ultrapure—Sodium Hyaluronates

Sodium Hyaluronate Pharma Grade is a highly purified and well-characterized sodium hyaluronate developed for use in biomedical and pharmaceutical applications.

Sodium Hyaluronate Pharma Grade is manufactured at the NovaMatrix/FMC BioPolymer strategic partner Kibun's state-of-the-art production facility in Kamogawa, Japan.

The fermentation of ultra-pure sodium hyaluronate is made from non-animal origin material and based on more than 10 years of experience and the plant in constructed and run in compliance with cGMP guidelines for bulk pharmaceuticals. The ultra pure sodium hyaluronates from Kibun are described in a DMF submitted to the US FDA.

9.4 Developments in Medical Polymers for Biomaterials Applications

As materials designers confront the fundamental challenges of medical science, the discovery of new biocompatible polymers is creating unprecedented excitement.

by Jon Katz

Like other important scientific concepts that change over time, the notion of biocompatibility has evolved in conjunction with the continuing development of materials used in medical devices. Until recently, a biocompatible material was essentially thought of as one that would "do no harm." The operative principle was that of inertness—as reflected, for example, in the definition of *biocompatibility* as "the quality of not having toxic or injurious effects on biological systems."

A cartilage repair unit injection molded from biodegradable polylactide (PLA).

When more-recent devices began to be designed with materials that were more responsive to local biological conditions, the salient principle became one of interactivity, with biocompatibility regarded as "the ability of a material to perform with an appropriate host response in a specific application." Of course, this conceptual shift was predicated on the ability to determine just what constituted "an appropriate host response"—the result of insights gained through tremendous advances in molecular biology and biological surface sciences. Such progress also lies behind the current definition of *biomaterial* as "a material intended to interface with biological systems to evaluate, treat, augment, or replace any tissue, organ, or function of the body." In particular, the design approach regarding implantable devices—and especially long-term implants—has moved away from attempts to develop inert biomaterials in favour of biomaterials that interact with and in time are integrated into the biological environment.

The biocompatibility of a medical implant will be influenced by a number of factors, including the toxicity of the materials employed, the form and design of the implant, the skill of the surgeon inserting the device, the dynamics or movement of the device in situ, the resistance of the device to chemical or structural degradation (biostability), and the nature of the reactions that occur at the biological interface. These factors vary significantly depending on whether the implant is deployed, for example, in soft tissue, hard tissue, or the cardiovascular system—to the extent that "biocompatibility may have to be uniquely defined for each application." Among the prominent applications for biomaterials are:

- *Orthopedics*—joint replacements (hip, knee), bone cements, bone defect fillers, fracture fixation plates, and artificial tendons and ligaments.
- *Cardiovascular*—vascular grafts, heart valves, pacemakers, artificial heart and ventricular assist device components, stents, balloons, and blood substitutes.
- *Ophthalmics*—contact lenses, corneal implants and artificial corneas, and intraocular lenses.
- *Other applications*—dental implants, cochlear implants, tissue screws and tacks, burn and wound dressings and artificial skin, tissue adhesives and sealants, drug-delivery systems, matrices for cell encapsulation and tissue engineering, and sutures.

The types of materials featured in the above uses include metals (stainless steel, titanium, cobalt chrome, nitinol), ceramics and glasses (alumina, calcium phosphate, hydroxyapatite), and a wide range of synthetic and natural polymers. This article focuses on polymers, and presents a brief overview of some of the more exciting recent developments that are radically expanding the capabilities of polymeric biomaterials. These include:

- New approaches to biodegradable polymers.
- "Combinatorial" and "supramolecular" chemistry.
- So-called intelligent materials.
- Other new formulations, including phospholipids, polymers for gene therapy, enhanced polyurethanes, and protein-based polymers.

It should be kept in mind that the examples presented run the gamut from newly reported research to products in clinical trials or awaiting regulatory approval to devices that are commercially available.

Novel Biodegradable Polymers

As for other biomaterials, the basic design criteria for polymers used in the body call for compounds that are biocompatible (new definition), processable, sterilizable, and capable of controlled stability or degradation in response to biological conditions. The reasons for designing an implant that degrades over time often go beyond the obvious desire to eliminate the need for retrieval. For example, the very strength of a rigid metallic implant used in bone fixation can lead to problems with "stress shielding," whereas a bioabsorbable implant can increase ultimate bone strength by slowly transferring load to the bone as it heals. For drug delivery, the specific properties of various degradable systems can be precisely tailored to achieve optimal release kinetics of the drug or active agent.

An ideal biodegradable polymer for medical applications would have adequate mechanical properties to match the application (strong enough but not too strong), would not induce inflammation or other toxic response, would be fully metabolized once it degrades, and would be sterilizable and easily processed into a final end product with an acceptable shelf life. In general, polymer degradation is accelerated by greater hydrophilicity in the backbone or end groups, greater reactivity among hydrolytic groups in the backbone, less crystallinity, greater porosity, and smaller finished device size.

Beginning in the 1960s, a range of synthetic biodegradable polymers have been developed, including polylactide (PLA), polyglycolide (PGA), poly (lactide-co-glycolide) (PLGA), poly (e-caprolactone), polydioxanone, polyanhydride, trimethylene carbonate, poly (ß-hydroxybutyrate), poly (g-ethyl glutamate), poly (DTH iminocarbonate), poly (bisphenol A iminocarbonate), poly (ortho ester), polycyanoacrylate, and polyphosphazene. There are also a number of biodegradable polymers derived from natural sources such as modified polysaccharides (cellulose, chitin, dextran) or modified proteins (fibrin, casein).

To date, the compounds that have been employed most widely in commercial applications are PGA and PLA, followed by PLGA, poly (e-caprolactone), polydioxanone, trimethylene carbonate, and polyanhydride. Some of the common PLA products include tissue screws, tacks, and suture anchors, as well as systems for meniscus and cartilage repair. The first FDA-cleared PLGA product was the Lupron Depot drug-delivery system (TAP Pharmaceutical Products Inc.; Lake Forest, IL), a controlled release device for the treatment of advanced prostate cancer that used biodegradable microspheres of 75:25 lactide/glycolide to administer leuprolide acetate over periods as long as 4 months (replacing daily injections). Another drug-delivery device, the Gliadel Wafer (Guilford Pharmaceuticals Inc.; Baltimore, MD), is used to prolong the life of patients suffering from a particularly deadly form of brain cancer, glioblastoma multiforme. In this case, dime-sized wafers of a biodegradable polyanhydride copolymer—poly[bis (p-carboxyphenoxy) propane:sebacic acid] in a 20:80 molar ratio—are implanted directly into the brain to deliver a powerful chemotherapeutic agent (BCNU) that has deleterious side effects when administered systemically.

Biodegradable Polymers for Tissue Engineering

One area of intense research activity is the use of biodegradable polymers for tissue engineering, which can be defined as "the application of engineering principles to create devices for the study, restoration, modification, and assembly of functional tissues from native or synthetic sources." Candidate materials include natural polymers (fibrin, collagen, gelatin, hyaluronan), synthetic polymers (e.g., PLA, PGA, PLGA, ethylene oxide block copolymers), and inorganic materials (tricalcium phosphate, calcium carbonate, nonsintered hydroxyapatite).

A recent project investigated the possibility of manufacturing biodegradable composites for use as bioactive matrices to guide and support tissue ingrowth.[5] Composites were prepared using polyhydroxybutyrate (PHB), a naturally occurring ß-hydroxyacid linear polyester, and as much as 30 per cent by volume of either hydroxyapatite (HA) or tricalcium phosphate (TCP) (Figure 9.1). One of the goals was to achieve a reasonably homogeneous distribution of the HA/TCP particles in the PHB matrix, as this uniformity would provide an anchoring mechanism when the materials would be employed as part of an implant. The composites were successfully manufactured through a compounding and compression molding process. It was observed that microhardness increased with an increase in bioceramic content for both the HA ad TCP compounds.

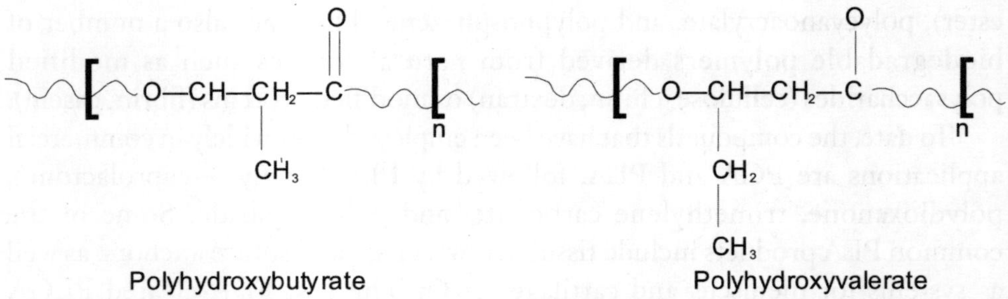

$$\sim\!\!\left[\!\!\sim O{-}\underset{\displaystyle CH_3}{CH}{-}CH_2{-}\overset{\displaystyle O}{\overset{\|}{C}}\!\!\sim\right]_n\!\!\sim \qquad \sim\!\!\left[\!\!\sim O{-}\underset{\displaystyle CH_2CH_3}{CH}{-}CH_2{-}\overset{\displaystyle O}{\overset{\|}{C}}\!\!\sim\right]_n\!\!\sim$$

Polyhydroxybutyrate Polyhydroxyvalerate

Fig. 9.1: Polyhydroxybutyrate (PHB) typically requires the presence of enzymes for biodegradation. It is often copolymerized with polyhydroxyvalerate (PHV)

Another team of researchers has addressed the problem of fabricating open-pore, biodegradable polymer scaffolds for cell seeding or other tissue engineering applications. The material selected was the tyrosine-derived polycarbonate poly (DTE-co-DT carbonate), in which the pendant group via the tyrosine—an amino acid—is either an ethyl ester (DTE) or free carboxylate (DT). Through alteration of the ratio of DTE to DT, the material's hydrophobic/hydrophilic balance and rate of in vivo degradation can be manipulated. It was shown that, as DT content increases, pore size decreases, the polymers become more hydrophilic and anionic, and cells attach more readily. Previously encountered problems with maintaining sufficient interconnectivity between pores in the structure were avoided through a novel phase-separation technique. Tyrosine-derived polyarylates under study as resorbable coatings for other biomaterials have been reported to significantly decrease blood-coagulation activation and bacterial adhesion without modifying the structure of the substrate.

Multiblock copolymers of poly (ethylene oxide) (PEO) and poly (butylene terephthalate) (PBT) are also under development as prosthetic devices and artificial skin and as scaffolds for tissue engineering. These materials are subject to both hydrolysis (via ester bonds) and oxidation (via ether bonds). Degradation rate is influenced by PEO molecular weight and content, and the copolymer with the highest water uptake degrades most rapidly.

Among the important series of physiological reactions or "cascades" is the fibrinolytic sequence, in which blood clots are removed from the circulation in part through the breakdown of the protein fibrin by the enzyme plasmin. Neurite-associated plasmin activity has also been shown to play a role in nerve growth, and a recent study describes the creation of a new biosynthetic material that imitates fibrin as a vehicle for promoting peripheral nerve regeneration. The material incorporates a recombinantly expressed fragment of human fibrin within a photo-cross-linked, polyethylene glycol (PEG)—based hydrogel matrix. According to the study, the resulting system degrades completely upon exposure to plasmin but is otherwise stable, demonstrating that it is possible to design biosynthetic materials with specific enzymatic degradability.

Combinatorial and Supramolecular Chemistry

A product of revolutionary advances in molecular biology, microfabrication, and information technology, combinatorial chemistry is an emerging discipline of tremendous potential for pharmacological design, biomaterials development, and the entire realm of polymer science. This new approach to the synthesis of materials and characterization of their properties uses multicomponent screening, high-throughput chemical synthesis, and advanced computational techniques to produce and analyze a large number of novel monomeric and polymeric entities.

In a combinatorial synthesis, automated methods are used to process a relatively small number of "ingredients" in a parallel fashion so as to generate a large "library" of elemental combinations on a microscopic scale. Such incrementally controllable, permutationally designed systems hold out the promise of precise structure/property correlations to determine which specific materials will fulfill specific performance needs. One of the first reported examples of a combinatorially prepared library of biomaterials involved A-B type copolymers in which one monomer was a diphenol and the second a diacid. A total of 14 different diphenols were copolymerized in all combinations with eight different diacids to produce 112 (14 × 8) structurally related polyarylate copolymers. The characteristics of this series of new materials—properties such as wettability, glass-transition temperature, and cellular response—were then analyzed in a systematic manner to identify the relationship between polymer structure, properties, and performance.

Another intriguing new field of great promise is supramolecular chemistry, which is concerned with developing molecular assemblies for biological applications based on macromolecular architectures that mimic nanoscale systems or mechanisms in nature. Novel synthesis methods based on supramolecular chemistry have been used to create branched or graft, cyclic, cross-linked, star, and dendritic polymer structures.

An excellent example of the ability of supramolecular polymer systems to meet complex performance requirements and function like natural chemomechanical materials can be seen in two recent studies using polyrotaxanes—polymers comprising cyclic compounds that are threaded onto linear polymeric chains capped with bulky end groups (Figure 9.2). The first study prepared a series of biodegradable polyrotaxanes in which a-cyclodextrins (a-CDs) were threaded onto a PEG chain capped with amino acids. The resulting structure could be adapted to accomplish a two-stage, controlled release of drugs bonded to the a-CDs: hydrolytic enzymes could first attack the peptide bonds of the macrostructure, degrading the terminal moieties and releasing the drug-immobilized a-CDs, and a second enzyme could then attack the a-CDs and release the drugs. The very rapid and complete degradation of the polyrotaxane prevents problems like that posed by residual crystalline oligomers that can result from the incomplete hydrolysis of highly crystalline materials like PLA.

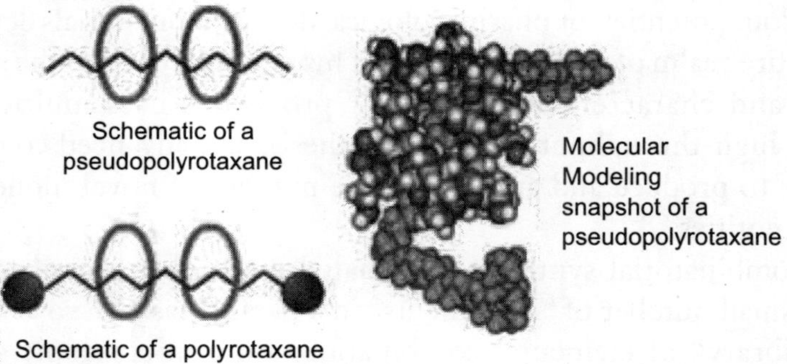

Fig. 9.2: Polyrotaxanes and other supramolecular polymer structures can be designed to mimic nanoscale systems in nature

Polyrotaxanes can also be designed to effect dynamic molecular functions similar to those of natural tissues through movement of the cyclic compounds along the polymer's linear chain. The second study used polyrotaxanes with b-cyclodextrins (b-CDs) threaded onto a triblock copolymer of PEG and poly (propylene glycol) (PPG) bounded by fluorescein-4-isothiocyanate end groups. It was observed that the majority of the b-CDs migrated toward the PPG segment with increasing temperature—a phenomenon that, with controlled temperature variation, could suggest the action of a molecular-scale piston. This stimulus response was seen to resemble the action of myosin molecules sliding along actin filaments in the muscle-contraction process.

Intelligent Materials

The polyrotaxane polymers described above can be considered intelligent biomaterials insofar as they can function in a manner similar to molecular structures in the body. Other intelligent materials currently under development include hydrogels exhibiting critical behaviour, anionic and cationic hydrogels, controlled porous structures, ultrapure biomaterials, tailored copolymers with desirable functional groups, biomimetic hydrogels, biodegradable polymers responding to specific biological conditions, and polymers precisely replicating selected properties.

One particularly fascinating model of an intelligent material has as its ultimate goal one of the most critical issues in modern medicine—the controlled delivery of insulin for the treatment of diabetes. This hydrogel system features an insulin-containing reservoir within a membrane of poly (methacrylic acid-g-poly[ethylene glycol]) copolymer in which glucose oxidase has been immobilized. The surface of the porous membrane contains a series of molecular "gates," which open and release insulin when the hydrogel shrinks at low pH values as a result of the interaction of glucose with glucose oxidase (Figure 9.3). In addition, the cross-linked polyethylene glycol graft in the decoupled state has the ability to adhere to a specific region in the upper intestine that is a preferred location for the delivery of insulin.

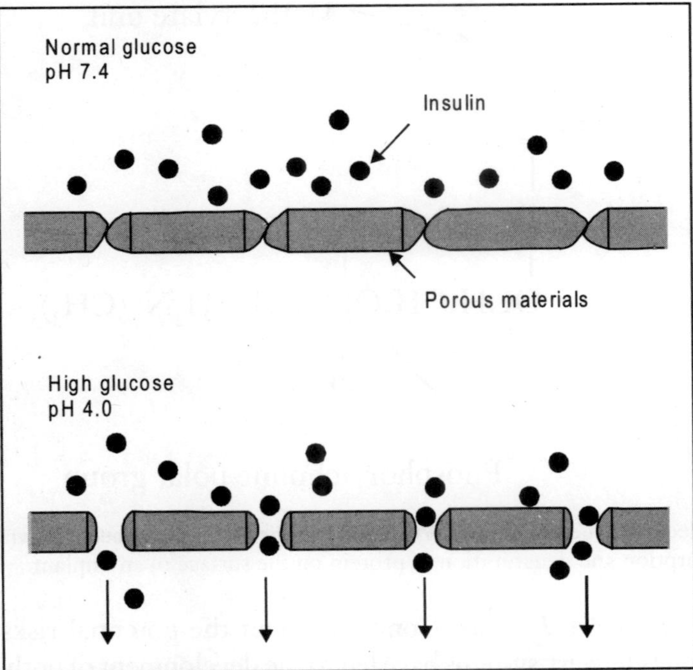

Fig. 9.3: The porous surface of the P (MAA-g-EG) hydrogel system responds to the presence of glucose

Other New Formulations

The tremendous range of current biomaterials research is proposing innovative new polymers for applications ranging from cardiovascular devices to gene therapy. Several of the more interesting formulations are highlighted below:

Phospholipids: Among the materials receiving a great deal of attention for its hemocompatible characteristics is 2-methacryloyloxyethyl phosphorylcholine, or MPC (Figure 9.4). Created in Japan in the mid-1970s, this polymer has been shown to inhibit to a significant degree the almost-instantaneous protein adsorption and subsequent denaturation that is the initial event affecting practically every material used in the body, and which can lead to thrombus formation. For example, it has been reported that a small-diameter (2-mm) vascular graft prepared from a blend of MPC polymer and segmented polyurethane (SPU) did not occlude for more than 8 months after implantation in a rabbit, whereas an identical SPU graft occluded within 90 minutes.[14] The polymer has also been tested as a coating for implantable glucose sensors and hemodialyzer filters and as a rinsing agent to protect contact lenses from protein deposition.

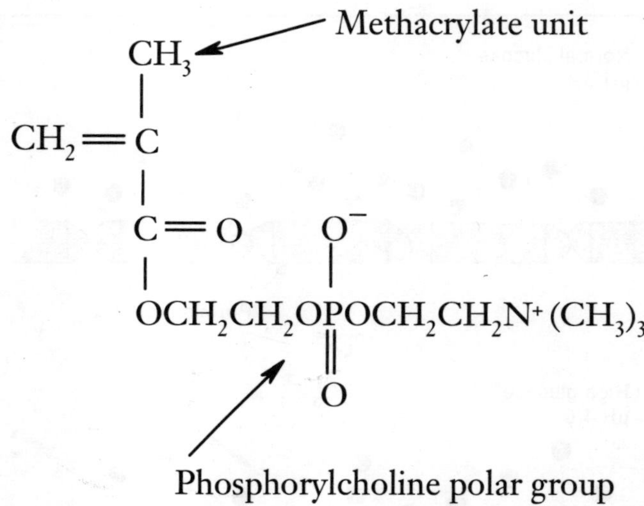

Fig. 9.4: Molecular structure of the MPC polymer. The material has been shown to inhibit the absorption and denaturation of protein on the surface of an implant.

Polymers for Gene Therapy: Concerns about the potential risks associated with viral gene-delivery systems have led to the development of both degradable and nondegradable, targeted and nontargeted polymeric gene carriers. Examples include PLL-PEG-lactose as a carrier for the transfection of plasmid DNA at hepatocytes; a biodegradable cationic polymer, poly (a-[4-aminobutyl]-L-glycolic acid), as a carrier for mouse plasmid DNA to prevent insulitis; and biodegradable gelatin-alginate microspheres as a carrier of adenovirus (Ad5-p53) for intracranial delivery.

Silicone-Urethane Copolymers: Novel families of silicone-urethane copolymers have been developed that, compared with traditional polyurethane biomaterials, offer advantages in biostability, thromboresistance, abrasion resistance, thermal stability, and surface lubricity, among other properties. Copolymer synthesis is performed via two methods: incorporation of silicone into the polymer backbone together with organic soft segments, and the use of surface-modifying end groups to terminate the copolymer chains. The organic soft block can be either polytetramethyleneoxide (PTMO) or an aliphatic polycarbonate used together with polydimethylsiloxane (PSX). Applications for the new materials include balloons, ventricular assist devices, vascular grafts, pacemaker leads, and orthopedic and urologic implants.

Protein-Based Polymers: A series of recently introduced casein- and soy-based biodegradable thermoplastics have recently joined collagen as a source of natural protein-based biomaterials. In comparison with collagen, however, these polymers are less susceptible to thermal degradation, can be easily processed via melt-based technologies, and can be reinforced with inert or bioactive ceramics. Temporary replacement implants, scaffolds for tissue engineering, and drug-delivery vehicles are among the potential biomaterials uses under investigation.

Conclusion

The discovery of novel polymeric biomaterials—and the refinement of traditional ones—is creating a thoroughly unprecedented excitement in the field as polymer chemists and other materials designers increasingly confront many of the fundamental challenges of medical science. As the biomaterials discipline itself evolves, the startling advances of the last few years in genomics and proteomics, in various high-throughput cell-processing techniques, in supramolecular and permutational chemistry, and in information technology and bioinformatics promise to support the quest for new materials with powerful analytic tools and insights of boundless energy and sophistication.

9.5 Overview

Biopolymer™ provides the tools needed for building and manipulating peptide, protein, DNA/RNA, and carbohydrate structures. Biopolymer is fully integrated with SYBYL® for structure building, refinement, visualization and analysis. Protein structures can be analyzed in detail using the SYBYL Molecular Spreadsheet.

The methyl glycoside of the Lewis b human blood group determinant complexed with Lectin IV of Griffonia simplicifolia. The protein is represented as tubes colour-coded by secondary structure type. The tetrasaccharides are shown as space-filled atoms.

Key Benefits

- ❖ Create models of peptides, proteins, RNA, DNA, and polysaccharides
- ❖ Virtual mutagenesis experiments
- ❖ Pattern searching of the Protein Data Bank
- ❖ Refine structures derived from crystallography, NMR, or modeling
- ❖ Create graphical displays that reveal structure-activity relationships
- ❖ Identify receptor-bound conformations of ligands

10

Smart Polymers and Polymeric Materials

10.1 Smart Polymers

Smart polymers are materials composed of polymers that respond in a *dramatic* way to very *slight* changes in their environment. Scientists studying natural polymers have learned how they behave in biological systems, and are now using that information to develop similar man-made polymeric substances with specific properties. These **synthetic polymers** are potentially very useful for a variety of applications including some related to biotechnology and biomedicine.

Smart polymers are becoming increasingly more prevalent as scientists learn about the chemistry and triggers that induce conformational changes in polymer structures and devise ways to take advantage of, and control them. New polymeric materials are being chemically formulated that **sense specific environmental changes** in biological systems, and adjust in a *predictable* manner, making them useful tools for drug delivery or other metabolic control mechanisms.

In this relatively new area of biotechnology, the potential biomedical applications and environmental uses for smart polymers appear to be limitless.

Also Known As:

Stimulus-Responsive Polymers

Examples:

The rapid conformational change of the smart polymer, at pH 8, made it a useful medium for drug delivery.

10.2 Biomedical Applications for Smart Polymers

Mimicking the Stimulus-Responsiveness of Natural Polymers

Scientists studying the natural polymers found in living organisms (*proteins, carbohydrates and nucleic acids*) have learned how they behave in biological systems as they perform their structural and physiological roles. That information is being put to use to develop similar man-made polymeric substances with specific properties and the ability to **respond to changes** in their environment. These synthetic polymers are potentially very useful for a variety of applications including some related to biotechnology and biomedicine.

Smart polymers are becoming increasingly more prevalent as scientists learn about the chemistry and triggers that induce conformational changes in polymer structures and devise ways to take advantage of, and control them. New polymeric materials are being chemically formulated that sense specific environmental changes in biological systems, and adjust *in a predictable manner* making them useful tools for drug delivery or other metabolic control mechanisms.

The **nonlinear response** of smart polymers is what makes them so unique and effective. A significant change in structure and properties can be induced by a **very small stimulus**. Once that change occurs, there is no further change, meaning a predictable all-or-nothing response occurs, with complete uniformity throughout the polymer. Smart polymers may change conformation, adhesiveness or water retention properties, due to slight changes in pH, ionic strength, temperature or other triggers.

Another factor in the effectiveness of smart polymers lies in the inherent nature of polymers in general. The strength of each molecule's response to changes in stimuli is the composite of changes of individual monomer units which, alone, would be weak. However, these weak responses, compounded *hundreds or thousands of times*, create a considerable force for driving biological processes.

Classification and Chemistry

Currently, the most prevalent use for smart polymers in biomedicine is for **specifically targeted drug delivery**. Since the advent of *timed-release pharmaceuticals*, scientists have been faced with the problem of finding ways to deliver drugs to a particular site in the body *without having them first degrade* in the highly acidic stomach environment. Prevention of adverse effects to healthy bone and tissue is also an important consideration. Researchers have devised ways to use smart polymers to control the release of drugs until the delivery system has reached the desired target. This release is controlled by either a chemical or physiological trigger.

Linear and matrix smart polymers exist with a variety of properties depending on reactive functional groups and side chains. These groups might be responsive to pH, temperature, ionic strength, electric or magnetic fields, and light. Some polymers are reversibly cross-linked by noncovalent bonds that can break and reform depending on external conditions. Nanotechnology has been fundamental in the development of certain nanoparticle polymers such as dendrimers and fullerenes, that have been applied for drug delivery. Traditional drug encapsulation has been done using lactic acid polymers. More recent developments have seen the formation of lattice-like matrices that hold the drug of interest integrated or entrapped between the polymer strands.

Smart polymer matrices release drugs by a chemical or physiological structure-altering reaction, often a hydrolysis reaction resulting in cleavage of bonds and release of drug as the matrix breaks down into biodegradable components. The use of natural polymers has given way to artificially synthesized polymers such as polyanhydrides, polyesters, polyacrylic acids, poly (methyl methacrylates), and polyurethanes. Hydrophilic, amorphous, low-molecular-weight polymers containing heteroatoms (i.e., atoms other than carbon) have been found to degrade fastest. Scientists control the rate of drug delivery by varying these properties thus adjusting the rate of degradation.

10.3 Polymeric Medical Materials

Polymeric materials are used in medical devices because of their unique properties, ease of manufacture, flexibility in design and low cost. Unique properties include:

- ❖ Flexibility (tubing, seals, vascular...)
- ❖ Sorption/diffusion (drug delivery, contact lens...)
- ❖ Formability (bone cement, dental fillings...)
- ❖ Wear resistance (prosthetic joints...)
- ❖ Specific stiffness of composites (crutches, wheelchairs...)

Polymers have been successfully used to aid repair and modification of the human body. The applications for polymers in the medical field cover an enormous range. Some example application areas are: Cardiovascular (heart valves, heart assist devices, blood vessel prosthesis, stents etc.), Internal Artificial Organs (kidney, lung, muscles etc.) Skin and Soft Tissue (skin, wound care) Ophthalmic and Craniofacial (contact lenses, cornea implants, facial surgery), Orthopaedic and Dental (joint replacement, limb replacement, dental implants) and mobility aids (crutches, wheelchairs).

Materials Selection

Only a small number of polymers are available as medical grades for medical applications and an even smaller number are used for implants. The choice of polymer depends on the extent of contact with body fluids, internal and external tissue as prescribed by the regulatory framework. In addition, thermal and mechanical properties and possible degradation over time must also be taken into consideration. Typical polymers and their uses are listed below.

Materials Application

Materials	*Application*
PVC, PA, PE, PS, Epoxy Resins	Non-contact: e.g. syringes, blood storage bags
Silicone Rubber, Natural Rubber, PVC, Polyurethane, PE, PP, Polyester	Short-term contact: e.g. catheters, feeding tubes, drainage tubes
Nylon, PP, Polyester	Medium-term contact: e.g. sutures and ligatures
PE, PET, Silicone Rubber, Polyurethane, PMMA, Polysulphones, Polyphosphazenes, Hydrogels, Thermoplastic Elastomers Polydimethylsiloxane (PDMS)	Long term contact: Implants

10.4 Graft-and-Block

Graft-and-block copolymers are comprised of two different polymers grafted together. A number of patents already exist for different combinations of polymers with different reactive groups. The product exhibits properties of both individual components which adds a new dimension to a smart polymer structure, and may be useful for certain applications. Cross-linking hydrophobic and hydrophilic polymers results in formation of micelle-like structures that can protectively assist drug delivery through aqueous medium until conditions at the target location cause simultaneous breakdown of both polymers.

A graft-and-block approach might be useful for solving problems encountered by the use of a common bioadhesive polymer, polyacrylic acid (PAAc). PAAc adheres to mucosal surfaces but will swell and degrade rapidly at pH 7.4, resulting in rapid release of drugs entrapped in its matrix. A combination of PAAc with another polymer that is less sensitive to changes at neutral pH might increase the residence time and slow the release of the drug, thus improving bioavailability and effectiveness.

Hydrogels are polymer networks that do not dissolve in water but *swell or collapse* in changing aqueous environments. They are useful in biotechnology for phase separation because they are reusable or recyclable. New ways to control the flow, or catch and release of target compounds, in hydrogels, are being investigated. Highly specialized hydrogels have been developed for the delivery and release of drugs into specific tissues. Hydrogels made from PAAc are especially common because of their bioadhesive properties and tremendous absorbency.

Enzyme immobilization in hydrogels is a fairly well-established process. Reversibly cross-linked polymer networks and hydrogels can be similarly applied to a biological system where the response and release of a drug is triggered by the target molecule itself. Alternatively, the response might be turned on or off by the product of an enzyme reaction. This is often done by incorporating an enzyme, receptor or antibody, that binds to the molecule of interest, into the hydrogel. Once bound, a chemical reaction takes place that triggers a reaction from the hydrogel. The trigger can be oxygen, sensed using oxidoreductase enzymes, or a pH-sensing response. An example of the latter is combined entrapment of glucose oxidase and insulin in a pH-responsive hydrogel. In the presence of glucose, the formation of gluconic acid by the enzyme triggers release of insulin from the hydrogel.

Two criteria for this technology to work effectively are *enzyme stability* and *rapid kinetics* (quick response to the trigger and recovery after removal of the trigger). Several strategies have been tested in type 1 diabetes research, involving the use of similar types of smart polymers that can detect changes in blood glucose levels and trigger production or release of insulin. Likewise, there are many possible applications of similar hydrogels as drug delivery agents for other conditions and diseases.

Smart polymers are not just for drug delivery. Their properties make them especially suited for **bioseparations**. The time and costs involved in purifying proteins might be reduced significantly by using smart polymers that undergo rapid reversible changes in response to a change in medium properties. Conjugated systems have been used for many years in physical and affinity separations and immunoassays. Microscopic changes in the polymer structure are manifested as precipitate formation, which may be used to aid separation of trapped proteins from solution.

These systems work when a protein or other molecule that is to be separated from a mix, forms a bioconjugate with the polymer, and precipitates with the polymer when its environment undergoes a change. The precipitate is removed from the media, thus separating the desired component of the conjugate from the rest of the mixture. Removal of this component from the conjugate depends on recovery of the polymer and a return to its original state, thus hydrogels are very useful for such processes.

Another approach to controlling biological reactions using smart polymers is to prepare recombinant proteins with built-in polymer binding sites close to ligand or cell binding sites. This technique has been used to control ligand and cell binding activity, based on a variety of triggers including temperature and light.

Future Applications

It has been suggested that polymers might be developed that can *learn and self-correct behaviour* over time. Although this might be a far distant possibility, there are other more feasible applications that appear to be coming in the near future. One of these is the idea of smart toilettes that analyze urine and help identify health problems. In environmental biotechnology, smart irrigation systems have been also been proposed. It would be incredibly useful to have a system that turns on and off, and controls fertilizer concentrations, based on soil moisture, pH and nutrient levels. Many creative approaches to targeted drug delivery systems that self-regulate based on their unique cellular surroundings, are also under investigation.

There are obvious possible problems associated with the use of smart polymers in biomedicine. The most worrisome is the possibility of toxicity or incompatibility of artificial substances in the body, including degradation products and byproducts. However, smart polymers have enormous potential in biotechnology and biomedical applications if these obstacles can be overcome.

10.5 Polymeric Materials Novel Polymerization Technology

Brief Summary: Dr. Caneba has discovered a free-radical polymerization process that has the following features:

1. Polymer radical trapping mechanism
2. Capability of producing effective amounts of block copolymers
3. Polymer products with relatively narrow molecular weight distributions

Specific Materials and Applications

Hydrophilic-Hydrophobic Copolymers

We have used styrene-based hydrophilic-hydrophobic block copolymers as coupling agent for wood-polystyrene composites. The block copolymer can be produced in large quantities and delivered as a water-based self-emulsion. When wood fibers were are immersed or sprayed with the emulsion, the resulting dried wood material is mixed with polystyrene and blended to form a composite. The composite was demonstrated to have a dramatically higher level of toughness without sacrificing yield stress compared to the composite without the block copolymer coupling agent. The block copolymer could be used to produce composites from recycled polystyrene and wood or paper. It also has the potential as a surfactant.

We are currently developing a procedure that can be used to produce a poly (methyl methacrylate)-based hydrophilic-hydrophobic block copolymers. This material can be used as coupling agent for PVC-based composites.

We have developed a new type of hydrophilic-hydrophobic material that is being produced from inexpensive raw materials in a very efficient process operation. The material is delivered in a no-VOC water-based self-emulsion. It has a unique chemical makeup and it can be marketed as a low-cost surfactant that has a wide-range of HLB numbers. Since the hydrophilic portion of this material comes from an acid polymer that can be distributed along the chain or at the ends (one or two ends of the linear chain), then this material can be used in a formulation where the acid can be reacted with a suitable base material, such as in adhesives, epoxy systems, polyurethane formulations, etc.

Silicone-Containing Copolymers

From the above-mentioned hydrophilic-hydrophobic block copolymers, we have been able to generate silicone-containing blocks by reacting the hydrophilic parts with amino-silicones. These copolymers can be used to impart a very thin semi-permanent water-repellent and lubricating layer to polystyrene, poly (methyl methacrylate), polyvinylchloride, etc.

Emulsion-Based Organic Block Copolymers

We have been able to generate emulsions from polystyrene and poly (methyl methacrylate) radicals that can be reacted with other monomers. For example, emulsions containing very high molecular weight polystyrene-poly (butyl acrylate) and poly (methyl methacrylate)-poly (butyl acrylate) have been generated to produce a thermoplastic elastomer from the dried solid.

Poly (methyl methacrylate)-based emulsions have been generated and formulated as interior eggshell very low-VOC paints. Most of the paint properties compare to conventional interior eggshell paints, although wet scrub resistance indicates failure after about 600-1,000 cycles. Neat very low VOC emulsions were demonstrated to be useful as clear coatings for wood.

10.6 Polymeric Materials for Photo-chromic Applications

This application is a continuation-in-part of U.S. Serial No. 08/435,126, entitled "Adhesive Photochromic Matrix Layers for use in Optical Articles," filed May 5, 1995, naming Amitava Gupta, Ronald D. Blum and Ven atramani S. Iyer as inventors. This application is also a continuation-in-part of Application Serial No. 08,167,103, entitled "Method and Apparatus for Manufacturing

Photochromic Lenses," filed December 15, 1993, naming Amitava Gupta and Ronald Blum as inventors, which is a continuation-in-part of Application Serial No. 08/165,056, entitled "Method and Apparatus for Manufacturing Photochromic Lenses," filed December 10, 1993, also naming Amitava Gupta and Ronald Blum as inventors.

Field of the Invention

This invention relates to thin layers of materials which function as effective carriers for photochromic additives, can be attached to many types of optical substrates (including ophthalmic lenses, semifinished lens blanks, optical preforms, goggles, safety glasses, windows, and windshields), and can be encapsulated between two layers of such optical products, at least one of which is transparent to ultraviolet radiation.

Background of the Invention

It is often desirable to incorporate photochromic properties into optical products which are used both in sunlight and in darkness. This allows such products to develop a dark tint that serves to reduce outdoor glare during exposure to sunlight, while also allowing them to turn clear when sunlight is not present. A combination of three or more photochromic additives, each with a coloured state absorbing in a particular part of the visible wavelength range (400- 750 NM) is usually used to obtain a neutral gray or brown tint.

While photochromic ophthalmic lenses have been successfully commercialized for many years, photochromic versions of other optical products, such as windshields, windows, goggles and safety glasses, are not yet in common use, because it is difficult and expensive to manufacture photochromic versions of these products.

Certain constraints also have to be imposed on the physical properties of materials, particularly plastic materials, in order to enable photochromic additives incorporated in them to perform their intended function (i.e., switch from a clear state to a dark state in presence of sunlight, and return to a clear state when exposure to sunlight ends). These constraints compromise the structural and optical performance of these materials, and can render them unsuitable for their intended applications.

It is useful to review the mechanism by which photochromic additives are believed to perform, in order to further understand the deficiencies of currently available photochromic materials, and to understand the design rationale for layers bearing photochromic additives.

Photochromic molecules exist in two ground state configurations ("doublet ground states"), one of which does not absorb visible radiation at any significant level, the other of which is able to absorb visible radiation strongly. Each of these ground state configurations has a corresponding electronically excited state. The four state of a typical organic photochromic molecule includes two ground states: a colourless state, 1, and a coloured state, 2, which are thermally interconvertible. For certain photochromic materials, state 2 can be converted to state 1 through absorption of visible light. An excited state 3 formed by absorption of ultraviolet light by ground state 1 decays to form an excited state 4 and from excited state 4 decays to ground state 2. The excited state 3 also directly decays to ground state 2 (the activating process). The ground state 2 forms the excited state 4 which returns to both ground state 1 and ground state 2 on deactivation (the bleaching process). The activated conversion of ground state 2 into ground state 1 introduces a temperature dependence into the dynamic range of the photochromic response of organic photochromic materials. Thus, the higher the temperature, the faster the conversion of ground state 2 into ground state 1, and the equilibrium population of ground state 2 decreases relative to the population of ground state 1. Accordingly, upon activation, the higher the temperature, the more photochromic additive exists in the colourless form 1.

Raising the temperature barrier between ground state 2 and ground state 1 slows down the rate of conversion of ground state 2 into ground state 1, and hence lessens the reduction of the photochromic response ("Activation Level") on raising the temperature. At the same time, however, it slows down the speed of switching of the photochromic additive to the transparent state once it is removed from sunlight exposure. The photochromic materials available in the state of the art, therefore, represent a compromise between the switching speed (measured as time required for the photochromic material to return to the colourless form when solar exposure is terminated, also referred to as the rate of the bleaching process) and the temperature dependence of the dynamic range of the photochromic response (measured by the population of the coloured form relative to the colourless form at thermal equilibrium when being exposed to solar or ultraviolet radiation at any particular temperature, also referred to as the level of activation). For example, U.S. Patent No. 5,110,881, issued to McBain, describes a plastic material in which this trade-off is shifted towards a better level of activation at a higher temperature, but the resulting switching (bleaching) rate is more than 5 minutes at 72°F, measured as $T_{0\ s}$, the time needed to regain 50 per cent of the optical transmission lost when exposed to sunlight or ultraviolet radiation ("Plastic Photochromic Eyewear: A Revolution", CN. Welch and J.C. Crano, PPG Industries Inc.) A third deficiency of currently available photochromic materials is that such materials are generally made by diffusing or infusing ("imbibing") photochromic additives into bulk plastic materials, or by incorporating the photochromic additives into a monomer or monomers, and subsequently polymerizing the monomer formulation to form the bulk plastic material (see, e.g., U.S. Patent No. 4,909,963 issued to Kwak and Chen, U.S. Patent No. 4,637,698 issued to Kwak and 8Hurditch, and U.S. Patent No. 5,185,390 issued to Fischer).

Summary of the Invention

In view of the above, it is an object of this invention to develop plastic materials which incorporate photochromic additives in which the speed of switching is maintained at a high level, while the level of photochromic response is rendered relatively insensitive to temperature changes. It is a further object of this invention to develop optical products (including but not limited to ophthalmic lenses, goggles, safety glasses, windows, and windshields) incorporating these materials that are simple and inexpensive to make and whose structural and optical performance are not significantly compromised.

According to an embodiment of the invention, a polymeric material comprising main chain ether groups or side chain alkoxylated groups is provided that is effective as a carrier of photochromic additives. The polymeric material preferably has a glass transition temperature of at least 125°F and a cross link density of 2 to 8 moles per liter, more preferably 2.5 to 6 moles per liter. The polymeric material preferably comprises one or more polymerized monomers selected from the group consisting of one or more mono- or multi¬ functional acrylates or methacrylates. Examples include bisphenol A derivatives of mono- or multi-functional acrylates or methacrylates; aromatic carbonates of mono- or multi-functional acrylates or methacrylates; aliphatic carbonates of mono- or multi-functional acrylates or methacrylates; and allyl and vinyl derivatives of mono- or multi-functional acrylates or methacrylates. Further examples include polyethylene glycol diacrylates, ethoxylated bisphenol A diacrylates, 2-phenoxyethyl methacrylates, alkoxylated trifunctional aliphatic acrylates, alkoxylated aliphatic diacrylate esters, tetrahydrofurfuryl acrylates, alkoxylated trifunctional acrylates, and alkoxylated difunctional acrylate esters.

According to another embodiment of the present invention, resin formulations containing monomers which can be cured to form polymeric materials, such as those discussed above, are provided. Examples of such monomers are discussed above. The resin formulations of the present invention preferably comprise either a photoinitiator or a thermal initiator. Preferred photoinitiators include bisdimethoxybenzoyl trimethylpentyl phosphine oxide, 1-hydroxy cyclohexyl phenyl ketone, and 2-hydroxy, 2-methyl, 1-phenyl propane. Preferred thermal initiators include organic peroxides, hydroperoxides, percarbonates, peracetates and azo derivatives. The resin formulations of the present invention can also include additional additives, such as antioxidants. Preferred antioxidants include thiodiethylene bis (hydrocinnamide) and bis (1,2,2,6,6,- pentamethyl-4-piperidyl sebacate). One or more photochromic additives can also be included in the resin formulations of the present invention. Preferred photochromic additives include spiro (indolino) aphthoxazines, spiro (indolino) pyridobenzoxazines, spiro (benzindolino) naphthoxazines, and octamethyl stilbene.

Optical products can be made from either the above polymeric material or the above resin. Optical products appropriate for the practice of the present invention include uncorrected lenses, ophthalmic lenses, semifinished lens blanks, optical preforms, goggles, safety glasses, windows, and windshields. Other embodiments of the present invention relate to methods of forming a polymeric material that is impregnated with a photochromic additive. According to a first embodiment, the photochromic additive is incorporated into the polymeric material by adding the photochromic additive to a monomer formulation before it is polymerized.

According to a second embodiment, the photochromic additive is incorporated into the polymeric material by diffusing the photochromic additive material into the polymeric material. This can be accomplished in a variety of ways. For example, the photochromic additive can incorporated into the polymeric material by

immersing it into a solution of the photochromic additives in a chemically inert solvent, by spraying or spin coating the polymeric material with a solution of the photochromic additives in a chemically inert solvent, by repeatedly providing a surface layer of polymeric material and subsequently incorporating the photochromic additive into the surface layer of polymeric material, and so forth.

The polymeric material, either with or without the incorporated photochromic additive, can be adhered to an optical product, for example, by an interfacial region composed of an interpenetrating network, by an adhesive layer, and so forth.

Another way by which the polymeric material incorporating the photochromic additive can be included in an optical product is by forming the polymeric material incorporating the photochromic additive as a solid powder, suspending the solid powder in a liquid, and applying the suspended powder to an optical product. Yet another way of including the polymeric material incorporating the photochromic additive into an optical product is by partially polymerizing the polymeric material; loading the partially polymerized polymeric material with the photochromic additives; dispersing the loaded, partially polymerized polymeric material in a carrier fluid; and applying the carrier fluid with the dispersed, loaded, partially polymerized polymeric material onto an optical product.

Still other objects and advantages of the invention and alternative embodiments will readily become apparent to those skilled in the art, particularly after reading the detailed description and claims set forth below.

Detailed Discussion of the Invention

From the above discussion, it is clear that the rate of conversion of ground state 2 into ground state 1 controls the level of activation as well as the switching (bleaching) speed. Thus, the faster the conversion of ground state 2 into ground state 1, the lower the level of activation, but the faster the bleaching rate. For slow bleaching systems, the level of activation is controlled by the second process of conversion of ground state 2 into ground state l, i.e., through absorption of visible light by ground state 2, and subsequent deactivation to form 1. For such slow switching materials, the level of activation approaches a maximum, but the bleaching rate is slow.

In several photochromic molecules, the bleaching rate provided by the photochromic molecule itself is much faster than the bleaching rate permitted by the segmental rotational rates of the plastic matrix constituting the optical product. Therefore, the bleaching rate is very largely determined by the nature of the matrix, in particular its segmental rotational rate. As a result, the level of activation is also largely determined by the nature of the matrix for these photochromic molecules. Such photochromic molecules are disclosed, for example, in U.S. Patent No. 5,349,065, issued to Tanaka, Tanaka and Kida, or U.S. Patent No. 5,021,196, issued to Crano, Kwiatkowski and Hurditch. Plastic products incorporating such molecules exhibit matrix dependent bleaching rates and activation levels.

A series of polymeric matrices were developed by the present inventors for use in conjunction with such photochromic molecules that are capable of achieving high levels of activation at temperatures up to and

exceeding, 100°F, and still retain bleaching rates (T_{05}) of less than 1 minute at room temperature (70°-90°F).

A key feature of the polymeric matrices of the present invention is that the barrier to conversion of ground state 2 into ground state 1 (Figure 10.1), presented by the polymeric segments, contains a component which is weakly dependent on temperature, i.e., it remains nearly constant over the temperature range of 70°F-100°F. This is accomplished by selecting appropriate monomers which, after curing, contain constituents (either at side chains or at main chains) capable of undergoing rotation with a very low activation energy (less than 40 kJ/Mol). A combination of such constituents are provided so that the polymeric network has several degrees of freedom of rotational motion available to it. At the same time the network is cross linked to ensure that the absolute magnitude of the rotational frequency remains restricted. This is done to ensure that the rate of conversion of ground state 2 into ground state 1 does not become excessively fast. Otherwise, the level of activation will become unacceptably low at the upper limit of the target temperature range (105°F). Generally speaking, the segmental mobility of polymeric matrices increases 5 fold or more when the ambient temperature is at or above its glass transition temperature. Accordingly, all matrices are formulated to have glass transition temperature well above the target temperature range (20°F or more), so that only local group or side chain motions are available to stimulate the conversion of ground state 2 into ground state 1.

It was found that when monomers containing main or side chain ether groups (embodied in certain ethoxylated or propoxylated derivatives or methoxy substituents present as side chains) were used to formulate the polymeric matrix for incorporation of photochromic additives, the matrices permitted rapid conversion of ground state 2 into ground state 1 in the absence of cross linking, or at low levels of cross link density. Such a matrix provided rapid bleaching rates, but low levels of activation at or near the upper limit of the target temperature range. As the cross link density was increased, the conversion rate of ground state 2 into ground state 1 was slowed down, simultaneously increasing the activation level at the temperature range 90°-100°F. A range of cross link density was found within which the activation level remained approximately constant (varying less than 30%) over a relatively wide temperature range (80°-100°F), leaving the switching speed (bleaching rate) also approximately unchanged at 30-45 seconds. It is believed that the polymeric matrices within this range of cross link densities and concentrations of ether groups present as alkoxy groups or side chains provide an optimum number of rotational degrees of freedom with low temperature dependence to a photochromic molecule incorporated in the matrix. Thus, a combination of polyethylene glycol (400) diacrylate, ethoxylated bisphenol A diacrylate, 2- phenoxyethyl methacrylate, either used as a three component mixture or mixed with bisallyl ethylene carbonate, provide independent control of crosslink density and glass transition temperature of the resulting polymeric matrices, in addition to providing both main chain ether groups and side chain alkoxylated groups in the resulting network.

In addition to the monomers listed above, it is possible to use a series of polyethylene glycol diacrylates of different molecular weights, alkoxylated trifunctional aliphatic acrylates (such as SR9008, available from Sartomer Corp), and alkoxylated aliphatic diacrylate esters (such as SR9209, available from

Sartomer Corp). In addition, other oligomeric mono- or multi-functional acrylates or methacrylates may also be used, provided that the range of cross link density does

not fall outside of 2-8 moles per liter, preferably 2.5- 6 moles per liter. Preferable candidates among acrylates and methacrylates are those which bear either aliphatic or aromatic ether linkages on the main chain, such as Bisphenol A derivatives, or aromatic carbonates derived thereof, aliphatic carbonates, as well as side chain substitutions bearing alkoxy linkages. In addition, allyl and vinyl derivatives may be used, such as styrene, substituted styrenes, or bisallyl carbonate, or derivatives thereof.

The monomer formulation was typically photopolymerized using photointiators such as bisdimethoxybenzoyl trimethylpentyl phosphine oxide (BAPO, available from Ciba Geigy Corp), 1-Hydroxy cyclohexyl phenyl ketone (Irgacure 184, available from Ciba Geigy Corp), or 2-hydroxy, 2-methyl, 1-phenyl propane (Durcure 1173, available from Radcure Corp). Alternatively the monomer formulation may be thermally polymerized using an organic peroxide, hydroperoxide, percarbonate, peracetate, or an azo derivative as a thermal polymerization initiator. Examples are benzoyl peroxide, 2,2' azobisisobutyronitrile, or diisopropyl percarbonate.

Antioxidants, such as thiodiethylene bis (hydrocinnamide) (Irganox 1035, available from Ciba Geigy Corp) or bis (1,2,2,6,6, -pentamethyl-4-piperidyl sebacate) (Tinuvin 292, available from Ciba Geigy Corp), may also be included in the monomer formulation.

Table 10.1 shows some typical formulations, and their glass transition temperatures. The formulations are custom developed for a given substrate material (e.g., an optical product made of CR-39 (Trademark of PPG Corp), polycarbonate of bisphenol A, or a polyurethane). In each case, the monomers are selected to develop strong bonding and compatibility with the substrate to which this matrix is to be applied.

Definition of terms: DEG-BAC: Diethylene Glycol Bisallyl Carbonate; PEGDA: Polyethylene Glycol Diacrylate; THFA: Tetrahydrofurfuryl Acrylate, EBDA: Ethoxylated Bisphenol A Diacrylate; PEMA: 2- Phenoxyethyl Methacrylate, 9008: Alkoxylated Trifunctional Acrylate, 9209: Alkoxylated Difunctional Acrylate Ester, 184: 1-Hydroxycyclohexyl Phenyl Ketone, TIN 1130: 2-Hydroxyphenyl Benzotriazole, TIN 292: Bis (1,2,2,6,6-Pentamethyl-4-Piperidinyl) Sebacate.

These polymeric matrices can be loaded with the photochromic additives. Preferred photochromic molecules for the practice of the present invention include spiro (indolino) naphthoxazines, spiro (indolino) pyridobenzoxazines, spiro (benzindolino) naphthoxazines, and octamethyl stilbene.

Several means are available for loading the photochromic additives into the polymeric matrices. For example, the photochromic additives can be added to the monomer formulation prior to polymerization, provided that the additives survive the conditions of polymerization. The matrix can be imbibed with the photochromic additives, either by slow diffusion, or by taking the matrix to a temperature above its glass transition temperature, then immersing it for a short time into a solution of the photochromic additives in a chemically inert solvent.

Alternatively, the matrix may be sprayed or spin coated with a solution of the photochromic additive, the solvent evaporated off, then the polymeric matrix may be heated to allow the photochromic additive thus deposited on the surface to diffuse inward, and eventually develop a gradient of concentration. The process of forming the matrix layer and addition of photochromic additives may be repeated several times in order to build up an overall thickness appropriate for the contemplated application.

The polymeric matrices described herein can be applied to optical products in many ways. For example, the polymeric matrix can be developed as a conformal layer bonded to the optical product via an interface region composed of an interpenetrating network. In another method of application of the polymeric matrix to the optical product, the polymeric matrix is produced in the form of a thin sheet which is adhesively bonded to the optical product of interest. In yet another method of application of the polymeric matrix to the optical product, the polymeric matrix material bearing the photochromic additive may be formed as a solid powder, then suspended in a liquid in order to dip, spray,-spin or brush coat the optical product. The matrix material will not dissolve in any solvent because it is cross linked. In yet another method of application of the polymeric matrix to the optical product, the matrix material may be partially polymerized to form a sticky, viscous material, loaded with the photochromic additives, dispersed in a carrier fluid, then dip, spray, spin or brush coated onto the optical product. Still other embodiments will become immediately apparent to those skilled in the art upon reading this specification and the claims below. All patents, patent applications and other references cited herein, including priority applications, are incorporated by reference in their entireties.

10.7 Molded Polymeric Material

MOLDED POLYMERIC MATERIAL INCLUDING MICROCELLULAR, INJECTION-MOLDED, AND LOW-DENSITY POLYMERIC MATERIAL Field of the Invention The present invention relates generally to polymeric structural foam processing, and more particularly to microcellular structural foams and systems and methods of manufacture.

Background of the Invention Structural foamed materials are known, and can be produced by injecting a physical blowing agent into a molten polymeric stream, dispersing the blowing agent in the polymer to form a two-phase mixture of blowing agent cells in polymer, injecting the mixture into a mold having a desired shape, and allowing the mixture to solidify therein.

A pressure drop in the mixture can cause the cells in the polymer to grow. As an alternative to a physical blowing agent, a chemical blowing agent can be used which undergoes a chemical reaction in the polymer material causing formation of a gas.

Chemical blowing agents generally are low molecular weight organic compounds that decompose at a critical temperature and release a gas such as nitrogen, carbon dioxide, or carbon monoxide. Under some conditions the cells can be made to remain isolated, and a closed-cell foamed material results. Under other, typically more violent foaming conditions, the cells rupture or become interconnected and an open-cell material results.

A sample of standard injection molding techniques described in the patent literature follow.

U.S. Patent No. 3,436,446 (Angell) describes a method and apparatus for molding foamed plastic articles with a solid skin by controlling the pressure and temperature of the mold.

U.S. Patent No. 4,479,914 (Baumrucker) describes a foamed article forming method in which a mold cavity is pressurized with gas to prevent premature diffusion of blowing gas from the material injected into the cavity. Pre-pressurization gas is vented during injection of material to be foamed, finally to a vacuum chamber creating a vacuum that draws the material throughout the mold cavity.

Particularly low-density (high void volume) molded polymeric foam materials include expanded polypropylene (EPP) and expanded polystyrene (EPS). Precursors to EPP or EPS can be provided as beads of already partially or fully-foamed polypropylene or polystyrene, respectively. These beads typically are injected into a steam chest mold and fused together to form a structural foam product from the beads. In some cases, further expansion and foaming of the beads occurs within the steam chest (typically with EPS). While EPS and EPP are useful products for many applications, they can have a less than ideal appearance. Lines of fusion between beads can easily be observed by the naked eye in products such as styrene foam cups and other EPP or EPS products.

Microcellular material typically is defined by polymeric foam of very small cell size and various microcellular material is described in U.S. Patent Nos. 5,158,986 and 4,473,665. These patents describe subjecting a single-phase solution of polymeric material and physical blowing agent to thermodynamic instability required to create sites of nucleation of very high density, followed by controlled cell growth to produce microcellular material. U.S. Patent No. 4,473,665 (Martini-Vvedenslcy) describes a molding system and method for producing microcellular parts. Polymeric pellets are pre-pressurized with a gaseous blowing agent and melted in a conventional extruder to form a solution of blowing agent and molten polymer, which then is extruded into a pressurized mold cavity. The pressure in the mold is maintained above the solubility pressure of the gaseous blowing agent at melt temperatures for given initial saturation.

When the molded part temperature drops to the appropriate critical nucleation temperature, the pressure on the mold is dropped, typically to ambient, and the part is allowed to foam.

U.S. Patent No. 5,158,986 (Cha *et al.*) describes an alternative molding system and method for producing microcellular parts. Polymeric pellets are introduced into a conventional extruder and melted. A blowing agent of carbon dioxide in its supercritical state is established in the extrusion barrel and mixed to form a homogenous solution of blowing agent and polymeric material. A portion of the extrusion barrel is heated so that as the mixture flows through the barrel, a thermodynamic instability is created, thereby creating sites of nucleation in the molten polymeric material. The nucleated material is extruded into a pressurized mold cavity. Pressure within the mold is maintained by counter pressure of air. Cell growth occurs inside the mold cavity when the mold cavity is expanded and the pressure therein is reduced rapidly; expansion of the mold provides a molded and foamed article having small cell sizes and high cell densities. Nucleation and cell growth occur separately according to the technique; thermally-induced nucleation takes place in the barrel of the extruder, and cell growth takes place in the mold.

While the above and other reports represent several techniques associated with the manufacture of microcellular material and the manufacture of material via injection molding, a need exists in the art for improved microcellular injection molding processes.

It is, therefore, an object of the invention to provide injection molding systems and methods effective in producing microcellular structural foams and, in particular, very thin articles. It is another object to provide systems and methods useful in injection molding of microcellular structural foams, but also useful in injection molding of conventional foams and continuous extrusion of microcellular or conventional foams.

Summary of the Invention The present invention is directed to injection molding systems and methods capable of forming molded articles, and systems and methods for forming a variety of foamed materials. In each of the various aspects of the invention described herein, in one set of embodiments the material is not microcellular foamed material as defined by cell size and densities, and in another set of embodiments the material is microcellular material.

The present invention involves, according to one aspect, an injection molding system constructed to produce microcellular structural foamed articles. The system includes an extruder having an inlet at an inlet end thereof designed to receive a precursor of microcellular material, a molding chamber, and an enclosed passageway connecting the inlet with the molding chamber. The molding chamber is constructed and arranged to receive a non-nucleated, homogeneous, fluid, single-phase solution of a polymeric material and a blowing agent, to contain the non-nucleated, homogeneous fluid single-phase solution of the polymeric material and the blowing agent in a fluid state at an elevated pressure within the passageway and to advance the solution as a fluid stream within the passageway in a downstream direction from the inlet end toward the molding chamber. The enclosed passageway includes a nucleating pathway in which blowing agent in the single-phase solution passing therethrough is nucleated. The nucleating pathway is constructed to include a polymer receiving end which receives a homogeneous fluid, single-phase solution of a polymeric material and a non-nucleated blowing agent, a nucleated polymer releasing end constructed and arranged to release nucleated polymeric material, and a fluid pathway connecting the receiving end to the releasing end. Optionally, the polymer releasing end can define an orifice of the molding chamber, or can be in fluid communication with the molding chamber. The nucleating pathway is constructed to have length and cross-sectional dimensions such that, the system is capable of subjecting fluid polymer admixed homogeneously with blowing agent to a pressure drop rate while passing through the pathway of at least about 0.1 GPa/sec, or at least about 0.3 GPa/sec, or at least about 1.0 GPa/sec, or at least about 3 GPa/sec, or at least about 10 GPa/sec, or at least about 100 GPa/sec. The nucleating pathway can also be constructed to have a variable cross-sectional dimension such that a fluid polymer flowing through the pathway is subjected to a variable pressure drop rate and/or temperature rise.

According to another aspect of the invention a system is provided having a molding chamber constructed and arranged to contain nucleated polymeric material at an elevated pressure in order to prevent cell growth at the elevated pressure. The pressurized molding chamber can be fluidly or mechanically pressurized in order to contain the nucleated polymeric material at such an elevated pressure. After reduction of the pressure on the pressurized molding chamber, the polymeric material can solidify the shape of a desired microcellular polymeric article as the molding chamber is constructed and arranged to have such an interior shape.

According to yet another aspect of the invention, the system is provided having a barrel with an inlet designed to receive a precursor of extruded material, an outlet designed to release a fluid non-nucleated mixture of blowing agent and foamed polymeric article precursor to the precursor, an orifice connectable to a source of the blowing agent, and a screw mounted for reciprocation within the barrel. The extrusion system can also have at least two orifices connectable to a source of the blowing agent and the orifice can be arranged longitudinally along the axis of the barrel in order to sequentially introduce the non-nucleated mixture through at least the two orifices into the barrel as the screw reciprocates. The system can also include a second extrusion barrel connected in tandem with the first barrel where the second barrel has an inlet designed to receive the fluid non-nucleated mixture and has a screw mounted for reciprocation within the barrel.

According to another aspect, the invention provides a method for establishing a continuous stream of the non-nucleated, fluid, single-phase solution of polymeric precursor and blowing agent, nucleating the stream to create a nucleated stream of the mixture, passing the nucleated stream into the enclosure, and allowing the mixture to solidify in the shape of the enclosure. Optionally, the stream can be continuously nucleated by continuously subjecting it to a pressure drop of a rate of at least about 0.1 GPa/sec while passing the stream into the enclosure, to create a continuous stream of nucleated material. Alternatively, the method involves intermittently nucleating the stream by subjecting it to a pressure drop at a rate of at least about 0.1 GPa/sec, while passing the stream into the enclosure so that non-nucleated material passes into the enclosure first, followed by the nucleated material. Conversely, the nucleated stream may be passed into the enclosure so that nucleated material passes into the enclosure, first followed by non-nucleated material. The method also involves removing a solidified microcellular article from the enclosure, and in a period of less than about 10 minutes providing a second nucleated mixture in the enclosure, allowing the second mixture to solidify in the shape of the enclosure, and removing a second solidified microcellular article from the enclosure.

The invention also provides a method involving accumulating a charge of a precursor of foamed polymeric material and a blowing agent, heating a first portion of the charge defining at least about 2 per cent of the charge to a temperature at least about 10 °C higher than the average temperature of the charge, and injecting the charge into a molding chamber.

Also provided is a method involving accumulating, in an accumulator fluidly connected to a molding chamber, a charge including a first portion comprising a fluid polymeric material essentially free of blowing agent and a second portion comprising a fluid polymeric material mixed with a blowing agent, and injecting the charge from the accumulator into a molding chamber.

Also provided is a method involving injecting a fluid, single-phase solution of a precursor of foamed polymeric material and a blowing agent into a molding chamber from an accumulator in fluid communication with extrusion apparatus while nucleating the solution to create a nucleated mixture, and allowing the mixture to solidify as a polymeric microcellular article in the molding chamber.

Also provided is a method involving injecting a blowing agent into an extruder barrel of polymer extrusion apparatus while an extrusion screw is moving axially within the barrel.

Also provided is a method involving injecting a blowing agent from an extrusion screw into a barrel of polymer extrusion apparatus.

Also provided is a method involving establishing in a barrel of extrusion apparatus a fluid polymeric article precursor, withdrawing a portion of the fluid precursor from the barrel, mixing the portion of the fluid precursor with blowing agent to form a mixture of the blowing agent and the portion of the fluid precursor, and introducing the mixture into the barrel.

Also provided is a method involving introducing polymeric material admixed with supercritical fluid into a mold including a portion having an interior dimension of less than about 0.125 inch and allowing the polymeric material to solidify in the mold.

Also provided is a method involving establishing a mixture of at least two, dissimilar, molten polymeric components with a super critical fluid blowing agent, and extruding the mixture to form a non-delaminated foam of the at least two components.

Also provided is a method involving injecting a single phase solution of polymeric material and blowing agent into an open mold, then closing the mold and forming a microcellular article in the shape of the mold.

Also provided is a method involving establishing a single-phase, non-nucleated solution of a polymeric material and blowing agent, introducing the solution into a molding chamber while nucleating the solution, cracking the mold thereby allowing cell growth to occur, and recovering a microcellular polymeric article having a shape similar to that of the molding chamber but being larger than the molding chamber.

Also provided is a method involving forming in an extruder a non-nucleated, homogeneous, fluid, single-phase solution of a precursor of microcellular polymeric material and a blowing agent, filling a molding chamber with the solution while nucleating the solution to form within the molding chamber a nucleated microcellular polymeric material precursor.

Also provided is a method involving injecting a fluid polymeric material/

blowing agent mixture into a molding chamber and allowing the mixture to solidify as a microcellular polymeric article in the molding chamber and removing a solidified microcellular polymeric article from the molding chamber, and in a period of less than about ten minutes, providing a second polymeric/ blowing agent mixture into the chamber and allowing the mixture to solidify as a second microcellular polymeric article in the chamber and removing the second microcellular polymeric article from the chamber.

Also provided is a method involving injecting a polymeric/blowing agent mixture into a molding chamber at a melt temperature of less than about 400°F, and molding in the chamber a solid foam polymeric article having a void volume of at least about 5 per cent and a length-to-thickness ratio of at least about 50: 1. In certain embodiments of this method, the melt temperature is less than about 380°F, in some embodiments less than about 300°F, and in other embodiments less than about 200°F.

Also provided is a method that involves injecting non-foamed polymeric material into a molding chamber and allowing the polymeric material to form a microcellular polymeric article having a shape essentially identical to that of the molding chamber.

The article includes at least one portion having cross-sectional dimensions of at least about 1/2 inch in each in each of three perpendicular intersecting cross-sectional axes and a void volume of at least 50 per cent.

Another method provided by the invention involves injecting a fluid precursor of foamed polymeric material into a molding chamber at a molding chamber temperature of less than about 100°C, and allowing the mixture to solidify in the molding chamber as a polymeric microcellular article. The article includes at least one portion having cross- sectional dimensions of at least l/2 inch in each of three perpendicular intersecting cross- sectional axes and a void volume of at least about 50 per cent. The molding chamber temperature can be less than about 75°C, 50°C, or 30°C, and the foamed polymeric material can be polyolefin.

Another method involves injecting a non-foamed polymeric material into a molding chamber and allowing the mixture to solidify in the molding chamber as a polymeric microcellular article having a void volume of at least about 50 per cent, and repeating the injecting and allowing steps at a cycle time of less than about one minute.

Another method provided by the invention involves injecting a fluid, single-phase solution of polymeric material and blowing agent into a molding chamber while subjecting the solution to a rapid pressure drop at a first pressure drop rate that is sufficient to cause microcellular nucleation. Essentially immediately thereafter cell growth is allowed and controlled by subjecting the material to a second pressure drop that is less than the first pressure drop and at a decreasing rate.

The systems of the invention include one including an accumulator having an inlet for receiving a precursor of foamed polymeric material and a blowing agent, and an outlet, a molding chamber having an inlet in fluid communication with the outlet of the accumulator, and heating apparatus associated with the accumulator constructed and arranged to heat, during operation of the system, a first section of the accumulator proximate the molding chamber to a temperature at least about 10 °C higher than the average temperature of the accumulator.

Also provided is a system including an extruder having an inlet for receiving a precursor of foamed polymeric material and being constructed and arranged to produce fluid polymeric material from the precursor, a first outlet positioned to deliver fluid polymeric material from the extruder, a blowing agent inlet downstream of the first outlet connectable to a source of a physical blowing agent, a mixing region downstream of the blowing agent inlet constructed and arranged to produce a mixture of fluid polymeric precursor and blowing agent, and a second outlet downstream of the mixing region positioned to deliver the mixture of fluid polymeric precursor and blowing agent, and an accumulator having a first inlet fluidly connected to the first outlet of the extruder and a second inlet fluidly connected to the second outlet of the extruder.

Also provided is a system for producing injection-molded microcellular material, including an extruder having an outlet at an outlet end thereof designed to release a non- nucleated, homogeneous, fluid, single-phase solution of a polymeric material and a blowing agent, and a molding chamber having an inlet in fluid communication with the outlet of the extruder. The system is constructed and arranged to deliver from the extruder outlet to the molding chamber inlet the single-phase solution and, during filling of the molding chamber, to nucleate the single-phase solution to form within the chamber a nucleated microcellular polymeric material precursor.

Also provided is an extrusion system including a barrel having an inlet designed to receive a precursor of extruded material, an outlet designed to release a fluid mixture of non-nucleated blowing agent and the precursor, an orifice connectable to a source of blowing agent, and a screw mounted for reciprocation within the barrel.

Also provided is a system for producing injection-molded microcellular material including an extruder having an outlet at an outlet end thereof designed to release a precursor of microcellular polymeric material and a blowing agent, and a molding chamber having an inlet in fluid communication with the outlet of the extruder. The system is constructed and arranged to cyclically inject the precursor of microcellular polymeric material and the blowing agent into the molding chamber.

The invention also provides an extrusion system including a barrel having an inlet designed to receive a precursor of extruded material, and outlet designed to release a fluid mixture of non-nucleated blowing agent and precursor, and an orifice connected to a source of blowing agent. A screw is mounted for reciprocation within the barrel.

Another system provided by the invention for producing molten polymeric microcellular material includes an inlet instructed and arranged to receive a precursor of molten polymeric microcellular material, a molding chamber, and a channel connecting the inlet with the molding chamber. The channel includes a divergent portion between the inlet and the molding chamber that increases in width by at least about 100 per cent while maintaining a cross-sectional area changing by no more than about 25 per cent.

Another system of the invention includes an inlet constructed and arranged to receive a precursor of molten polymeric microcellular material, a molding chamber, and a channel connecting the inlet with the molding chamber. The channel includes a nucleating pathway having length and cross-sectional dimensions that, when a fluid, single-phase solution of polymeric material and blowing agent is passed through the pathway at rates for which the system is constructed, creates a pressure drop in the fluid pathway at a pressure drop rate sufficient to cause microcellular nucleation. The channel includes a cell growth region between the nucleating pathway and the molding chamber that increases in cross-sectional dimension in the direction of the molding chamber.

Another system of the invention is as described immediately above but, while not necessarily including the cell growth region that increases in cross-sectional dimension, includes a nucleating pathway having a width to height ratio of at least about 1.5: 1.

Another system of the invention is similar to that described immediately above but, while the nucleating pathway need not necessarily have a width to height ration of at least 1.5: 1, has a width equal to one dimension of the molding chamber.

In another aspect, the invention provides a method that involves injecting a blowing agent into an extruder barrel of polymer extrusion apparatus while an extrusion screw is moving axially within the barrel. In one embodiment, the method involves injecting a blowing agent from an extrusion screw into a barrel of polymer extrusion apparatus. This injection technique can be used with any of a wide variety of microcellular and conventional techniques. In another aspect, the invention involves an extrusion screw constructed and arranged for rotation within a barrel of polymer extrusion apparatus that includes, within the screw, a lumen communicating with an orifice in a surface of the screw. The lumen can be used to inject blowing agent into the extrusion barrel.

In another aspect the invention provides a system for producing injection-molded articles. The system includes an extruder, a molding chamber, a runner fluidly connecting the extruder and the molding chamber, and a temperature control device in thermal communication with the runner. In another aspect, the invention involves establishing a fluid mixture blowing agent and injection-molded material precursor in an extruder, passing the mixture through a runner into a molding chamber, solidifying the portion of the fluid mixture in the chamber while maintaining a portion of the mixture in the runner in a fluid state, and injecting additional fluid mixture into the runner thereby urging the portion of the fluid mixture and the runner into the chamber.

The invention also provides a method that involves withdrawing a portion of a fluid polymeric article precursor from an extrusion barrel, mixing the portion of the fluid precursor with blowing agent to form a mixture, and re-introducing the mixture into the barrel.

The invention also provides a system including an extruder with an extruder barrel, a molding chamber, and a mixing chamber in fluid communication with a first, upstream orifice in the barrel, a second, downstream orifice in the barrel, and a source of a blowing agent.

In another aspect, the invention provides a molded foam article having a shape essentially identical to that of a molding chamber, including at least one portion having a cross-sectional dimension of no more than about 0.125 inch.

Another aspect involves a three-dimensional polymeric foam article having three intersecting, principal axes corresponding to the three dimensions, one of the dimensions associated with a first axis varying as a function of position along a second, perpendicular axis. The article includes at least one portion having a cross-sectional dimension of no more than about 0.125 inch and has a void volume of at least about 20 per cent.

Another aspect involves a three-dimensional polymeric foam article having three intersecting, principal axes corresponding to the three dimensions, one of the dimensions associated with a first axis varying as a function of position along a second, perpendicular axis. The article includes at least one portion having a cross-sectional dimension of no more than about 0.125 inch.

In another aspect, the invention provides an injection molded polymeric part having a length-to-thickness ratio of at least about 50: 1, the polymer having a melt index of less than about 10.

In another aspect, the invention provides an injection molded polymeric part having a length-to-thickness ratio of at least about 120: 1, the polymer having a melt flow rate of less than about 40.

In another aspect, the invention provides an injection molded polymeric foam having a void volume of at least about 5 per cent, and having a surface that is free of splay and swirl visible to the naked human eye.

In another embodiment, the invention provides an article having a thickness of less than about 0.125 inch at a void volume of at least about 20 per cent. A method of making such an article is provided as well, that can involve introducing polymeric material admixed with a supercritical fluid into a mold including a portion having an interior dimension of less than about 0.125 inch, and allowing the polymeric material to solidify in the mold, the introducing and allowing steps taking place within a period of time of less than 10 seconds.

The invention, in another embodiment, provides a molded polymeric article having a shape essentially identical to that of a molding chamber and including at least one portion having a cross-sectional dimension of at least 1/2 inch in each of three perpendicular intersecting cross-sectional axes. The article has a void volume of at least about 50 per cent and is defined by cells including cell walls of average cell wall thickness.

The article is free of periodic solid boundaries of thickness greater than about five time the average cell wall thickness.

In another embodiment, the invention provides a molded polymeric foam article including at least one portion having a cross-sectional dimension of no more than about 0.075 inch and a void volume of at least about 5 per cent.

The invention, in another embodiment, provides a molded polymeric foam article including at least one portion having a cross-sectional dimension of between about 0.075 inch and about 0.125 inch and a void volume of at least about 10 per cent.

In another embodiment, the invention provides a molded polymeric foam article including at least one portion having a cross-sectional dimension of between about 0.125 inch and about 0.150 inch and a void volume of at least about 15 per cent.

In another embodiment, the invention provides a molded polymeric foam article including at least one portion having a cross-sectional dimension of between about 0.150 inch and about 0.350 inch and a void volume of at least about 20 per cent.

The invention, in another embodiment, a molded polymeric article including a plurality of cells wherein at least 70 per cent of the total number of cells have a cell size of less than 150 microns.

In another embodiment, the invention provides a system. The system includes a barrel having an inlet, at an upstream end, designed to receive a polymeric article precursor, and an outlet at a downstream end. The barrel includes a blowing agent port, between the upstream end and the downstream end, fluidly connectable to a blowing agent source for introducing blowing agent from the source into the precursor in the barrel to form a mixture of precursor material and blowing agent in the barrel. The system also includes a metering device having an inlet connected to the blowing agent source and an outlet connected to the barrel. The metering device constructed and arranged to meter the mass flow rate of the blowing agent from the blowing agent source to the blowing agent port. The system further includes a molding chamber having an inlet in fluid communication with the outlet of the barrel to receive the mixture of precursor material and blowing agent from the barrel.

The invention, in another embodiment, provides a method of forming a polymeric foam article. The method includes urging a stream of polymeric article precursor flowing in a downstream direction within a barrel of an extrusion apparatus. The method further includes introducing a blowing agent into the stream at a rate metered by the mass flow of the blowing agent to form a mixture of fluid polymeric article precursor and blowing agent. The method further includes injecting the mixture of fluid polymeric article precursor into a molding chamber fluidly connected to the barrel.

The invention, in another embodiment, provides a system. The system includes a barrel having an inlet, at an upstream end, designed to receive a polymeric article precursor, and an outlet, at a downstream end. The barrel includes, between the upstream end and the downstream end, a blowing agent port having a plurality of orifices. The blowing agent port is fluidly connectable to a blowing agent source for introducing blowing agent from the source into the precursor in the barrel through respective orifices to form a mixture of precursor material and blowing agent in the barrel. The system further includes a molding chamber having an inlet in fluid communication with the outlet of the barrel to receive the mixture of precursor material and blowing agent from the barrel.

The invention, in another embodiment, provides a method for forming a polymeric article. The method includes urging a stream of polymeric article precursor flowing in a downstream direction within a barrel of an extrusion apparatus. The method further includes introducing a blowing agent from a blowing agent source into the stream through a plurality of orifices in a blowing agent port fluidly connecting the barrel with the blowing agent source to form a mixture of precursor material and blowing agent, and injecting the mixture of precursor material into a molding chamber fluidly connected to the barrel.

In another embodiment, the invention provides a system for producing injection- molded microcellular material The system includes an accumulator constructed and arranged to accumulate a precursor of microcellular material and a blowing agent, and including an outlet. The system further includes an injector constructed and arranged to cyclically inject the precursor of microcellular material through the outlet of the accumulator. The system further includes a molding chamber having an inlet in fluid communication with the outlet of the accumulator. The molding chamber constructed and arranged to receive the precursor of microcellular material.

In another embodiment, the invention provides a method. The method includes accumulating a charge of a precursor of microcellular polymeric material and a blowing agent, and injecting the charge into a molding chamber.

Other advantages, novel features, and objects of the invention will become apparent from the following detailed description of the invention when considered in conjunction with the accompanying drawings, which are schematic and which are not intended to be drawn to scale. In the figures, each identical or nearly identical component that is illustrated in various figures is represented by a single numeral. For purposes of clarity, not every component is labeled in every figure, nor is every component of each embodiment of the invention shown where illustration is not necessary to allow those of ordinary skill in the art to understand the invention.

Detailed Description of the Invention Commonly-owned, co-pending U.S. patent application serial no. 08/777,709, entitled"Method and Apparatus for Microcellular Polymer Extrusion", filed December 20,1996, and commonly-owned, co-pending international patent publication nos. WO 98/08667, published March 5,1998 and WO 98/31521, published July 23,1998, are incorporated herein by reference.

The various embodiments and aspects of the invention will be better understood from the following definitions. As used herein,"nucleation"defines a process by which a homogeneous, single-phase solution of polymeric material, in which is dissolved molecules of a species that is a gas under ambient conditions, undergoes formations of clusters of molecules of the species that define"nucleation sites", from which cells will grow. That is,"nucleation"means a change from a homogeneous, single-phase solution to a mixture in which sites of aggregation of at least several molecules of blowing agent are formed. Nucleation defines that transitory state when gas, in solution in a polymer melt, comes out of solution to form a suspension of bubbles within the polymer melt.

Generally this transition state is forced to occur by changing the solubility of the polymer melt from a state of sufficient solubility to contain a certain quantity of gas in solution to a state of insufficient solubility to contain that same quantity of gas in solution.

Nucleation can be effected by subjecting the homogeneous, single-phase solution to rapid thermodynamic instability, such as rapid temperature change, rapid pressure drop, or both. Rapid pressure drop can be created using a nucleating pathway, defined below.

Rapid temperature change can be created using a heated portion of an extruder, a hot glycerine bath, or the like."Microcellular nucleation", as used herein, means nucleation at a cell density high enough to create microcellular material upon controlled expansion.

A"nucleating agent"is a dispersed agent, such as talc or other filler particles, added to a polymer and able to promote formation of nucleation sites from a single-phase, homogeneous solution. Thus"nucleation sites"do not define locations, within a polymer, at which nucleating agent particles reside."Nucleated"refers to a state of a fluid polymeric material that had contained a single-phase, homogeneous solution including a dissolved species that is a gas under ambient conditions, following an event (typically thermodynamic instability) leading to the formation of nucleation sites."Non-nucleated"refers to a state defined by a homogeneous, single-phase solution of polymeric material and dissolved species that is a gas under ambient conditions, absent nucleation sites. A"non-nucleated"material can include nucleating agent such as talc.

A"polymeric material/blowing agent mixture"can be a single-phase, non-nucleated solution of at least the two, a nucleated solution of at least the two, or a mixture in which blowing agent cells have grown."Nucleating pathway"is meant to define a pathway that forms part of microcellular polymeric foam extrusion apparatus and in which, under conditions in which the apparatus is designed to operate (typically at pressures of from about 1500 to about 30,000 psi upstream of the nucleator and at flow rates of greater than about 10 pounds polymeric material per hour), the pressure of a single-phase solution of polymeric material admixed with blowing agent in the system drops below the saturation pressure for the particular blowing agent concentration at a rate or rates facilitating rapid nucleation. A nucleating pathway defines, optionally with other nucleating pathways, a nucleation or nucleating region of a device of the invention."Reinforcing agent", as used herein, refers to auxiliary, essentially solid material constructed and arranged to add dimensional stability, or strength or toughness, to material. Such agents are typified by fibrous material as described in U.S. Patent Nos. 4,643,940 and 4,426,470."Reinforcing agent"does not, by definition, necessarily include filler or other additives that are not constructed and arranged to add dimensional stability. Those of ordinary skill in the art can test an additive to determine whether it is a reinforcing agent in connection with a particular material.

The present invention provides systems and methods for the intrusion and injection molding of polymeric material, including microcellular polymeric material, and systems and methods useful in intrusion and injection molding and also useful in connection with other techniques. For example, although injection and intrusion molding are primarily described, the invention can be modified readily by those of ordinary skill in the art for use in other molding methods such as, without limitation, low-pressure molding, co-injection molding, laminar molding, injection compression, and the like. For purposes of the present invention, microcellular material is defined as foamed material having an average cell size of less than about 100 microns in diameter, or material of cell density of generally greater than at least about 106 cells per cubic centimeter, or preferably both. Non-microcellular foams have cell sizes and cell densities outside of these ranges. The void fraction of microcellular material generally varies from 5 per cent to 98 per cent. Supermicrocellular material is defined for purposes of the invention by cell sizes smaller than 1 um and cell densities greater than 1012 cells per cubic centimeter.

In preferred embodiments, microcellular material of the invention is produced having average cell size of less than about 50 microns. In some embodiments particularly small cell size is desired, and in these embodiments material of the invention has average cell size of less than about 20 microns, more preferably less than about 10 microns, and more preferably still less than about 5 microns. The microcellular material preferably has a maximum cell size of about 100 microns. In embodiments where particularly small cell size is desired, the material can have maximum cell size of about 50 microns, more preferably about 25 microns, more preferably about 15 microns, more preferably about 8 microns, and more preferably still about 5 microns. A set of embodiments includes all combinations of these noted average cell sizes and maximum cell sizes. For example, one embodiment in this set of embodiments includes microcellular material having an average cell size of less than about 30 microns with a maximum cell size of about 50 microns, and as another example an average cell size of less than about 30 microns with a maximum cell size of about 35 microns, etc. That is, microcellular material designed for a variety of purposes can be produced having a particular combination of average cell size and a maximum cell size preferable for that purpose. Control of cell size is described in greater detail below.

In one embodiment, essentially closed-cell microcellular material is produced in accordance with the techniques of the present invention. As used herein,"essentially closed-cell"is meant to define material that, at a thickness of about 100 microns, contains no connected cell pathway through the material.

Although not shown in detail, screw 38 includes feed, transition, gas injection, mixing, and metering sections.

Positioned along barrel 32, optionally, are temperature control units 42.

Control units 42 can be electrical heaters, can include passageways for temperature control fluid, and or the like. Units 42 can be used to heat a stream of pelletized or fluid polymeric material within the barrel to facilitate melting, and/or to cool the stream to control viscosity and, in some cases, blowing agent solubility. The temperature control units can operate differently at different locations along the barrel, that is, to heat at one or more locations, and to cool at one or more different locations. Any number of temperature control units can be provided.

Barrel 32 is constructed and arranged to receive a precursor of polymeric material. As used herein,"precursor of polymeric material"is meant to include all materials that are fluid, or can form a fluid and that subsequently can harden to form a microcellular polymeric article. Typically, the precursor is defined by thermoplastic polymer pellets, but can include other species. For example, in one embodiment the precursor can be defined by species that will react to form microcellular polymeric material as described, under a variety of conditions. The invention is meant to embrace production of microcellular material from any combination of species that together can react to form a polymer, typically monomers or low-molecular-weight polymeric precursors which are mixed and foamed as the reaction takes place. In general, species embraced by the invention include thermosetting polymers in which a significant increase in molecular weight of the polymer occurs during reaction, and during foaming, due to crosslinking of polymeric components. For example, polyamides of the condensation and addition type, including aliphatic and aromatic polyamides such as polyhexamethyleneadipamide, poly (e-caprolactam), polyenes such as cycloaromatic polymers including polydicyclopentadiene, acrylic polymers such as polyacrylamide, polyacrylamate, acrylic ester polymers such as 2-cyanoacrylic ester polymers, acrylonitrile polymers, and combinations.

Preferably, a thermoplastic polymer or combination of thermoplastic polymers is selected from among amorphous, semicrystalline, and crystalline material including polyolefins such as polyethylene and polypropylene, fluoropolymers, cross-linkable polyolefins, polyamides, polyvinyl chloride, and polyaromatics such as styrenic polymers including polystyrene. Thermoplastic elastomers can be used as well, especially metallocene-catalyzed polyethylene.

Typically, introduction of the precursor of polymeric material utilizes a standard hopper 44 for containing pelletized polymeric material to be fed into the extruder barrel through orifice 46, although a precursor can be a fluid prepolymeric material injected through an orifice and polymerized within the barrel via, for example, auxiliary polymerization agents. In connection with the present invention, it is important only that a fluid stream of polymeric material be established in the system.

Immediately downstream of downstream end 48 of screw 38 is a region 50 which can be a temperature adjustment and control region, auxiliary mixing region, auxiliary pumping region, or the like. For example, region 50 can include temperature control units to adjust the temperature of a fluid polymeric stream prior to nucleation, as described below. Region 50 can include instead, or in addition, additional, standard mixing units, or a flow-control unit such as a gear pump. In another embodiment, region 50 can be replaced by a second screw in tandem which can include a cooling region. In an embodiment in which screw 38 is a reciprocating screw in an injection molding system, described more fully below, region 50 can define an accumulation region in which a single-phase, non-nucleated solution of polymeric material and a blowing agent is accumulated prior to injection into mold 37.

Microcellular material production according to the present invention preferably uses a physical blowing agent, that is, an agent that is a gas under ambient conditions (described more fully below). However, chemical blowing agents can be used and can be formulated with polymeric pellets introduced into hopper 44. Suitable chemical blowing agents include those typically relatively low molecular weight organic compounds that decompose at a critical temperature or another condition achievable in extrusion and release a gas or gases such as nitrogen, carbon dioxide, or carbon monoxide. Examples include azo compounds such as azo dicarbonamide.

As mentioned, in preferred embodiments a physical blowing agent is used. One advantage of embodiments in which a physical blowing agent, rather than a chemical blowing agent, is used is that recyclability of product is maximized. Use of a chemical blowing agent typically reduces the attractiveness of a polymer to recycling since residual chemical blowing agent and blowing agent by-products contribute to an overall non-uniform recyclable material pool. Since foams blown with chemical blowing agents inherently include a residual, unreacted chemical blowing agent after a final foam product has been produced, as well as chemical by-products of the reaction that forms a blowing agent, material of the present invention in this set of embodiments includes residual chemical blowing agent, or reaction by-product of chemical blowing agent, in an amount less than that inherently found in articles blown with 0.1 per cent by weight chemical blowing agent or more, preferably in an amount less than that inherently found in articles blown with 0.05 per cent by weight chemical blowing agent or more. In particularly preferred embodiments, the material is characterized by being essentially free of residual chemical blowing agent or free of reaction by-products of chemical blowing agent. That is, they include less residual chemical blowing agent or by-product that is inherently found in articles blown with any chemical blowing agent. In this embodiment, along barrel 32 of system 30 is at least one port 54 in fluid communication with a source 56 of a physical blowing agent. Any of a wide variety of physical blowing agents known to those of ordinary skill in the art such as helium, hydrocarbons, chlorofluorocarbons, nitrogen, carbon dioxide, and the like can be used in connection with the invention, or mixtures thereof, and, according to a preferred embodiment, source 56 provides carbon dioxide as a blowing agent. Supercritical fluid blowing agents are especially preferred, in particular supercritical carbon dioxide. In one embodiment solely supercritical carbon dioxide is used as blowing agent. Supercritical carbon dioxide can be introduced into the extruder and made to form rapidly a single-phase solution with the polymeric material either by injecting carbon dioxide as a supercritical fluid, or injecting carbon dioxide as a gas or liquid and allowing conditions within the extruder to render the carbon dioxide supercritical in many cases within seconds. Injection of carbon dioxide into the extruder in a supercritical state is preferred. The single-phase solution of supercritical carbon dioxide and polymeric material formed in this manner has a very low viscosity which advantageously allows lower temperature molding, as well as rapid filling of molds having close tolerances to form very thin molded parts, which is discussed in greater detail below.

A pressure and metering device 58 typically is provided between blowing agent source 56 and that at least one port 54. Device 58 can be used to meter the mass of the blowing agent between 0.01 lbs/hour and 70 lbs/hour, or between 0.04 lbs/hour and 70 lbs/hour, and more preferably between 0.2 lbs/hour and 12 lbs/liour so as to control the amount of the blowing agent in the polymeric stream within the extruder to maintain blowing agent at a desired level. According to one set of embodiments, the amount of blowing agent in the polymeric stream is between about 0.1 per cent and 25 per cent by weight of the mixture of polymeric material and blowing agent, preferably between about 1.0 per cent and 25 per cent by weight, more preferably between about 6 per cent and 20 per cent by weight, more preferably between about 8 per cent and 15 per cent by weight, more preferably still between about 10 per cent and 12 per cent by weight, based on the weight of the polymeric stream and blowing agent. The particular blowing agent used (carbon dioxide, nitrogen, etc.) and the amount of blowing agent used is often dependent upon the polymer, the density reduction, cell size and physical properties desired. In embodiments where nitrogen is used as blowing agent, blowing agent is present in an amount between 0.1 per cent and 2.5 per cent, preferably between 0.1 per cent and 1.0 per cent in some cases, and where carbon dioxide is used as blowing agent the mass flow of the blowing agent can be between 0.1 per cent and 12 per cent in some cases, between 0.5 per cent and 6.0 per cent in preferred embodiments.

The pressure and metering device can be connected to a controller (not shown) that also is connected to drive motor 40 to control metering of blowing agent in relationship to flow of polymeric material to very precisely control the weight percent blowing agent in the fluid polymeric mixture. For example, the mass flow rate of the blowing agent can be controlled so that it varies by no more than +/-0.3 per cent in preferred cases.

Although port 54 can be located at any of a variety of locations along the barrel, according to a preferred embodiment it is located just upstream from a mixing section 60 of the screw and at a location 62 of the screw where the screw includes unbroken flights.

A preferred embodiment, port 54 is located at a region upstream from mixing section 60 of screw 38 (including highly-broken flights) at a distance upstream of the mixing section of no more than about 4 full flights, preferably no more than about 2 full flights, or no more than 1 full flight. Positioned as such, injected blowing agent is very rapidly and evenly mixed into a fluid polymeric stream to quickly produce a single-phase solution of the foamed material precursor and the blowing agent.

Port 54, in the preferred embodiment, is a multi-hole port including a plurality of orifices 64 connecting the blowing agent source with the extruder barrel. In preferred embodiments a plurality of ports 54 are provided about the extruder barrel at various positions radially and can be in alignment longitudinally with each other. For example, a plurality of ports 54 can be placed at the 12 o'clock, 3 o'clock, 6 o'clock, and 9 o'clock positions about the extruder barrel, each including multiple orifices 64. In this manner, where each orifice 64 is considered a blowing agent orifice, the invention includes extrusion apparatus having at least about 10, preferably at least about 40, more preferably at least about 100, more preferably at least about 300, more preferably at least about 500, and more preferably still at least about 700 blowing agent orifices in fluid communication with the extruder barrel, fluidly connecting the barrel with a source of blowing agent.

Also in preferred embodiments is an arrangement in which the blowing agent orifice or orifices are positioned along the extruder barrel at a location where, when a preferred screw is mounted in the barrel, the orifice or orifices are adjacent full, unbroken flights 65. In this manner, as the screw rotates, each flight, passes, or"wipes"each orifice periodically. This wiping increases rapid mixing of blowing agent and fluid foamed material precursor by, in one embodiment, essentially rapidly opening and closing each orifice by periodically blocking each orifice, when the flight is large enough relative to the orifice to completely block the orifice when in alignment therewith. The result is a distribution of relatively finely-divided, isolated regions of blowing agent in the fluid polymeric material immediately upon injection and prior to any mixing.

In this arrangement, at a standard screw revolution speed of about 30 rpm, each orifice is passed by a flight at a rate of at least about 0.5 passes per second, more preferably at least about 1 pass per second, more preferably at least about 1.5 passes per second, and more preferably still at least about 2 passes per second. In preferred embodiments, orifices 54 are positioned at a distance of from about 15 to about 30 barrel diameters from the beginning of the screw (at upstream end 34).

The described arrangement facilitates a method of the invention that is practiced according to one set of embodiments. The method involves introducing, into fluid polymeric material flowing at a rate of at least about 40 lbs/hr., a blowing agent that is a gas under ambient conditions and, in a period of less than about 1 minute, creating a single-phase solution of the blowing agent fluid in the polymer. The blowing agent fluid is present in the solution in an amount of at least about 2.5 per cent by weight based on the weight of the solution in this arrangement. In preferred embodiments, the rate of flow of the fluid polymeric material is at least about 60 lbs/hr., more preferably at least about 80 lbs/hr., and in a particularly preferred embodiment greater than at least about 100 lbs/hr., and the blowing agent fluid is added and a single-phase solution formed within one minute with blowing agent present in the solution in an amount of at least about 3 per cent by weight, more preferably at least about 5 per cent by weight, more preferably at least about 7 per cent, and more preferably still at least about 10 per cent (although, as mentioned, in a another set of preferred embodiments lower levels of blowing agent are used). In these arrangements, at least about 2.4 lbs per hour blowing agent, preferably CO2, is introduced into the fluid stream and admixed therein to form a single-phase solution. The rate of introduction of blowing agent is matched with the rate of flow of polymer to achieve the optimum blowing agent concentration.

A supercritical fluid blowing agent also provides an advantage in that it facilitates the rapid, intimate mixing of dissimilar polymeric materials, thereby providing a method for mixing and molding dissimilar polymeric materials without post-mold delamination.

Dissimilar materials include, for example, polystyrene and polypropylene, or polystyrene and polyethylene. These dissimilar materials typically have significantly different viscosity, polarity, or chemical functionality that, in most systems, precludes formation of a well-mixed, homogeneous combination, leading to delamination or other physical property reduction or physical property degradation. Preferably, in this embodiment, at least two dissimilar components are present with the minority component in an amount of at least about 1 per cent by weight, preferably at least about 5 per cent, more preferably at least about 10 per cent, more preferably still at least about 20 per cent.

Typical prior art techniques for forming combinations of dissimilar polymeric materials involves extruding and pelletizing dissimilar polymeric materials which then are provided, as pellets, in hopper 44 of a system. Using a supercritical fluid blowing agent, in accordance with this aspect of the invention, eliminates the necessity for using pre-mixed pellets or compounding equipment. In this aspect, a mixture of different polymer pellets, for example a mixture of polystyrene pellets and polypropylene pellets, can be provided in hopper 44, melted, intimately mixed with a supercritical fluid blowing agent, and extruded as a well-mixed homogeneous mixture. In this aspect of the invention a single-phase solution of blowing agent and multi-component polymeric material, including dissimilar materials, can be formed at flow rates and within time periods specified below. This aspect of the invention can be used to form polymeric articles composed of at least two dissimilar polymeric materials that resist delamination via extrusion, molding as described herein, or other techniques.

The described arrangement facilitates a method that is practiced according to several embodiments of the invention, in combination with injection or intrusion molding. The method involves introducing, into fluid polymeric material flowing at a rate of from about 0.4 to about 1.4 lbs/la., a blowing agent that is a gas under ambient conditions and, in a period of less than about 1 minute, creating a single-phase solution of the blowing agent fluid in the polymer. The blowing agent fluid is present in the solution in an amount of at least about 2.5 per cent by weight based on the weight of the solution in this arrangement. In some embodiments, the rate of flow of the fluid polymeric material is from about 6 to 12 lbs/hr. In these arrangements, the blowing agent fluid is added and a single-phase solution formed within one minute with blowing agent present in the solution in an amount of at least about 3 per cent by weight, more preferably at least about 5 per cent by weight, more preferably at least about 7 per cent, and more preferably still at least about 10 per cent (although, as mentioned, in a another set of preferred embodiments lower levels of blowing agent are used). In these arrangements, at least about 2.4 lbs per hour blowing agent, preferably CO2, is introduced into the fluid stream and admixed therein to form a single-phase solution. The rate of introduction of blowing agent is matched with the rate of flow of polymer to achieve the optimum blowing agent concentration.

Downstream of region 50 is a nucleator 66 constructed to include a pressure-drop nucleating pathway 67. As used herein,"nucleating pathway"in the context of rapid pressure drop is meant to define a pathway that forms part of microcellular polymer foam extrusion apparatus and in which, under conditions in which the apparatus is designed to operate (typically at pressures of from about 1500 to about 30,000 psi upstream of the nucleator and at flow rates of greater than about 5 lbs polymeric material per hour), the pressure of a single-phase solution of polymeric material admixed with blowing agent in the system drops below the saturation pressure for the particular blowing agent concentration at a rate or rates facilitating nucleation. Nucleating pathway 67 includes an inlet end 69 for receiving a single-phase solution of polymeric material precursor and blowing agent as a fluid polymeric stream, and a nucleated polymer releasing end 70 for delivering nucleated polymeric material to molding chamber, or mold, 37. Nucleator 66 can be located in a variety of locations downstream of region 50 and upstream of mold 37. In a preferred embodiment, nucleator 66 is located in direct fluid communication with mold 37, such that the nucleator defines a nozzle connecting the extruder to the molding chamber and the nucleated polymer releasing end 70 defines an orifice of molding chamber 37. According to one set of embodiments, the invention lies in placing a nucleator upstream of a mold. Although not illustrated, another embodiment of nucleator 66 includes a nucleating pathway 67 constructed and arranged to have a variable cross-sectional dimension, that is, a pathway that can be adjusted in cross-section. A variable cross-section nucleating pathway allows the pressure drop rate in a stream of fluid polymeric material passing therethrough to be varied in order to achieve a desired nucleation density.

In one embodiment, a nucleating pathway that changes in cross-sectional dimension along its length is used. In particular, a nucleating pathway that decreases in cross-sectional dimension in a downstream direction can significantly increase pressure drop rate thereby allowing formation of microcellular material of very high cell density using relatively low levels of blowing agent. These and other exemplary and preferred nucleators are described in co-pending U.S. patent application serial no. 08/777,709 entitled "Method and Apparatus for Microcellular Extrusion" and International patent application serial no. PCT/US97/15088, entitled "Method and Apparatus for Microcellular Polymer Extrusion" of Anderson, *et al.*, both referenced above.

While pathway 67 defines a nucleating pathway, some nucleation also may take place in the mold itself as pressure on the polymeric material drops at a very high rate during filling of the mold.

The system illustrates one general embodiment of the present invention in which a single-phase, non-nucleated solution of polymeric material and blowing agent is nucleated, via rapid pressure drop, while being urged into molding chamber 37 via the rotation action of screw 38. This embodiment illustrates an intrusion molding technique and, in this embodiment, only one blowing agent injection port 54 need be utilized.

In another embodiment, screw 38 of system 30 is a reciprocating screw and a system defines an injection molding system. In this embodiment screw 38 is mounted for reciprocation within barrel 32, and includes a plurality of blowing agent inlets or injection ports 54,55,57,59, and 61 arranged axially along barrel 32 and each connecting barrel 32 fluidly to pressure and metering device 58 and a blowing agent source 56. Each of injection ports 54, 55, 57, 59, and 61 can include a mechanical shut- off valve 154,155,157,159, and 161 respectively, which allow the flow of blowing agent into extruder barrel 38 to be controlled as a function of axial position of reciprocating screw 38 within the barrel. In operation, according to this embodiment, a charge of fluid polymeric material and blowing agent (which can be a single-phase, non- nucleated charge in some embodiments) is accumulated in region 50 downstream of the downstream end 48 of screw 38. Screw 38 is forced distally (downstream) in barrel 32 causing the charge in region 50 to be injected into mold 37. A mechanical shut-off valve 64, located near orifice 70 of mold 37, then can be closed and mold 37 can be opened to release an injection-molded part. Screw 38 then rotates while retracting proximally (toward the upstream end 34 of the barrel), and shut-off valve 161 is opened while shut- off valves 155,157,154, and 159 all are closed, allowing blowing agent to be injected into the barrel through distal-most port 61 only. As the barrel retracts while rotating, shut-off valve 161 is closed while shut-off valve 159 is opened, then valve 159 is closed while valve 154 is opened, etc. That is, the shut-off valves which control injection of blowing agent from source 56 into barrel 32 are controlled so that the location of injection of blowing agent moves proximally (in an upstream direction) along the barrel as screw 38 retracts proximally. The result is injection of blowing agent at a position along screw 38 that remains essentially constant. Thus, blowing agent is added to fluid polymeric material and mixed with the polymeric material to a degree and for a period of time that is consistent independent of the position of screw 38 within the barrel. Toward this end, more than one of shut-off valves 155,157, etc. can be open or at least partially open simultaneously to achieve smooth transition between injection ports that are open and to maintain essentially constant location of injection of blowing agent along barrel 38.

Once barrel 38 is fully retracted (with blowing agent having been most recently introduced through injection port 55 only), all of the blowing agent shut-off valves are closed. At this point, within distal region 50 of the barrel is an essentially uniform fluid polymeric material/blowing agent mixture. Shut-off valve 64 then is opened and screw 38 is urged distally to inject the charge of polymeric material and blowing agent into mold 37.

The embodiment of the invention involving a reciprocating screw can be used to produce non-microcellular foams or microcellular foam. Where non-microcellular foam is to be produced, the charge that is accumulated in distal region 50 can be a multi-phase mixture including cells of blowing agent in polymeric material, at a relatively low pressure. Injection of such a mixture into mold 37 results in cell growth and production of conventional foam. Where microcellular material is to be produced, a single-phase, non-nucleated solution is accumulated in region 50 and is injected into mold 37 while nucleation takes place.

The described arrangement facilitates a method of the invention that is practiced according to another set of embodiments in which varying concentrations of blowing agent in fluid polymeric material is created at different locations in a charge accumulated in distal portion 50 of the barrel. This can be achieved by control of shut-off valves 155, 157,154,159, and 161 in order to achieve non-uniform blowing agent concentration. In this technique, articles having varying densities may be produced, such as, for example, an article having a solid exterior and a foamed interior. One technique for forming articles having portions that vary in density is described more fully below.

Although, molding chamber 37 can include vents to allow air within the mold to escape during injection. The vents can be sized to provide sufficient back pressure during injection to control cell growth so that uniform microcellular foaming occurs. In another embodiment, a single-phase, non-nucleated solution of polymeric material and blowing agent is nucleated while being introduced into an open mold, then the mold is closed to shape a microcellular article.

According to another embodiment an injection molding system utilizing a separate accumulator is provided. An injection molding system 31 includes an extruder. The extruder can include a reciprocating screw. At least one accumulator 78 is provided for accumulating molten polymeric material prior to injection into molding chamber 37.

The extruder includes an outlet 51 fluidly connected to an inlet 79 of the accumulator via a conduit 53 for delivering a non-nucleated, single-phase solution of polymeric material and blowing agent to the accumulator.

Accumulator 78 includes, within a housing 81, a plunger 83 constructed and arranged to move axially (proximally and distally) within the accumulator housing. The plunger can retract proximally and allow the accumulator to be filled with polymeric material/blowing agent through inlet 79 and then can be urged distally to force the polymeric material/blowing agent mixture into mold 37. When in a retracted position, a charge defined by single-phase solution of molten polymeric material and blowing agent is allowed to accumulate in accumulator 78. When accumulator 78 is full, a system such as, for example, a hydraulically controlled retractable injection cylinder forces the accumulated charge through nucleator 66 and the resulting nucleated mixture into molding chamber 37. This arrangement illustrates another embodiment in which a non-nucleated, single-phase solution of polymeric material and blowing agent is nucleated as a result of the process of filling the molding chamber. Alternatively, a pressure drop nucleator can be positioned downstream of region 50 and upstream of accumulator 78, so that nucleated polymeric material is accumulated, rather than non- nucleated material, which then is injected into mold 37.

In another arrangement, a reciprocating screw extruder can be used with system so as to successively inject charges of polymeric material and blowing agent (which can remain non-nucleated or can be nucleated while being urged from the extruder into the accumulator) while pressure on plunger 83 remains high enough so that nucleation is prevented within the accumulator (or, if nucleated material is provided in the accumulator cell growth is prevented). When a plurality of charges have been introduced into the accumulator, shut-off valve 64 can be opened and plunger 83 driven distally to transfer the charge within the accumulator into mold 37. This can be advantageous for production of very large parts.

A ball check valve 85 is located near the inlet 79 of the accumulator to regulate the flow of material into the accumulator and to prevent backflow into the extruder, and to maintain a system pressure required to maintain the single-phase solution of non- nucleated blowing agent and molten polymeric material or, alternatively, to prevent cell growth of nucleated material introduced therein. Optionally, injection molding system 31 can include more than one accumulator in fluid communication with extruder 30 and molding chamber 37 in order to increase rates of production.

System 31 includes several additional components that will be described more fully below.

Molding chambers according to alternative embodiments for use with injection molding systems of the invention. A movable wall molding chamber 71 is illustrated schematically, including mold cavity 84, temperature control elements 82, moveable wall 80, pressurizing means (not shown) and in the preferred embodiment illustrated, at least one nucleator 66 including a nucleating pathway 67 having an inlet end 69 and an releasing end 70 which defines an orifice of mold cavity 84. In one embodiment, movable wall molding chamber 71 includes a plurality of nucleators 66. Movable wall 80 can be adjusted to increase the volume of the mold as the mold is filled with a nucleated mixture of polymeric precursor and blowing agent, thus maintaining a constant pressure within the mold. In this way, cell growth can be restricted, or controlled, appropriately.

A gas counter-pressure molding chamber 73 is illustrated schematically, including mold cavity 84, temperature control elements 82, pressure controller 86, seals 92, and in the preferred embodiment illustrated, at least one nucleator 66 including a nucleating pathway 67 including defining an orifice of the molding chamber 73. As described previously, the nucleating pathway 67 has an inlet end 69 and an releasing end 70 which defines an orifice of chamber 84. The pressure within the mold can be maintained, via pressure controller 86, to restrict or control cell growth in the nucleated mixture introduced into the mold.

Any combination of a movable-wall mold, a mold having a gas pressure controller, and temperature control elements in a mold can be used for a variety of purposes. As discussed, conditions can be controlled so as to restrict or control cell growth in a nucleated mixture within the mold. Another use for temperature control measurements is that a portion of the mold wall, or the entire mold wall, can be maintained at a relatively high or relatively low temperature, which can cause relatively greater or lesser cell growth at regions near the wall (regions at and near the skin of the microcellular mold and product) relative to regions near the center of the article formed in the mold.

In one embodiment of the invention, relatively thick microcellular polymeric material is molded, for example material including at least one portion have a thickness of at least 0.500 inch by establishing a nucleated, microcellular polymeric precursor in a mold and rapidly"cracking", or opening the mold to allow a part larger than the interior of the closed mold itself to form. When the mold is cracked, cell growth occurs due to a corresponding pressure drop. The nucleated mixture is allowed to partially solidify in the shape of the mold, or enclosure, to form a first microcellular polymeric article in the shape of the enclosure, is removed from the enclosure, and allowed to expand further to form a second microcellular polymeric article having a shape that is larger than the shape of the enclosure. In some aspects, the injection or protrusion may continue after cracking of the mold, to control density and cell structure. That is, a solution can be introduced into the mold while being nucleated and, simultaneously, the mold can be cracked and then further opened to control back pressure in the mold and to control the size of the final part and cell density and structure. This can be accomplished, as well, with an analogous moveable wall mold, described herein.

The invention allows for rapid, cyclic, polymeric foam molding. After injection and molding, in a period of less than about 10 minutes, a second nucleated mixture can be created by injection into the molding chamber and allowed to foam and solidify in the shape of the enclosure, and to be removed. Preferably, the cycle time is less than about one minute, more preferably less than about 20 seconds. The time between introduction of the material into the mold and solidification is typically less than about 10 seconds.

Low cycle times are provided due to reduced weight in foam material (less mass to cool) and low melt temperatures made possible by reduced viscosity of a supercritical fluid blowing agent. With lower melt temperatures less heat absorption is required before part ejection.

Another embodiment of the invention that makes use of system 31 is illustrated, and system 31 now will be described more fully. System 31 also includes a blowing agent-free conduit 88 connecting an outlet 90 of the extruder with an accumulator inlet 91. Inlet 91 of the accumulator is positioned at the face of plunger 83 of the accumulator. A mechanical shut-off valve 99 is positioned along conduit 88, preferably near outlet 90. Extruder outlet 90 is located in the extruder upstream of blowing agent inlet 54 (or multiple blowing agent inlets, as in the extrusion arrangement, where that arrangement is used in the system) but far enough downstream in the extruder that it can deliver fluid polymeric material 94. The fluid polymeric material 94 delivered by conduit 88 is blowing-agent-poor material, and can be essentially free of blowing agent. Thus, the system includes a first outlet 90 of the extruder positioned to deliver fluid polymeric material essentially free of blowing agent, or at reduced blowing agent concentration, from the extruder to a first inlet 91 of the accumulator, and a second outlet 51 downstream of the mixing region of the extruder positioned to deliver a mixture of fluid polymeric material and blowing agent (a higher blowing agent concentration than is delivered from outlet 90, i.e. blowing-agent-rich material) to a second inlet 79 of the accumulator. The accumulator can include heating units 96 to control the temperature of polymeric material therein. The accumulator includes an outlet that is the inlet 69 of nucleator 66. A passage (or nozzle) defining nucleating pathway 67 connects accumulator 78 to the molding chamber 37.

A series of valves, including ball check valves 98 and 85 located at the first and second inlets to the accumulator, and mechanical valves 64 and 99, respectively, control flow of material from the extruder to the accumulator and from the accumulator to the mold as desired, as described below according to some embodiments.

The invention involves, in all embodiments, the ability to maintain pressure throughout the system adequate to prevent premature nucleation where nucleation is not desirable (upstream of the nucleator), or cell growth where nucleation has occurred but cell growth is not desired or is desirably controlled.

Practicing the method according to one embodiment of the present invention involves injecting blowing agent-poor material into a mold to form a nearly solid skin, followed by injecting blowing agent-rich material into the mold to form a foamed core.

Although, with proper synchronization this method can also be used to form articles having a foamed exterior and a solid interior.

A situation in which polymeric material that does not contain blowing agent, or contains blowing agent only to a limited extent (material 94) is provided both at the distal end of the accumulator and the proximal end of the accumulator. That is, blowing agent-poor material 94 is provided just in front of plunger 83 and in nucleating pathway 67 and just upstream of nucleating pathway 67. Between these regions of blowing agent-poor material 94 is a region of blowing agent-rich material 101 in the accumulator. At this point, mechanical valve 64 connecting to mold 37 is opened and plunger 83 is driven downstream to force the material in accumulator 78 into mold 37. This is illustrated in Fig. 10.6. The first section of blowing agent-poor material lines the exterior of the mold, forming an essentially solid exterior wall, then the blowing agent-rich material 101 fills the center of the mold and is nucleated while entering the mold. The distal limit of motion of the plunger stops short of the end of accumulator and the region of blowing agent-poor material that had been located just in front of the plunger is now positioned at the distal end of the accumulator and filling the nucleating pathway of the accumulator. Valve 64 then is closed and the resultant part is removed from mold 37. With mechanical valve 99 closed, the extruder is driven to introduce blowing agent-rich material, preferably as a single-phase, non-nucleated solution of polymeric material and blowing agent, into the accumulator as the plunger retracts proximally. The plunger applies an essentially constant pressure to material in the accumulator, maintaining material 101 in a non-nucleated state. When the plunger has reached nearly its proximal limit, mechanical valve 99 is opened and blowing agent-poor material 94 is allowed to fill a section of the accumulator just in front of the plunge.

An injection-molded microcellular article having a blowing agent-poor exterior wall and a blowing agent-rich, microcellular foamed interior can be formed without the necessity of filling accumulator 78 with blowing agent-rich material sandwiched between blowing agent- poor material, as illustrated. In this embodiment blowing agent-rich material fills the mold but the distal-most part of the accumulator, defined by the nucleating pathway 67, is heated to a greater extent than is the remainder of the accumulator. This can be accomplished using heating units 103 positioned adjacent the nucleator. If needed, additional heating units can be provided to heat material in the accumulator upstream of the nucleating pathway. Material in the distal-most portion of the accumulator is heated to a great enough extent that, when the charge in the accumulator is injected into the mold, blowing agent in the highly-heated section very quickly diffuses out of the polymer and through vents (not shown) in the mold. In the polymeric material upstream of the distal-most, more highly-heated charge section, cell growth occurs to form microcellular material faster than blowing agent can diffuse out of the polymer. The distal most portion of the charge that is heated can define at least about 2 per cent of the charge, or at least about 5 per cent, or at least about 10 per cent, or at least about 20 per cent of the charge, and can be heated to a temperature at least about 10 °C higher than the average temperature of the charge, or at least about 20 °C, 40 °C, or 80 °C higher than the average temperature of the charge, prior to injecting the charge into a molding chamber.

In another embodiment of the invention a single-phase, homogeneous solution of polymeric material and blowing agent can be injected into a mold while maintaining pressure in the mold high enough to prevent nucleation. That is, injection occurs without nucleation. The homogeneous, single-phase solution then can be frozen into a solid state in the mold, and the mold opened. At this point nucleation and foaming do not occur.

The molded article then can be heated to cause nucleation and foaming, for example by placement in a glycerine bath.

A variety of articles can be produced according to the invention, for example, consumer goods and industrial goods such as polymeric cutlery, automotive components, and a wide variety of other injection molded parts.

An injection molding system 100 according to another embodiment of the invention. Injection molding system 100 includes an extruder, including a barrel 102 having a first, upstream end 104 and a second, downstream end 106 connected to a molding chamber 108. Mounted for reciprocation and rotation within barrel 102 is a screw 110 operably connected, at its upstream end, to a drive motor (not illustrated). A sidestream 114, connecting an intake 113 and a port 115 of the barrel, the port downstream of the intake, includes a melt pump 116 and mixer 118 fluidly connected in sequence. Melt pump 116 can be a gear pump or a small extruder, which are known in the art.

Mixer 118 includes a blowing agent injection port 120 for introducing a blowing agent therein. Mixer 118 can be a static mixer or a cavity transfer mixer, which are also known in the art. The arrangement illustrated in Fig. 10.8 facilitates another method of the invention that is useful for forming injection molded microcellular parts having varying material densities, as described previously. The method involves introducing into extruder barrel 102 a precursor of polymeric material, melting the precursor of polymeric material, and advancing molten polymeric material 124 towards the downstream end 106 of extruder 100. As the molten polymeric material 124 advances through extruder barrel 102, a portion is diverted and advanced through intake 113 into sidestream 114 by melt pump 116 (for example, after the distal end of screw 110 is retracted proximally of intake 113 of side arm 114). As the molten polymeric material in sidestream 114 advances through mixer 118, blowing agent from gas injection port 120 is introduced and mixed thoroughly therein to form a single-phase, non-nucleated solution of blowing agent and molten polymeric material which is advanced from sidestream 114 into the downstream end 106 of extruder barrel 102 through port 115, while reciprocating screw 110 fully retracts. This creates a blowing agent-rich region 122 at the distal most end of the barrel and a blowing agent-poor region proximal of the blowing agent-rich region. The relative amount of blowing agent-rich material and a blowing agent-poor material can be controlled by the rate at which material is passed through side arm 14 and enriched with blowing agent. Thereafter, the reciprocating action of screw 110 is used to inject the blowing agent-rich, single-phase solution of non- nucleated blowing agent and molten polymeric material 122 followed by a portion of the blowing agent-poor molten polymeric material 124 into molding chamber 108.

In another embodiment, the invention provides a technique for rapidly and efficiently introducing a blowing agent into a fluid polymeric precursor in injection molding apparatus as described herein, as well as in extrusion apparatus in accordance with essentially any arrangement. This embodiment includes an extrusion screw, having an orifice in a surface of the screw positionable within an extrusion barrel (not shown) that fluidly communicates with a source of blowing agent.

The orifice defines the terminus of a lumen passing from a location connectable to the source, such as a location at the proximal end of the screw. In a preferred embodiment, the lumen passes longitudinally along the rotational axis of the screw from the proximal end of the screw and connects with one or more orifices on the surface of the screw.

The one or more orifices preferably are located at outer surfaces of screw flights or can be slightly recessed from outer surfaces of flights, this positioning allowing introduction of blowing agent in a manner such that the blowing agent undergoes shearing/diffusion against the inner surface of the barrel. One or more orifices can be located in regions between flights as well, or a combination of orifices at a variety of locations can be used.

An extruder screw 130 includes a flight 132 and a lumen 134 that provides communication with an orifice 136 on an exterior surface 138 of flight 132.

Portion 140 of lumen 134 passes from the lumen at the central axis of the screw to orifice 136. One advantage in the introduction of blowing agent via an orifice within a screw is that uniformity of blowing agent level or distribution within a polymeric precursor can be provided in an arrangement using a reciprocating screw because of a fired injection point on the screw.

The techniques of the invention described above can be used also in gas-assist co-injection. In this technique a precursor of microcellular material is extruded and nucleated while being introduced into a mold, as described above, while gas is injected into the melt stream in such a way as to form, in the mold, an exterior layer against the mold walls of nucleated polymeric material and a central void filled with the co-injected gas. Cell growth can be made to occur as in other embodiments.

Microcellular polymeric articles or non-microcellular polymeric foam articles can be produced having thicknesses, or cross-sectional dimension, of less than 0.125 inch, preferably no more than 0.100 inch, more preferably no more than about 0.075 inch, more preferably no more than about 0.050 inch, more preferably no more than about 0.025 inch, more preferably still no more than about 0.010 inch, via injection molding, because a single-phase solution of polymer precursor and supercritical fluid has a particularly low viscosity and, in this manner, can be injected into a mold and formed as a foamed article therein. For example, a single-phase solution of supercritical fluid and polymer can be introduced into a mold and a conventionally-foamed or microcellular article can be produced thereby. The low viscosity of the fluid injected into the mold allows injection-mold cycle times, as described above, of less than 10 minutes, preferably less than 5 minutes, and more preferably less than 1 minute, preferably less than 30 seconds, more preferably less than 20 seconds, more preferably less than 10 seconds, and more preferably still less than 5 seconds.

The invention provides also for the production of molded microcellular polymeric articles or molded non-microcellular polymeric foam articles of a shape of a molding chamber, including at least one portion have a cross-sectional dimension of no more than about 0.125 inch or, in other embodiments, smaller dimensions noted above, the article having a void volume of at least about 5 per cent. Preferably, the void volume is at least about 10 per cent, more preferably at least about 15 per cent, more preferably at least about 20 per cent, more preferably at least about 25 per cent, and more preferably still at least about 30 per cent. In other embodiments the article has a void volume of at least about 50 per cent. This is a significant improvement in that it is a challenge in the art to provide weight reduction in polymeric material, via foam void volume, in articles having very small dimensions.

The articles of the invention include the above-noted void volumes in those sections that are of cross-sectional dimension of no more than about 0.125 inch, or other, smaller dimensions noted above.

The invention also provides for the production of molded microcellular polymeric articles or molded non-microcellular foam polymeric articles having a variety of thicknesses and void volumes.

In one set of embodiments, the molded articles include at least one portion having a cross-sectional dimension of between about 0.075 inch and about 0.125

inch and a void volume of at least about 10 per cent. The molded articles of this embodiment preferably have a void volume of at least about 20 per cent, more preferably at least about 30 per cent, more preferably at least about 40 per cent, and still more preferably at least about 50 per cent.

In another set of embodiments, the molded articles include at least one portion having a cross-sectional dimension of between about 0.125 inch and about 0.150 inch and a void volume of at least about 15 per cent. The molded articles of this embodiment preferably have a void volume of at least about 20 per cent, more preferably at least about 30 per cent, more preferably at least about 40 per cent, and still more preferably at least about 50 per cent.

In another set of embodiments, the molded articles include at least one portion having a cross-sectional dimension of between about 0.150 inch and about 0.350 inch and a void volume of at least about 20 per cent. The molded articles of this embodiment preferably have a void volume of at least about 30 per cent, more preferably at least about 40 per cent, more preferably at least about 50 per cent, more preferably at least about 60 per cent, and still more preferably at least about 70 per cent. In certain preferred embodiments of this set, the molded articles include at least one portion having a cross-sectional dimension of between about 0.150 and about 0.250 inch.

The methods of the invention also allow the production of higher weight reduction, as described herein, and smaller cells in injection molded parts having thicknesses greater than 0.125 inch, for example between 0.200 inch and about 0.500 inch thickness.

The invention also provides a system and method to produce thick and thin foam molded parts with surfaces replicating solid parts. At least a portion of the surface of these parts is free of splay and swirl visible to the naked human eye. In conjunction with Example 9 (below) demonstrate formation of polymeric parts having surfaces free of splay and a swirl visible to the naked human eye. Such molded parts can be produced when the temperature of the melt and mold temperature and a blowing agent concentration is optimized to allow blowing agent to diffuse away from the surface of the part so that the surface includes a skin layer essentially free of cells. This skin layer is essentially solid polymer, thus the part appears as a solid polymeric part appears to the naked human eye. Splay and a swirl, in foamed polymeric material, is caused by bubbles at the surface being dragged against a mold wall. Where bubbles at the surface are removed, due to temperature control, splay and a swirl is avoided. In these embodiments molded parts are produced having an outer skin of essentially solid polymeric material free of cells, having a thickness at least three times the average cell size of the foam material. Preferably, the outer skin thickness is at least about five times the average cell size of the material. Another reason that molded parts can be produced, according to the invention, that are free of visible splay and a swirl is that the diffusion rate of a supercritical fluid blowing agent is believed by the inventors to be more rapid than that of typical blowing agents, allowing diffusion at the surface of the article to occur, as described, to form a solid skin layer.

As mentioned, the invention provides for the production of molded foam polymeric material, preferably microcellular material having thin sections. In particular, articles having high length-to-thickness ratios can be produced. The invention provides injection molded polymeric materials having length-to-thickness ratios of at least about 50: 1 where the polymer has a melt index of less about 10. Preferably the length-to- thickness ratio is at least about 75: 1, more preferably at least about 100: 1, and more preferably still at least 150: 1. In another embodiment an article is provided having a length-to-thickness ratio of at least about 120: 1, the polymer having a melt flow rate of less than about 40. In this embodiment, the length-to-thickness is preferably at least about 150 : 1, more preferably at least 175: 1, more preferably at least about 200: 1, and more preferably still at least 250: 1. Length-to-thiclcness ratio, in this context, defines the ratio of the length of extension of a portion of a polymeric molded part extending away from the injection location in the mold (nozzle) and the thickness across that distance.

One particularly advantageous combination of features of the invention is a thin molded part at a relatively high void volume. In particular, the invention provides foam polymeric articles having a portion of thickness less than about 1.2 millimeters and a void volume of at least 30 per cent. In another embodiment a polymeric article having a thickness of less than about 0.7 millimeters is provided having a void volume of at least 15 per cent.

In another set of embodiments, a series of molded polymeric articles are provided. At least 70 per cent of the total number of cells in the polymeric articles of this set of embodiments have a cell size of less than 150 microns. Preferably, at least 30 per cent of the total number of cells have a cell size of less than 800 microns, more preferably less than 500 microns, and more preferably less than 200 microns. In some embodiments of this set, at least 80 per cent, preferably at least 90 per cent, more preferably at least 95 per cent, and still more preferably at least 99 per cent of the total number of cells have a cell size of less than 150 microns. In certain embodiments, at least 80 per cent, more preferably at least 90 per cent, more preferably at least 95 per cent, and still more preferably at least 99 per cent of the total number of cells have a cell size of less than 100 microns. The molded articles of this set of embodiments can have a variety of void volumes. For example, the molded articles can have a void volume of at least 10 per cent, at least 20 per cent, at least 30 per cent, at least 40 per cent, or at least 50 per cent.

In preferred embodiments, articles are provided having thicknesses as defined herein at void volumes defined herein, where the maximum thickness exists over at least about 25 per cent of the article, that is, a least about 25 per cent of the area of a thin molded part is of a thickness less than that described. In other embodiments more of the article can be of thickness less than the maximum defined, for example 50 per cent or 100 per cent.

Molding system 150 of these figures is designed to allow injection molding of microcellular polymeric material, particularly microcellular polyolefins such as polypropylene and polyethylene. System 150 allows for the production of relatively thick parts while avoiding EPP procedures of typical existing polypropylene foams.

System 150 includes an inlet 152, constructed and arranged to receive a precursor of molded polymeric microcellular material, such as can be provided by an extruder and/or accumulator. A channel 154 connects inlet 152 with a molding chamber 156. Channel 154 includes a nucleating pathway 158 that has length and cross-sectional dimensions that create a pressure drop in a fluid, single-phase solution of polymeric material and blowing agent at a pressure drop rate sufficient to cause microcellular nucleation, when the solution is passed through the nucleating pathway at rates for which the system is constructed. Since the design of a molding system and the rate of introduction of polymeric material into a mold typically are planned in conjunction with each other, those of ordinary skill in the art will understand the meaning of reference to rates for which the system is constructed. Specifically, the nucleating pathway has length and cross-sectional dimensions that can create a pressure drop at a rate of at least about 0.3 GPa/sec in fluid polymeric material and blowing agent, as a single phase solution, for example when passing through the pathway at a rate of greater than 40 pounds fluid per hour. Other flow rates and pressure drop rates suitable for microcellular nucleation are apparent from reading the present application.

Channel 154 includes a cell growth region 160 between nucleating pathway 158 and molding chamber 156 that increases in cross-sectional dimension in the direction of the molding chamber. Channel 154 also includes a divergent portion 162 between inlet 152 and the molding chamber, specifically between the inlet and the nucleating pathway.

Divergent portion 162 increases in width in a downstream direction (toward molding chamber 156) while decreasing in clearance (height). The result is an increase in width while maintaining a cross-sectional area that does not change significantly. Specifically, the divergent portion increases in width by at least about 100 per cent, preferably at least by about 200 per cent, and more preferably still by at least 300 per cent, while maintaining a cross-sectional area that changes by no more than about 25 per cent, preferably by no more than about 15 per cent, and more preferably still by no more than about 10 per cent. Divergent portion 162 allows for introduction of microcellular molded material precursor through inlet 152 and delivery of the precursor to nucleating pathway 158 while widening the pathway flow to a dimension equal to the width of molding chamber 156 while maintaining a relatively constant pressure profile in the material.

The arrangement of divergent section 162 and nucleating pathway 158 allows the nucleating pathway to have a width-to-height ratio of at least about 1.5: 1, more preferably at least about 2.0: 1, more preferably at least about 5.0: 1, more preferably at least about 10: 1, and more preferably still at least about 20: 1. This allows the nucleating pathway to have a width equal to one dimension of molding chamber 156, thus microcellular polymeric articles that are both wide and thick can easily be nucleated and molded within system 150. In addition, conventional (i.e., non-microcellular) foam polymeric material can be injection-molded using the system of 150, as well.

Specifically, non-foamed polymeric material can be injected into molding chamber 156 and allowed to foam, to have a shape essentially identical to that of the molding chamber (including by definition larger where an expanded, or"cracked"mold is used), the article having at least one portion having cross-sectional dimensions of at least one half inch in at least two perpendicular intersecting cross-sectional axes, and a void volume of at least about 50 per cent. Higher void volumes of 60 per cent, 70 per cent and 80 per cent also can be achieved using this system, in combination with any of higher thicknesses of at least about 0.75 inch, one inch, or 1.5 inch.

One specific advantage provided by the physical arrangement of divergent portion 162, nucleating pathway 158, and cell growth region 160 of system 150 allows for injection of the fluid, single-phase solution of polymeric material and blowing agent into the molding chamber system and, at a significant width dimension, subjecting the solution to a rapid pressure drop at the nucleating pathway to cause microcellular nucleation, and essentially immediately thereafter allowing and controlling cell growth in cell growth region 160 by subjecting the material to a pressure drop that is controlled, at a rate less than the pressure drop rate to which the solution is subjected in nucleating pathway 158, and that decreases during cell growth. That is, the pressure drop rate experienced by nucleated polymeric material in cell growth region 160 decreases during cell growth to provide uniform, controlled microcellular material.

The use of molding system 150 in conjunction with extrusion and/or accumulation apparatus described previously allows for the production of unique, thick and wide polymeric molded articles, including microcellular polymeric molded articles, that have a uniformity in cell structure much better than that of EPP and EPS foams. As noted above, in the production of EPP and EPS foams lines of fusion between beads, after molding, can be easily observed by the naked eye. The molded articles of the present invention, in contrast, are free of lines of fusion in the cell structure. That is, they are free of periodic solid boundaries (lines of fusion in molded EPP or EPS) of thickness greater than about 5 times the average cell wall thickness. Preferably, the articles are free of periodic solid boundaries of thickness greater than about 4 times average cell wall thickness, and more preferably still free of periodic boundaries greater than about 3 times average cell wall thickness.

System 150 also allows for production of thick and wide polymeric material, including microcellular polymeric material, at mold temperatures much lower than those typical of steam chest molding of EPP and EPS. In particular, a fluid precursor of foamed polymeric material can be injected into molding chamber 156 at a molding chamber temperature of less than about 100°C. The mixture then is allowed to solidify in the molding chamber as a polymeric article, preferably a polymeric microcellular article, including at least one portion having cross-sectional dimensions of at least one half inch in each of three perpendicular intersecting cross-sectional axes and a void volume of at least about 50 per cent (or higher values noted above). Preferably, the mold temperature is less than about 75°C in this technique, more preferably less than about 50°C, and more preferably still less than about 30°C.

The system also allows very rapid cycle times of injection molding of polymeric microcellular material of void volume of at least about 50 per cent (or higher values noted above). In particular, a cycle time (repeated injecting of non-foam material, allowing the mixture to solidify in the molding chamber as a polymeric microcellular article, and removing the article from the mold and repeating) can be carried out at cycle time of less than about 1 minute, more preferably less than about 45 seconds, more preferably less than about 30 seconds, and more preferably still less than about 25 seconds.

It is known in the art that molding of material can inherently give a skin-foam- skin structure, and that the skin-foam-skin structure can be controlled based on temperature and other injection conditions to give a thicker or thinner skin. It is also known that a skin-foam-skin structure has a higher strength-to-weight ratio than a similar part without a skin or with a relatively thin skin. It is common practice to calculate strength in a skin-foam-skin molded partly based upon "I-beam" formulations.

However, the applicants are unaware of any prior work that takes into account cell size in the prediction or calculation relating to the strength-to-weight ratio of any skin-foam- skin molded structure.

It is another feature of the present invention that very strong, thin parts can be made. In particular, due to the ability to form very thin foam parts with very small cells, that retain a skin-foam-skin structure, previously impossible with thin foam parts, unexpected tensile strength-to-weight ratios in molded materials is achieved. In particular, the present invention provides molded polymeric parts including at least one very thin section, having strength-to-weight ratios (represented as strength-to-density), of at least about 280,000 psi/g/cm3, more preferably at least about 290,000 psi/g/cm3, and more preferably still at least about 300,000 psi/g/cm3. The thin sections of these parts have a thickness of less than about 0.250 inch, or of less than about 0.150 inch, or of less than about 0.100 inch, and in each of these cases have each of the strength-to-weight ratios described above.

Although not wishing to be bound by any theory, the applicants believe that the unexpected strength-to-weight ratios observed in accordance with the invention are due to maximizing the number of cell walls across a thin section as cell size is minimized.

That is, looking at a cross-section of a thin skin-foam-skin structure with relatively larger cells, relatively fewer cell walls will exist across the structure, and the possibility of one cell bridging the entire foam structure exists. Such a bridge would create a very weak link in the structure. In contrast, in microcellular skin-foam-skin structure of the present invention, the number of cells (thus the number of cell walls) across the structure between skin sections is maximized, and a uniform cellular polymer network and uniform strength characteristic across the foam between the skin structures exists. Thus, while in thin parts of the invention the average strength throughout the part may be similar to that of the average strength with a structure having larger cells, articles of the present invention are stronger because the point of typical minimum strength representing a cell or void bridging the entire structure is eliminated.

It is another feature of the present invention that articles can be produced that are opaque without the use of opacifiers. This is because polymeric foam diffracts light, thus it is essentially opaque and has a white appearance. It is a feature of the invention that microcellular foams are more opaque, and uniformly so, than conventional foams.

This is a significant advantage in connection with articles constructed and arranged to contain material that is subject to destruction upon exposure to light, such as food containers. Such material can involve food consumable by animals such as humans, containing vitamins that can be destroyed upon exposure to light. While opacifiers such as pigments are typically added to articles, pigmented material is less amenable to recycling. The present invention provides thin, opaque articles that include less than about 1 per cent by weight auxiliary opacifer, preferably less than about 0.05 per cent by weight auxiliary opacifer, and more preferably still material that is essentially free of auxiliary opacifer. "Auxiliary opacifer", in the present invention, is meant to define pigments, dyes, or other species that are designed specifically to absorb light, or talc or other materials that can block or diffiact light. Those of ordinary skill in the art can test whether an additive is an opacifer. Microcellular blow molded articles of the invention have the appearance of essentially solid, white, plastic articles, which offers significant commercial appeal.

The systems of the invention can include a restriction element (not shown) as described in co-pending U.S. application serial no. 09/285,948, filed April 2,1999, entitled "Methods for Manufacturing Foam Material Including Systems With Pressure Restriction Element" which is incorporated herein by reference. The restriction element, such as a check valve, is positioned upstream of a blowing agent injection port to maintain the solution of polymer and blowing agent in the extruder above a minimum pressure throughout an injection cycle, and preferably above the critical pressure required for the maintenance of a single-phase solution of polymer and blowing agent.

The systems of the invention can include heated runners (not shown). The term "runners used herein, is meant to define a fluid pathway that fluidly connects the outlet end of the injection system (outlet of nucleator according to some embodiments) and the molding chamber, and/or fluidly connecting various portions of the molding cavity for example where complex molded shapes are desired. Runners are known in the art. In some conventional foam injection molding systems, material left in runners hardens, and is removed with the molded part. The present invention provides runners addressed by thermal control units, such as passageways for flowing heated fluid. This is useful in accordance with certain embodiments of the present invention in which it is advantageous to maintain the polymeric article precursor material in a fluid state within the runners in order to eliminate a pressure drop that can occur if a gap in material were to occur within the rumer, when, for example, hardened material has been removed.

The arrangement of the invention can involve, for example, an extruder for supplying a fluid, single-phase solution of polymeric material and blowing agent, a nucleating pathway, and downstream of the pathway a runner between the pathway and a molding chamber, the runner including a valve at its downstream end to be opened when the mold is to be filled and closed when the mold is to be opened and an article removed. If molten polymeric material is used, then if the runner is heated the nucleated material in the runner will remain fluid and suitable for injection into the mold. The embodiment of the invention including temperature-controlled runners can find use in any of a wide variety of injection moldings systems, involving any number of runners between various components, and valves positioned, if needed, appropriately to allow for filling molds or mold sections periodically without the need for removal and discarding of hardened material from the runners. The runner can be the nucleating pathway.

The function and advantage of these and other embodiments of the present invention will be more fully understood from the examples below. The following examples are intended to illustrate the benefits of the present invention, but do not exemplify the full scope of the invention. The examples below demonstrate advantages of injection molding of a charge of polymeric material and supercritical fluid blowing agent, in that articles are formed that have a surface, corresponding to an interior surface of a molding chamber, that is free of splay and swirl visible to the naked human eye.

Example 1: A two stage injection molder (Engel manufacture) was constructed with a 32: 1 l/d, 40 mm plasticizing unit which feeds melted polymer into a 40 mm diameter plunger.

The plunger and plasticizing units were connected by a spring loaded ball check joiner assembly. The plunger was able to inject into a mold through a typical pneumatically driven shut-off nozzle. Injection of supercritical C02 was accomplished by placing at approximately 16 to 20 diameters from the feed section an injection system that included one radially positioned port containing 176 orifices of. 02 inch diameter. The injection system included an actuated control valve to meter a mass flow rate of blowing agent at rates from 0.2 to 12 lbs/hr.

The plasticator was equipped with a two stage screw including a conventional first stage feed, barrier, transition, and metering section, followed by a multi-flighted mixing section for blowing agent homogenization. The barrel was fitted with heating/cooling bands. The design allowed homogenization and cooling of the homogeneous single phase solution of polymer and gas.

The hydraulic system used to move all parts of the molding machine was modified to have a melt pressurization pressure of at least 1000 psi, but not more than 28,000 psi at all times. This technique controls and maintains the single phase solution of polymer and gas at all times before plastic injection into the mold.

Example 2: Injection Molding Microcellular Polystyrene The molding machine as described in example 1 was used to mold microcellular polystyrene plaques. Polystyrene pellets (Novacor 2282,11 M.I.) were fed into the plasticator and, in most cases, mixed with blowing agent to form a single-phase solution, then nucleated by injection into a 5 × 11 × 0. 050 inch, center gated plaque mold.

Injection occurred through a cold sprue. Injection rate was varied to determine the relationship between the processing variables and cell size and weight reduction. Cell size was controlled to under 30 microns and weight reduction as high as 20 per cent. See Tables 10.1 and 10.2 and corresponding Figs. 10.10-10.15.

Table 10.1: Effect of Injection Speed on Cell Size and Weight Reduction Injection Blowing Weight Cell Size Fig. Speed Agent Reduction ("/sec) (microns) (%) (%) 11 No Cells 11. 9 11 10 5 100 11. 9 11 11 4 10 11.9 19 12 2 10 11.9 is 13 2 10 11.9 18 13 1 30 11.9 12 - Table 10.2: Effect of Gas Concentration on Cell Size and Weight Reduction Melt temperature = 160'C Mold Temperature = 66°C Injection Speed = 4.0"/sec Sprue =. 375' diameter Gas Weight Blowing Agent Cell Size Fig. Concentration Reduction (%) (microns) (lbs./hr.) (%) 0.9 20 to 150 21 13.4 14 1.4 1 to 5 23 21 15 **Example 3:** Injection Molding Microcellular Polyethylene Terephthalate The injection molding machine described in example 1 was used to mold PET (Eastman, 0.95 IV) into a 5 × 11 ×. 200 inch cavity after drying for four hours at 350 F.

The melt processing temperature was 550 F, the mold temperature was 151 F and was injected with 12 per cent C02. The melt pressurization pressure was maintained at 3000 psi and the injection speed was 5.0 inches per second.

The weight reduction was 30 per cent and the cell size was 30 to 40 microns in diameter.

Example 4: Injection Molding Polypropylene to high levels of Weight Reduction The injection molding machine described in example 1 was used to mold polypropylene (4 melt flow rate (MFR), copolymer, Montell 7523), polypropylene (20 MFR, copolymer, Montell SD-376) and a talc-filled polypropylene (4 MFR, 40% talc- filled, Montell 65f4-4) into a 5 × 11 × inch plaque with variable thickness. High weight reductions were accomplished by using the following conditions: Table 10.3: Material Part Weight Melt Gas Percent Mold Thickness Reduction Temperature (%) Temperature (inches) (%) (F) (F) 7523. 050 14. 6 310 12. 5 100 SD-376. 100 30 320 12 150 65f4-4. 100 15 330 15 200 **Example 5:** Injection molding Polystyrene parts with density reductions greater than 70 per cent The injection molding machine described in example 1 was used to mold polystyrene under conditions similar to those found in example 2, but with mold temperatures ranging from 150 F to 250 F and cooling times ranging from 3.2 to 22.8 seconds. Large density reductions were seen as follows: Table 10.4: Mold Melt Cooling Solid Part Foam Density Temperature Temperature Time Density Density Reduction (F) (F) (sec) 150 F 250 3. 2. 88 g/cc. 37 g/cc 58% 250 F 250 22. 8. 88 g/cc. 16 g/cc 82% **Example 6:** Post Mold Nucleation and Cell Growth of a Solidified Polymer/ Supercritical Fluid Part The injection molding machine described in Example

1 was used to mold polystyrene (Novacor 2282,11 M.I.). Polystyrene pellets were fed into the plasticator and injected as described in Example 2. The material injected into the mold was cooled in the mold to a temperature below the solidification temperature of the polystyrene. The mold was opened and the part was removed in a non-foamed state. The part then was subjected to an external heat source (glycerine bath) whereupon it foamed. A microcellular article resulted.

Example 7: Demonstration of Viscosity Reduction in Polymer Molding This example demonstrates the advantage of using supercritical fluid blowing agent to reduce viscosity for introduction of polymeric material into a mold, at relatively low melt temperatures, while realizing the benefits of microcellular foaming.

A molding machine was used to mold polystyrene as described in Example 2 with the following exception. The mold had dimensions of 5 × 11 × 0.020 inches. Under the same conditions of Example 2 polystyrene was injected with 0 per cent blowing agent. The maximum flow length obtainable was 1 inch resulting in a length-to-thickness ratio of 50. An identical experiment was run with 15 per cent supercritical carbon dioxide blowing agent. The maximum flow length was at least 5.5 inches with a length-to-thickness ratio of 270.

Example 8: Injection Molding of Polypropylene Below its Crystalline Melting Point The injection molding machine described in Example 1 was used to mold polypropylene (4 MFR, copolymer, Montell 7523) into a 5 × 11 ×. 050 inch mold. With 0 per cent blowing agent, minimum melt temperatures needed to fill such a mold is 430° F.

With 15 per cent supercritical carbon dioxide blowing agent it was possible to inject polypropylene below its crystalline melting point which is nominally 325° F. Actual melt temperature was 310°F.

Example 9: Demonstration of a Microcellularly Foamed Article with a Near Perfect Surface A molding machine as described in Example 2 was used to mold polystyrene (Novacor 2282 11 M.I.). Polystyrene pellets were fed into a plasticator and mixed with C02 blowing agent to form a single phase solution of supercritical C02 and polystyrene, then nucleated by injection into a 5 × 11 ×. 050 inch plaque mold. Processing conditions were optimized to identify the appropriate conditions to obtain a high nucleation density, as well as a solid, splay-free looking skin. Photocopies of photomicrographs are provided as Figs. 10.16-10.18 to demonstrate the effectiveness of this technique. Fig. 10.16 is provided for comparison, and shows solid non-foamed polystyrene, injection molded using standard, non-foam injection molding techniques: Fig. 10.17 is a photocopy of a photomicrograph of a surface of a microcellular injection molded article of the invention having a smooth surface free of splay and a swirl visible to the naked human eye.

As can be seen, ideal conditions involve a balance of melt temperature, mold temperature, and blowing agent concentration. The melt temperature must be high enough so that blowing agent diffusion rate in the melt is relatively rapid, and the mold temperature must be high enough so that blowing agent diffusion out of the melt occurs to a significant degree at the surface, but the mold temperature must be low enough to avoid warpage and other distortion of the product. The blowing agent diffusion rate is dependent upon melt temperature, blowing agent concentration, differential pressure, and mold temperature. The diffusion rate of the blowing agent out of the melt must be greater than the rate at which the polymer surface cools and solidifies.

Table 10.5: Temp. Temp. Injection Blowing Surface Characteristics Fig. Melt Mold Speed Agent (F°) (F°) (I.P.S.) (%) SolidLooking 17 400 175 5. 0 11. 50 Streaks 350 80 3. 0 5. 15 Bubbles 350 80 4. 4 23% Warped, Small Bubbles 410 180 2. 0 11. 50 on Surface Temp. Temp. Injection Blowing Surface Characteristics Fig. Melt Mold Speed Agent (F°) (F°) (I.P.S.) (%) Solid, But Cracked 385 160 3. 5 9. 75 Swirls 18 400 87° 10 13. 25% **Example 10:** The injection molding system as described in example 1 was used to mold Polypropylene (Montell 6823, Montell 6523, and blends of 6823 and Dow's metallocene catalyzed polyethylene) into a 1.25 × 4 × 0.600 deep cavity. The melt temperature and gas percentage could be varied to produce various densities with cell sizes ranging from 1 to 50 microns. The resulting parts were of densities as low as 1.8 pounds per cubic foot and as high as 20 PCF.

Higher densities could easily be made if desired. Additionally these same densities and cell structures can be made through a method of crack molding whereby the melted polymer is injected into a partially opened mold and then cracked completely open. Low density PP and low density blends of PE and PP, respectively, were produced with uniform cell structures. See Table 10.6 for parameters.

Table 10.6: Gas (lbs/hr) Cell Size Density Melt Material % Gas (PCF) Temperature (F). 6 20 12. 7 310 6823 14. 6 1-15 4.9 300 6823 14 Those skilled in the art would readily appreciate that all parameters listed herein are meant to be exemplary and that actual parameters will depend upon the specific application for which the methods and apparatus of the present invention are used. It is, therefore, to be understood that the foregoing embodiments are presented by way of example only and that, within the scope of the appended claims and equivalents thereto, the invention may be practiced otherwise than as specifically described.

10.8 Polymeric Material Properties

C-MOLD CAE software comprises a set of computer programs for plastics molding simulation. The analyses require accurate material-property data to generate the best predictions. Obtaining quality data is critical; simulation results can be only as good as the material-property data used. This chapter serves as a reference for material-property data producers, providing essential information about the generation of this data for C-MOLD analyses. This chapter also contains other useful information, such as polymer nomenclature and unit conversion charts specific to such applications. Much of the material presented in this chapter is based on the document *Characterizing Polymers for C-MOLD Simulations, Third Edition* (PL-E3-0994), available separately from AC Technology Polymer Laboratories.

In terms of requests for data requirements, loose terminology is common among users. For example, it is commonplace to hear a request for "flow" data, in reference to a unique set of data that includes polymer rheology, melt specific heat, density, and thermal conductivity. In a similar manner, "pvT" is used loosely to include all of the above, in addition to pvT data for the material.

In the following sections, each of the C-MOLD injection-molding simulations is described briefly, followed by its material-property data specification. These are followed by the details of test methods and the C-MOLD polymer data models. Appendix A contains the format of our master material data file. Such files contain complete information on a material, including raw data and fits, and can be read directly into C-MOLD Database.

Wherever possible, C-MOLD adheres to standard ISO definitions with respect to polymer nomenclature.

Why Use Standard Test Methods

Standards are nationally and internationally defined test procedures used to characterize a particular property of a material. A standard test method ensures that data is produced to internationally accepted norms. Such data can be verified or tested anywhere else in the world, using comparable equipment, to yield results within the stated accuracy of the method. Standard test methods go through an extensive review by experts in the field, resulting in robust and well-defined techniques. Round-robin tests are conducted to provide a sound level of confidence in the standard test methods. Widely recognized standards-setting organizations for the plastics industry include the ASTM (USA), DIN (Germany), JIS (Japan), and the international ISO.

Standardization is the key to quality data. AC Technology Polymer Laboratories began to implement standard test methods in 1992. First, we investigated the compatibility of the standard test methods with our data requirements for CAE. We then looked for differences between our existing test methods and the standards. Following this, we conducted experiments for each property to determine the advantage, if any, of adopting the standard test method. Then, wherever possible, we harmonized with the standard method. Because of the fundamental-properties approach used in C-MOLD analyses, the changes needed in our testing procedures were minimal. AC Technology Polymer Laboratories now generates data to these standards, so that the data meet industry goals of accuracy and conformance.

Models Used in C-MOLD

To use a computer simulation of any engineering process, it is important to provide accurate information about material behaviour under the conditions encountered during processing, such as temperature, shear rate, pressure, or cooling rate. Standard procedures can be provided to measure these properties under some ideal conditions. It is impractical (and, fortunately, unnecessary) to perform the measurements under all circumstances. Instead, it is more reasonable to find a good model that describes the material behaviour under the conditions of interest. Such models can be derived from scientific principles or from semi-empirical rules. The model constants can be determined from limited experiments, then used to describe material behaviour under other conditions.

Modeling material behaviour in the field of polymer processing has never been an easy task, for several reasons:

- Material properties often vary from batch to batch (or from time to time).
- Using regrind or recycled materials can affect behaviour during processing.
- It is difficult to measure properties at the high temperatures, shear rates, pressures and cooling rates typical of actual processes.
- Material behaviour varies with changing conditions; for example, the temperature sensitivity of melt viscosity depends on shear rate as well as temperature.

Typical properties required for thermoplastics injection molding process simulation are:

- In the melt state: thermodynamic properties (density, heat capacity) and transport properties (rheological properties and thermal conductivity of the molten state).
- In the solid state: mechanical properties.

More sophisticated calculations might require additional properties.

The models described below have been selected after intensive review. These are believed to be the most valid to represent the behaviour of the material. C-MOLD accepts several other models; however, their use is not recommended.

Rheological Properties

Polymer rheology is the most important property used in flow simulations. Most polymers exhibit two regimes of flow behaviour, Newtonian and shear-thinning. Newtonian flow occurs at low shear rates, but with increasing shear, the viscosity tends to fall away in what is termed shear-thinning behaviour. Viscosity also decreases with increasing temperature.

To incorporate the dependence of melt viscosity on shear rate, temperature, and pressure, the following 5-constant (n, τ*, B, Tb, β), Cross-exp model is adequate for simulating the filling stage in injection molding:

$$(T, \gamma, p) = \frac{\eta_0(T,p)}{1+\left(\frac{\eta_0\gamma}{\tau^*}\right)^{1-n}} \quad \ldots (10.1)$$

with

$$\eta_0(T,p) = B\exp\left(\frac{T_b}{T}\right)\exp(\beta p) \quad \ldots (10.2)$$

The Cross-exp model treats polymer viscosity as a function of temperature, T, and shear rate, τ. It handles both the Newtonian and the shear-thinning flow regions found in polymer rheology. Unlike many other models in present use, the constants of the Cross-exp model do have a physical significance. The transition between the two regimes is characterized by τ*, the shear stress at which shear thinning behaviour begins to manifest itself. The slope of the shear-thinning curve is characterized by $(1 - n)$. Depending on the material, it might not be possible to generate the complete curve. However, the best rheology data spans both of these regions. If there are insufficient points in the shear-thinning region, the value of n might be erroneous. Similarly, if the data set lies almost entirely in the power-law region, the values of the zero-shear viscosity, η_o and τ*, will tend to be ill-defined, even though the product, $\eta_o n\, x\, (\tau^*)^{1-}$n, will be well-defined, corresponding to the multiplicative factor in the power-law region.

The remaining three constants of the Cross-exp model are used to model the zero-shear-rate viscosity. Concerning T_b, which characterizes the temperature sensitivity of η_o, note that this quantity tends to depend on temperature, especially in the vicinity of the glass transition. As such, this modeling is adequate for the filling stage, where the bulk of the polymer is in the vicinity of the processing temperature, usually more than 100°C away from the glass transition, where Tb is relatively constant (see Figure 10.1). A more sophisticated model is adopted for post-filling simulations.

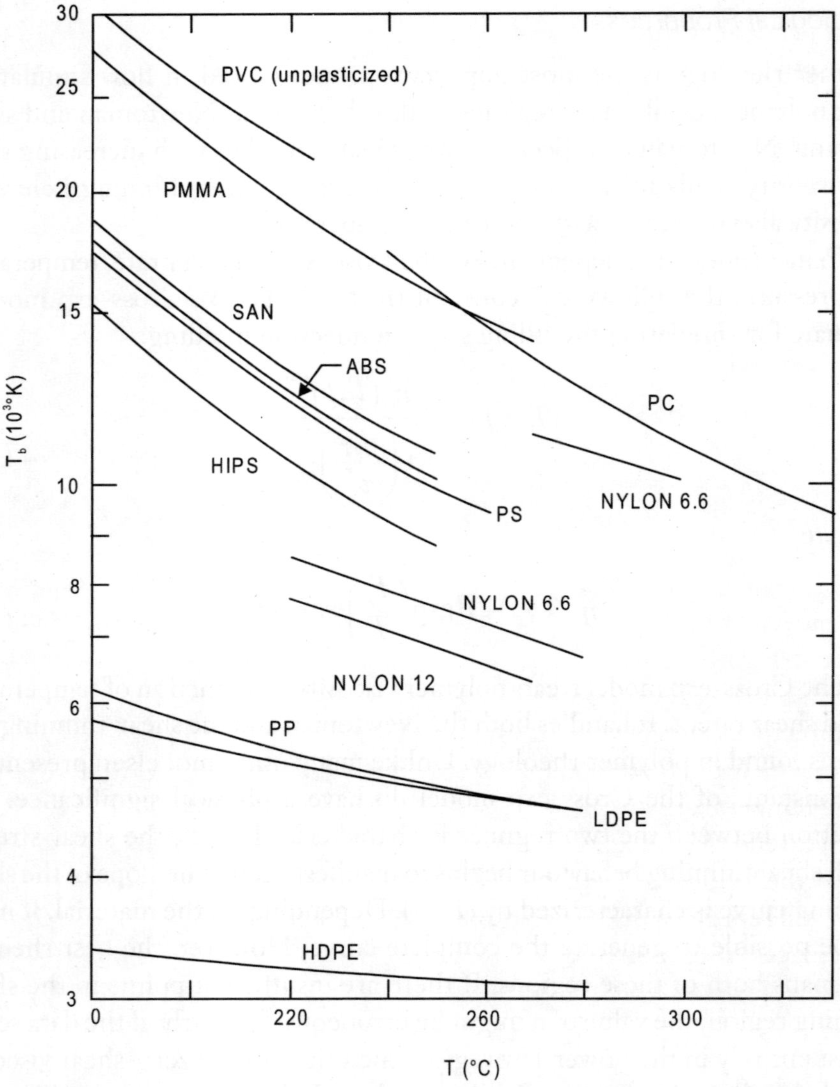

Fig. 10.1: Temperature-sensitivity factor, Tb, in Equation 2-2, versus temperature level based on results from, as discussed.

Concerning β, which characterizes the pressure dependence of η_o, note that this quantity tends to be more difficult to determine experimentally. It is not necessary to determine this constant if the analysis is used only for design purposes, such that β = 0. Since most of the viscosity measurement is performed in the processing pressure range (near 50 MPa), the measured data already includes the pressure effect. If the analysis is used for research purposes, a more elaborate procedure could be included to determine the value of β. There does tend to be a correlation between the pressure and the temperature sensitivity of η_o.

In particular, the following equation gives a reasonable representation of the results:

$$\beta \approx 4.5\times10^{-7}\frac{K}{Pa}\frac{T_b}{T^2} \qquad \text{... (10.3)}$$

That is, if T_b is known, as from Figure 10.3, then Equation 2-3 can be used to give a reasonable value for β.

Finally, concerning the constant B in Equation 2.2, which fixes the level of η_o, note that η_o at a fixed temperature and pressure strongly depends on the molecular weight, *Mw*, being approximately proportional to *Mw* raised to the 3.4 power.

$$\eta_0 = C_1 M_{14}^{\ 3.4} \qquad \text{... (10.4)}$$

Although the Cross-exp viscosity model in Equations 2.1 and 2.2 has been found adequate for simulating the filling stage, when the bulk of the polymer remains near the high injection temperature (i.e., a large hot core region with thin cold layers by the walls), it has been found to be inappropriate for simulating the post-filling stage, when the polymer undergoes substantial cooling throughout the cavity. In fact, the inadequacy of the 5-constant model can be seen directly from Figure 10.1, since this model corresponds to a constant value for Tb, which will be a poor approximation when modeling the behaviour over a large temperature range.

To extend the modeling into the post-filling stage, it is more appropriate to employ the following 7-constant (*n*, τ*, D_1, *D2*, *D3*, *A1*, *A*), Cross-WLF model, which still represents the shear-thinning behaviour according to Equation 10.1, but replaces Equation 10.2 with a more extensive model based on the WLF functional form:

$$\eta_0(T,p) = D_1 \exp\left[-\frac{A_1(T-T^*)}{A_2+(T-T^*)}\right] \qquad \text{... (10.5)}$$

where

$$T^* = D_2 + D_3 p \qquad \text{... (10.6)}$$

and

$$A_2 = A_2 + D_3 p \qquad \text{... (10.7)}$$

T* is a reference temperature and is typically taken as the glass-transition temperature of the material. That is, D_2 corresponds to the glass-transition temperature at low pressure (such as 1 atm), whereas D_3 characterizes the linear pressure dependence of *T*(p)*.

Thermal and Mechanical Properties

In addition to the shear viscosity, the simulation of polymer flow dynamics during the filling and post-filling stages requires other properties: mass density, ρ, or specific volume, υ = 1/ρ; specific heat, C_p; thermal conductivity, *k*; and transition temperature, Ttrans. To complete warpage analysis, mechanical properties are also required.

The fill time in injection molding processes is typically very short compared to the characteristic cooling time associated with the given cavity thickness. The assumptions of constant thermal properties and density of molten polymer are considered adequate for C-MOLD Filling and C-MOLD Cooling simulations.

However, a more accurate representation for the thermal properties is essential for C-MOLD Post-Filling. In particular, the compressibility of the polymer becomes a critical ingredient in modeling the material behaviour during the post-filling stage, when additional material is packed into the cavity under high holding pressure to compensate for shrinkage due to continuous cooling.

In dealing with filled polymers, it is appropriate to calculate the composite property values (subscript c below) as follows:

$$\rho_c = (1-\phi)\rho_\alpha + \phi\rho_f \qquad \text{... (10.8)}$$

where ϕ denotes the volume fraction of filler, and subscript f refers to property values of the filler; and

$$(C_p)_C = (1-\psi)(C_p)_m + \psi(Cp)_f \qquad \text{... (10.9)}$$

where ψ denotes the weight fraction of filler.

Further, the thermal conductivity of the composite is approximated as:

$$\frac{1}{k_c} = \frac{(1-\phi)}{k_\alpha} + \frac{\phi}{k_f} \qquad \text{... (10.10)}$$

although a more accurate treatment of k_c would account for the geometry and possible orientation of the filler, as detailed.

In particular, from the definitions of ϕ and ψ, it follows:

$$\psi = \frac{\phi}{\phi + (1-\phi)\dfrac{\rho_\alpha}{\rho_f}} \qquad \text{... (10.11)}$$

and

$$\phi = \frac{\psi\left(\dfrac{\rho_\alpha}{\rho_f}\right)}{1-\psi\left(1-\dfrac{\rho_\alpha}{\rho_f}\right)} \qquad \text{... (10.12)}$$

such that ψ can be determined from ϕ, or vice versa.

If the measured values of the thermal properties are not available, the suggested values of the thermal properties for various generic materials in the molten state listed in Table 10.1 can be used in the analysis.

Table 10.1: Suggested Values from [9] for Thermal Properties of Various Unfilled Generic Materials in the Molten State

Polymers	$\left(\frac{10^3}{\rho F\left(kg/m^3\right)}\right)$	$C_p\left(10^3\frac{f}{kg-{}^\circ K}\right)$	$k\left(\frac{W}{m-{}^\circ K}\right)$
ABS	1.02	2.4	0.18
POM	1.24	2.3	0.23
ASA	0.94	2.1	0.18
ESTER	1.23	2.7	0.16
HDPE	0.84	3.0	0.27
LDPE	0.77	3.4	0.31
NYLON	0.99	4.4	0.25
PC	1.06	1.9	0.24
PEI	1.08	2.1	0.22
PES	1.21	1.8	0.18
PMMA	1.04	2.3	0.21
PP	0.77	3.1	0.15
PPO	0.95	1.7	0.19
PS	0.94	2.1	0.15
PSF	1.09	1.9	0.26
PVC	1.32	1.8	0.18
SAN	0.95	2.1	0.17

Specific Volume (pvT Diagram)

This data is used to obtain information about the compressibility and volumetric expansion of polymeric materials. If the data are obtained under equilibrium, they are fundamental thermodynamic properties of the material. The data are seen to reflect transitions as the material moves from one physical state to another.

The pvT behaviour of an amorphous thermoplastic at atmospheric pressure can be summarized as in Figure 10.2(a). The kink in the curve characterizes the glass-transition temperature, which is a function of pressure. The slopes of the υ-T curve, b2m and b2s, denote the bulk thermal-expansion coefficients in the liquid and solid phases, respectively. On the other hand, a schematic diagram for the pvT behaviour of a semi-crystalline material under atmospheric pressure is shown in Figure 10.2(b). The abrupt transition in the υ-T curve is associated with the crystallization temperature, which is also a function of pressure.

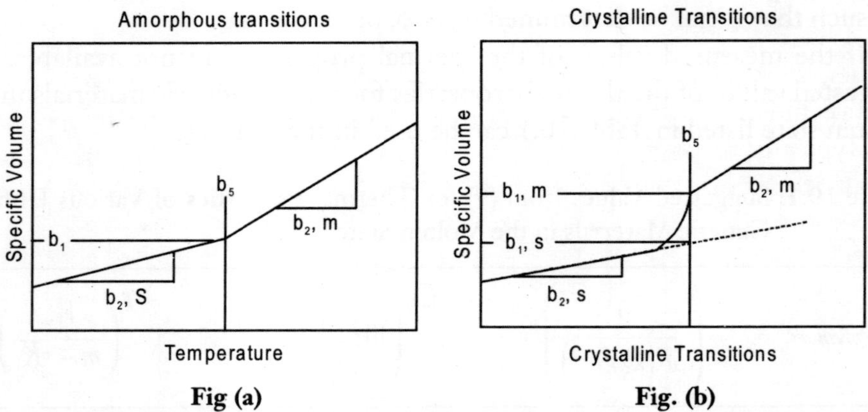

Fig. 10.2: (a) Schematic pvT diagram for amorphous polymers. (b) Schematic pvT diagram for crystalline polymers

To incorporate the dependence of specific volume (or mass density) on temperature and pressure, we have found the following 2-domain, modified Tait equation to be adequate:

$$v(T,p)=v\left(1-Cln\left(1+\frac{p}{B(T)}\right)\right)+v_t(T,p) \qquad \text{... (10.13)}$$

with

If $T > T_t(p)$

$$v_0 = b1_\alpha + b2_\alpha \overline{T}$$
$$E(T)=b3_\alpha \exp(-b4_\alpha \overline{T}) \qquad \text{... (10.14)}$$
$$v_t(T,p)=0$$

If $T > T_t(p)$

$$v_0 = b1_s + b2_s T$$
$$B(T)=b3_m \exp(-b4_s T) \qquad \text{... (10.15)}$$
$$v_t(T,p)=b7\exp(b8T-b9p)$$

where C = 0.894 (universal constant) and $\overline{T} = T - b_5$.

In addition, the transition temperature is assumed to be a linear function of pressure:

$$T_t(p) = b_5 + b_6 p \qquad \text{... (10.16)}$$

The transition temperature is the glass-transition temperature, *Tg*, for amorphous polymers and the crystallization temperature, *Tc*, for crystalline polymers.

Specific Heat, Cp

A schematic specific-heat diagram for amorphous polymers is shown in Figure 10.3(a). The inflection point corresponds to the glass-transition temperature, Tg. Typically, the variation in the specific-heat of amorphous materials can be about 50-70 per cent between the processing temperature and room temperature.

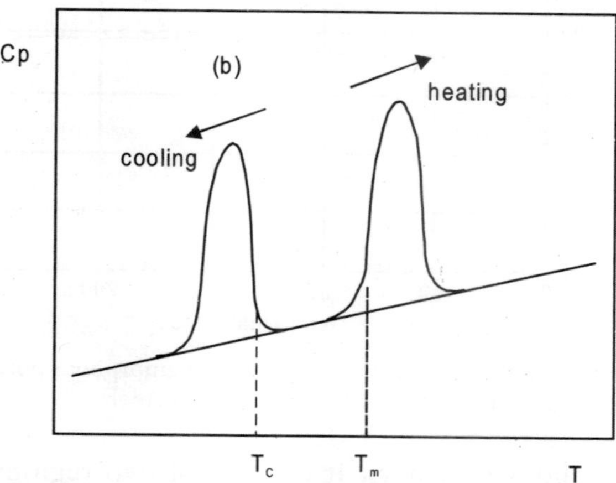

Fig. 10.3: (a) Schematic specific heat diagram for amorphous polymers. (b) Schematic specific heat diagram for crystalline polymers.

A schematic specific-heat diagram for crystalline polymers is shown in Figure 10.3(b). Note that the area under the peak and above the straight base-line represents the latent heat released during the crystallization process. A hysteresis is observed between the melting and the crystallization peaks, due to supercooling effects. Additionally, because the crystallization process depends on the cooling-rate, the crystallization peak shifts to lower temperatures at higher cooling rates.

Commonly available specific heat data are measured under a heating scan. The polymer, however, undergoes high cooling rate (quenching) during the injection molding process. While this will not affect the transitions of amorphous polymers significantly, the transition shifts for semi-crystalline materials can be dramatic. Accordingly, specific heat data measured under a high cooling rate is desirable for C-MOLD analyses.

Thermal Conductivity, k

Thermal conductivity is one of the most important properties that influence the injection molding pressure prediction. Similar to specific heat, thermal conductivity also exhibits variations from room temperature to processing temperature. Shown in Figure 10.4(a) is a schematic thermal conductivity

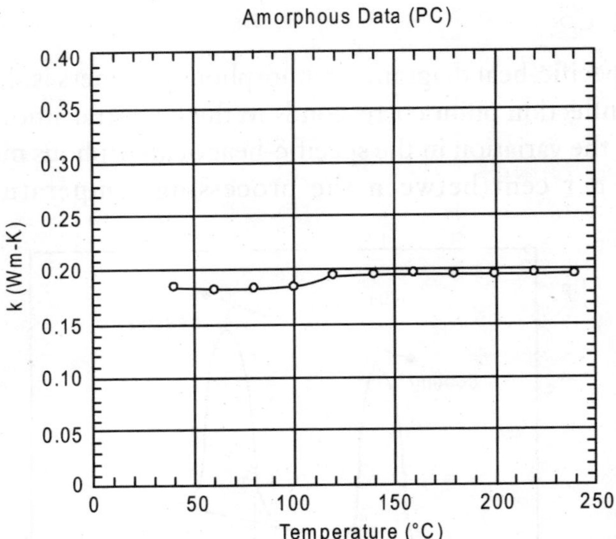

Fig. 10.4: (a) Schematic thermal conductivity diagram for amorphous polymers. (b) Schematic thermal conductivity diagram for crystalline polymers

diagram for amorphous polymers. It consists of two regions in a piece-wise linear manner. Thermal conductivity remains constant when temperature is above Tg and decreases linearly when temperature is below Tg. The slope of the line below Tg is about 0.04 W/m - K per 100 °C and is reasonably universal to all pure, amorphous polymers.

Thermal conductivity for semi-crystalline polymers, however, shows an abrupt increase when temperature drops below the crystallization temperature, Tc. This is because of the appearance of the crystalline phase, which creates regions of high thermal conductivity. Figure 10.4(b) gives a schematic thermal conductivity diagram for semi-crystalline polymers.

Transition Temperature, Ttrans

The transition temperature is used by C-MOLD analyses as the polymer freeze temperature. This temperature corresponds to the glass-transition temperature, Tg, for amorphous polymers and to the crystallization temperature, Tc, for crystalline polymers. In theory, the transition points determined from pvT, specific heat, thermal conductivity, and viscosity measurements should be identical. However, typical data for these transition points are not so close, due to the limitations of today's measurement techniques. Particularly for semi-crystalline materials, rate dependence tends to create a significant spread in the transitions measured by the various instruments. At this moment, the best way to determine transition temperature is by a DSC cooling scan.

Transition temperature, which can be observed in several measurements, is a physical property of a polymer. One misconception is that C-MOLD simulations require the no-flow temperature to calculate frozen-layer thickness and frozen-in stress. In fact, polymer (or stress) does not freeze at the no-flow temperature, but rather at the melt-to-solid transition temperature. Using the no-flow temperature, which is typically much higher than the transition temperature, instead of the polymer freeze temperature will over-predict residual stress and pressure in the cases of slow filling and post-filling of thin parts. As mentioned in the previous sections, the melt-to-solid transition is cooling-rate dependent. More complicated models with crystallization kinetics can be incorporated in C-MOLD simulations to determine the transition temperature as a function of cooling rate.

If the data is not readily available, there are correlations between the transition temperature and many other measurements. For example, Tg and the melting temperature, Tm, can be found easily in handbooks and are provided by resin suppliers. Heat deflection temperature (HDT, ASTM D648, at 66 psi) and Vicat temperature (ASTM D1525) are often used as measures of the temperature resistance of polymers. All of these data are determined under heating conditions. However, the data agrees reasonably well with the transition temperature measured under DSC heating scans. The transition temperature under DSC cooling scan, which is used by the C-MOLD simulations, can be estimated by a simple correlation:

$$\left.\begin{aligned} T_{trans} &= \text{handbook } T_g \text{ for amorphous materials} \\ T_{trans} &= \text{handbook } T_m - 30°\text{C for amorphous materials} \end{aligned}\right\} \quad \ldots (2.17)$$

or

$$\left.\begin{aligned} T_{trans} &= \text{Vicat temperature for amorphous materials} \\ T_{trans} &= \text{Vaicat temperature for amorphous materials} \end{aligned}\right\} \quad \ldots (2.18)$$

or

$$\left.\begin{aligned} T_{trans} &= \text{heat deflection temperature for amorphous materials} \\ T_{trans} &= \text{heat deflection temperature} -30°\text{C for crystalline materials} \end{aligned}\right\} \quad \ldots (2.19)$$

Note: One simple rule of thumb is to pick the lowest transition temperature from all sources of data available.

See Table 10.2 for typical transition temperatures of generic grades in handbook T_g and T_m.

A single transition temperature is easy to determine from the DSC cooling scan for homopolymers. Multiple transitions can be found in polymer blends,

Table 10.2: Typical transition temperatures for each generic grade from handbook Tg and Tm, and the correlation in Equation 2-17

Generic Type	*Class*[1]	*Tg* (°C)	*Tm* (°C)	*Ttrans* (°C)
ABS	A	105		105
ASA	A	104		104
HIPS	A	100		100
PC	A	144		144
PEI	A	220		220
PES	A	230		230
PMMA	A	100		100
PS	A	100		100
PVC	A	80		80
SAN	A	100		100
POM (ACETAL)	C		160	130
HDPE	C		135	105
LDPE	C		120	90
PA 6 (Nylon 6)	C		215	185
PA 66 (Nylon 66)	C		265	235
PA 6, 12 (Nylon 6, 12)	C		220	190
PBT	C		230	200
PET	C		250	220
PP	C		165	135
PPS	C		290	260
PU		120		120

1. A, amorphous; C, crystalline.

and it is difficult to determine which transition is to be used in the simulation. A classic case is the Xenoy resin, where the polymer flows below the crystallization temperature of the PBT component. An additional test for Vicat temperature is typically required. The closest DSC cooling scan transition below the Vicat temperature is the one to use. It is very helpful if you know the composition of the blend before the test.

Mechanical Properties

In warpage analysis, additional properties, such as the thermal expansion coefficient and mechanical properties, are required. The following models can be used, based on different degrees of complexity:

Thermal expansion coefficient:

- ❖ Isotropic thermal expansion coefficient
- ❖ Transversely-isotropic thermal expansion coefficient

Mechanical properties:

- Isotropic elastic model
- Transversely-isotropic elastic model
- Isotropic viscoelastic-elastic model with WLF form of shift function
- Isotropic viscoelastic model with WLF form of shift function
- Transversely-isotropic viscoelastic model with WLF form of shift function

A viscoelastic model has to be employed when the stress relaxation data are required as a function of time and temperature. For example, a part under load and subject to a heating cycle can deform as a stress relaxation. The changes are due to stress relaxation over the period of time and can only be modeled by viscoelastic models. However, it is very difficult to generate such stress relaxation (or creep) data, and limited data for model coefficients are available from any source. For most practical applications, it is found that the transversely-isotropic elastic model is adequate to predict the warpage of the part, and this is the model recommended for use. Transversely-isotropic means that properties transverse to the flow direction are isotropic. This model will account for variation in properties in the flow direction and transverse to the direction of flow. The following properties are required, based on the recommended model:

Thermal expansion coefficient in the flow direction, α_1, and in the transverse direction, α_2.

Modulus of elasticity in the flow direction, E_1, and in the transverse direction, E_2.

Poisson's ratio in the longitudinal (flow)-to-transverse direction, v_{12}, and in the transverse-to-thickness direction, v_{23}.

Shear modulus, G.

However, if the transversely-isotropic mechanical properties are not available, a very simple isotropic-elastic model can be used. The isotropic-elastic properties for many commercial grades can be easily found in many handbooks and are also readily available in the C-MOLD Database. A crude warpage analysis can be performed with this simple model, but this prediction is sometimes different from reality, since orientation effects are neglected. The procedure to derive model coefficients for such a simple model is described below:

Isotropic thermal expansion coefficient, α, can be obtained from the data sheet provided by resin suppliers.

Isotropic elastic modulus, E, and Poisson's ratio, v, can be obtained from the data sheet provided by resin suppliers.

If Poisson's ratio is not known, it can be derived from the isotropic elastic modulus and the bulk modulus determined from the 2-domain, modified Tait pvT coefficients, as given below. Note that this correlation relies on the accuracy of available data. Large error may be encountered in the estimation. For most polymers, Poisson's ratio is between 0.2 and 0.4. For most isotropic materials, Poisson's ratio is between zero (no lateral contraction) and 0.5 (constant volume deformation).

$$\text{Bulk Moduls } \kappa = \frac{E}{3(1-2v)}$$

$$= -v\frac{\partial p}{\partial v}|p=0, T=25°\text{C}$$

$$= \frac{b3_s \exp[-b4_s(298.16)-b4]}{0.0894} \quad \text{... (2.20)}$$

Suggested Polymer Testing Procedures

These procedures are guidelines to generating consistent data for CAE programs. The tests conform to ASTM procedures wherever such standards exist. Such tests are described in minimal detail so as not to duplicate information provided in the standard test methods (STMs). Non-standard tests are described in greater detail.

The equipment used for these tests is fairly common in modern plastics testing facilities. This section therefore assumes familiarity with such test equipment. No attempt is made to describe the operation or use of the instruments. Refer to instrument manuals or contact the manufacturer if such details are desired.

To ensure traceability of data, all transducers and measurement devices must be calibrated to a national standard such as NIST. We also recommend that regular verifications be carried out to ensure that the instruments maintain calibration. Round-robins are an excellent method for determining precision and bias of the laboratory instruments.

Specimen Preparation

It is recommended that the same batch of material be used for all of the measurements. If batch-to-batch variability is suspected, then a blend of statistically sampled lots should be used. Pre-processing (e.g., drying) is very important and should be performed according to the recommendations of the resin supplier.

For measurements of mechanical properties, the specimens should be taken from injection-molded plaques molded at normal processing conditions. High injection speeds are specified to maximize any anisotropic effects that may occur. The plaques must be nominally five inches (125 mm) square and one-eighth inch (3 mm) thick. Specimens must be cut from regions of the plaque where the flow direction is clearly defined, to characterize the effect of the anisotropy adequately (see Figure 10.2-5). Specimens should not be taken close to gate areas. Specimens should not be conditioned, annealed, irradiated, or subjected to any other treatment that will change their properties as molded.

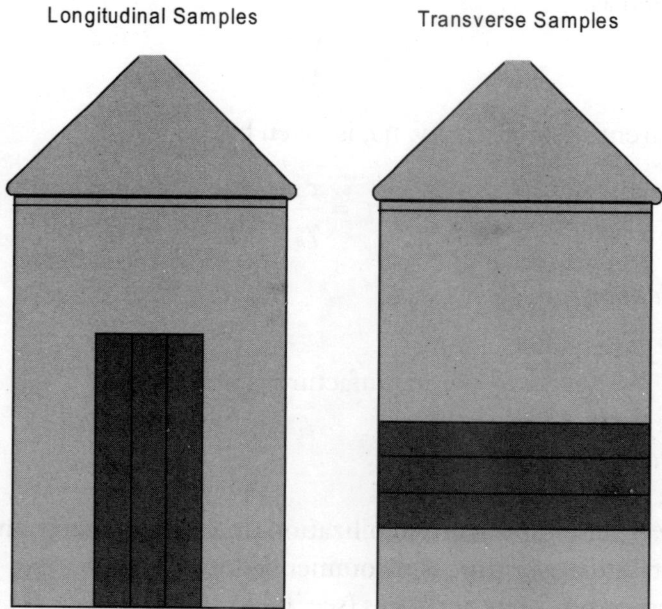

Fig. 10.5: Geometry of test specimen

Straight edge specimens are typically cut using machining operations such as high-speed milling. This is one of the most important steps in the measurement. The test specimen ideally must meet the dimensional requirements according to the ASTM specifications. It should be free from cold working or heat distortion, with no tool marks or incipient cracks. It should be cut in a manner that minimizes the formation of any stress in the part due to the cutting operation.

Rheology

Melt Viscosity

Method: ASTM D3835, Rheological Properties of Thermoplastics with a Capillary Rheometer.

Description of Method: Viscosity is measured using a capillary rheometer (see Figure 10.6). The rheometer piston pushes the material specimen at a constant temperature and flow rate, Q, through a cylindrical die of known length, L, and diameter, D. The apparent shear rate is defined as:

$$\gamma_a = \frac{32Q}{\pi D^3} \quad \text{... (2.21)}$$

By measuring the pressure drop, Δp, across the die, the wall-shear stress, τw, is calculated as:

$$\tau_w = \frac{\Delta pd}{4L} \quad \text{... (2.22)}$$

The apparent shear viscosity, ηa, is given by:

$$\eta_a = \frac{\tau_w}{\gamma_a} \quad \text{... (2.23)}$$

Test Specification

- *Specimen form*: pellets
- *Specimen pre-processing*: per manufacturer's instructions
- *Die entry angle*: 180°
- *Die diameter*: 1 mm
- *Die L/D*: 15 to 20
- *Dwell time*: per temperature stabilization time of rheometer; measurement of flow stability over time is recommended.
- *Test temperatures*: 3 temperatures (see Table 10.2-3)
- *Shear rates*: 3 per decade, 10-10,000/s (see Table 10.2-3)

General Guidelines

Use the maximum L/D that still produces measurements in the desired shear-rate range. The selection of capillary die might vary due to the limitations of the instrument and the viscosity of the polymers.

Obtain viscosity at three different temperatures: two to span the processing temperature range, and one below the minimum processing temperature, if possible. A wider temperature range will give a better prediction of the temperature sensitivity of polymers. It may be difficult to measure viscosity below the processing range for some polymers that have a very narrow processing temperature range, such as PA and PU. All measurements then have to be performed within the processing temperature range.

Make at least six measurements (two per decade change of shear rate) for each melt temperature. See Table 10.3 for suggested shear-rate ranges of different generic grades that cover the Newtonian and power-law viscosity regions.

Table 10.3: Suggested temperature and shear-rate ranges

Generic Name	*Temperature (°C)*	*Shear Rate (1/s)*
ABS	200-260	1-1,000
POM	180-220	10-10,000
ASA	200-240	1-1,000
HDPE	190-260	1-1,000
NYLON	230-280	10-10,000
PC	270-320	10-10,000
PEI	360-400	10-10,000
PES	310-400	10-10,000
LDPE	180-240	10-10,000
PMMA	230-300	1-1,000
PP	190-240	1-1,000
PPO	250-300	1-1,000
PS	180-260	1-1,000
PVC	180-210	1-1,000
SAN	220-260	1-1,000

Make Weissenberg-Rabinowitsch correction.

Bagley correction for juncture losses can be performed for capillary viscosity data if more than one die (L/D value) is used, as shown in Figure 10.6.

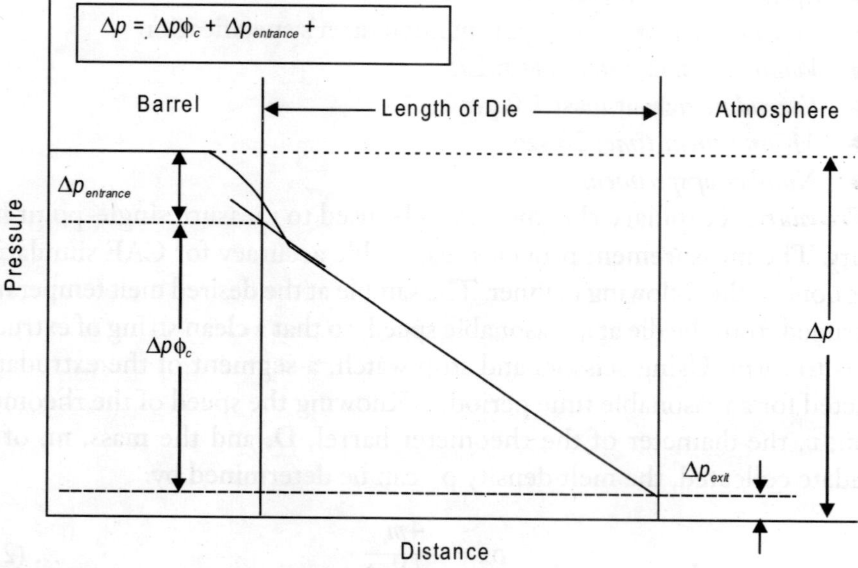

Fig. 10.6: A two-barrel capillary rheometer

Thermal Properties

Solid Density

Method: ASTM D792, Specific Gravity (Relative Density) and Density of Plastics by Displacement, Method A.

Description of Method: Solid density ρ is measured using a hydrostatic balance apparatus. The specimen is weighed in air, then immersed in a liquid, and its weight loss upon immersion is determined.

Test Specification:

- *Specimen form*: molded plaques
- *Specimen pre-processing*: maintain samples dry as molded
- *Immersion liquid*: water
- *Number of specimens*: 3

Melt Density

Method: Extrusion flow-rate measurement.

Description of Method: Melt density is measured using a capillary rheometer. The rheometer piston pushes the melt at processing temperature, at constant rate and temperature, through a die of known length and diameter. The mass of a sample of extrudate from a measured time interval is determined. The polymer melt pressure is recorded.

Test Specification:

- *Specimen form*: pellets
- *Specimen pre-processing*: per manufacturer's specification
- *Volumetric flow rate*: 0.04 mL/s
- *Extrudate size*: at least 1.0 gm
- *Measurement time*: 20 sec
- *Number of specimens*: 3

Procedure: A capillary rheometer can be used to measure single-point melt density. The measurement provides reasonable accuracy for CAE simulations and is done in the following manner. The sample at the desired melt temperature is extruded from the die at a reasonable speed, so that a clean string of extrudate begins to form. Using scissors and stop watch, a segment of the extrudate is collected for a reasonable time period, t. Knowing the speed of the rheometer piston, v, the diameter of the rheometer barrel, D, and the mass, m, of the extrudate collected, the melt density ρ_m can be determined by:

$$\rho_m = \frac{4m}{\pi D^2 \upsilon t} \quad \ldots (2.24)$$

pvT

Method: High-pressure dilatometry.

Description of Method: This test method covers measurement of the affect of temperature and pressure on the specific volume (pvT) of polymers. The specimen is heated in an enclosed cell and the change in its volume when subjected to a range of pressures is measured. Measurement techniques differ, mainly in the method used to apply the pressure. Both the direct method (piston and cylinder as shown in Figure 10.7) and the indirect method (high-pressure dilatometer, as shown in Figure 10.8) are found to yield pvT data of sufficient accuracy for the simulation programs.

Apparatus: Method A, also called the direct method, applies the pressure directly to the specimen, using a piston and cylinder set-up. Piston deflections are used to measure volume change. Since the volume of the cell is known, the absolute specific volume can be measured by this technique.

Method B, also called high-pressure dilatometry, applies the pressure to the specimen by means of a confining fluid. The specimen and confining fluid are enclosed in a chamber fitted with a bellows. The deflection of the bellows is used to measure the change in volume. Since this method measures volume changes, a reference density measured by an independent method such as ASTM D792 is needed.

Test Specimens: Specimens may be in the form of polymer pellets or may be cut from molded plaques. Since polymer pellets often contain voids within them, it is recommended that the specimens be cut from injection- or compression-molded samples. The quantity of specimen should not exceed 1-3 gms.

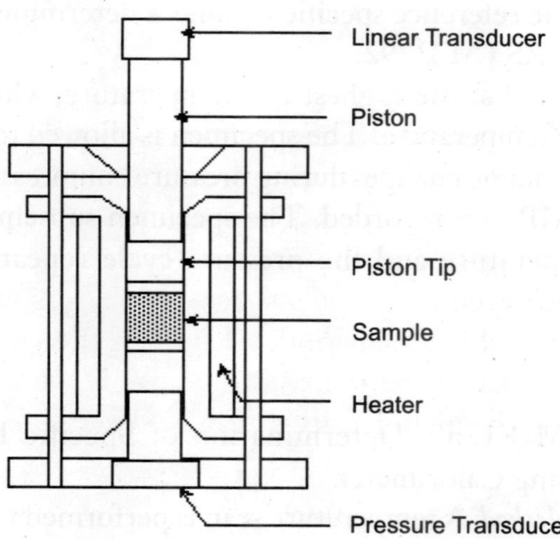

Fig. 10.7: Direct pvT measurement

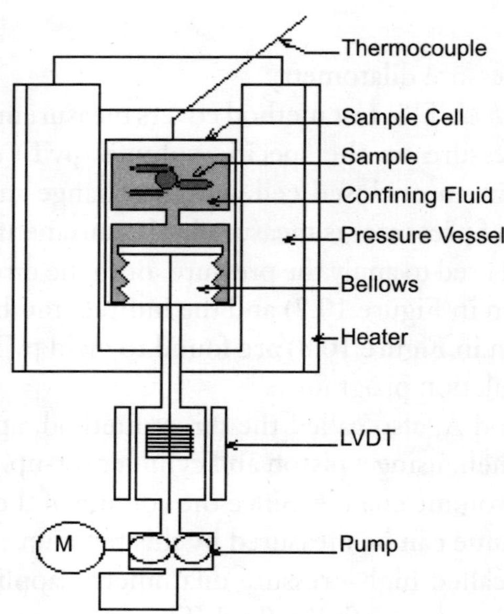

Fig. 10.8: Indirect pvT measurement

Conditioning: Specimens should be well dried prior to testing. A minimum of eight hours at 50 °C in a dessicant drier is recommended.

Procedure: The specific specimen loading procedures depend on the apparatus, and the manufacturer-recommended procedures should be used. In addition, the high-pressure dilatometry technique, being a relative method, requires the acquisition of a series of isothermal measurements as described below, at close to the ambient conditions, to which the reference specific volume will be mapped. The reference specific volume is determined by displacement methods following ASTM D792.

The test is started at the highest test temperature, which is typically the normal processing temperature. The specimen is allowed to equilibrate at the test temperature. Volume changes during pressure compression cycles between 10 MPa and 200 MPa are recorded. The specimen subsequently is cooled to the next test temperature and the pressure cycle repeated, until ambient temperatures are achieved.

Specific Heat

Method: ASTM E1269, Determination of Specific Heat Capacity by Differential Scanning Calorimeter.

Description of Method: A temperature scan is performed using empty pans in both chambers to establish a baseline. A specimen of mass, m, is then loaded into one of the pans, and the scan is repeated. The specific heat, C_p, is calculated from the difference in heat flow, ΔQ, between the baseline and the specimen needed to change the temperature by an amount, ΔT:

$$Cp = \frac{\Delta Q}{m\Delta T} \qquad \ldots (2.25)$$

Test Specifi7cation:

- *Specimen form*: pellets or plaques
- *Specimen pre-processing*: per resin suppliers instruction
- *Initial temperature*: mid-process temperature
- *Final temperature*: 50 °C
- *Cooling rate*: 20 °C/min
- *Equilibrium time*: 1 min
- *Sample pan material*: Aluminum
- *Sample pan type*: Standard or volatile
- *Purge gas*: 99.99 per cent pure N2
- *Purge gas flow rate*: 30 cm3/s

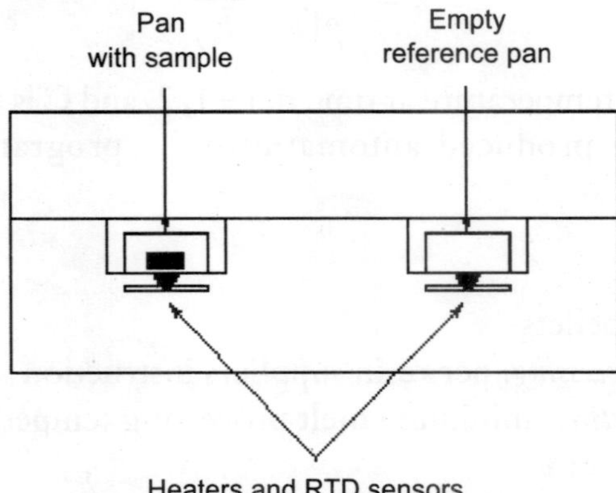

Fig. 10.9: Differential scanning calorimeter (DSC) measurement

General Guidelines:

❖ Specimens, sample pans and covers must not be touched by hand. Contamination can lead to spurious peaks. Moist specimens may show a peak at 100 °C.

❖ Temperature ranges for the measurement must cover the region between the processing temperature and the solid state. Avoid excessive high temperature at the beginning of the cooling scan, which may lead to material degradation and jeopardize the accuracy of the entire measurement.

❖ The melt specific heat is the average specific heat over the processing temperature range.

❖ In the presentation of tabulated specific heat data as a function of temperature, more data points must be included near transitions to adequately map these regions. In other areas, data points at intervals of 10 °C to 20 °C are adequate.

Thermal Conductivity

Method: Transient Line-Source Technique.

Description of Method: A probe is inserted into a molten specimen at its processing temperature. The probe contains a line-source heater running the length of the probe, and a temperature sensor in the middle of the probe. When thermal equilibrium is attained, a known amount of heat, Q, is supplied to the line-source heater, and the temperature rise in the sensor is recorded over a period of time. The thermal conductivity, k, is calculated from the following equation:

$$k = \frac{QC\,In\left(\frac{t_2}{t_1}\right)}{4p(T_2 - T_1)} \quad \ldots (2.26)$$

where Ti is the temperature at time ti, i = 1, 2, and C is the probe constant. Cooling scans are produced automatically by programming a range of temperatures.

Test Specification:

- *Specimen form*: pellets
- *Specimen pre-processing*: per resin suppliers instruction
- *Loading temperature*: minimum melt processing temperature
- *Probe length*: 50 mm
- *Settling time*: 45 s
- *Acquisition time*: 40-55 s

Apparatus: The K-System II from AC Technology is recommended to make measurements of thermal conductivity. This technique requires a very short measurement time and can significantly reduce the risk of thermal degradation.

Procedure: Load the specimen quickly, tamping frequently to eliminate entrapped air. Alternatively, use a molded rod of the resin.

Use well-cleaned probes, free from oils and residual polymer. Make measurements as soon as temperature stability is attained.

The melt thermal conductivity should be measured at the lowest temperature in the processing range.

For tabulated data, scan thermal conductivity from the processing temperature to the solid-state temperature. Measurements can be made at intervals of 20 °C to 40 °C, except near the transition temperature. More data

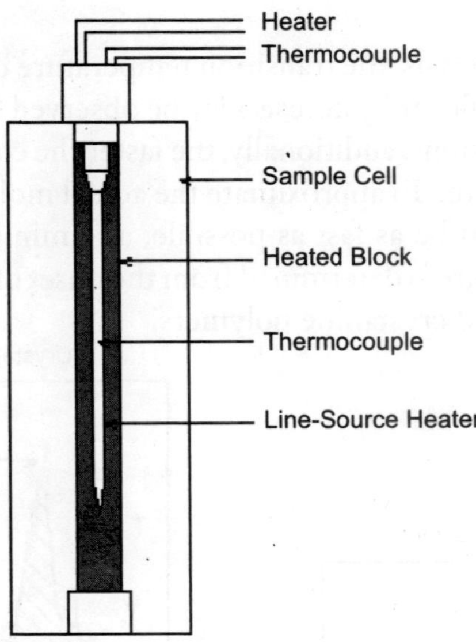

Fig. 10.10: Transient line-source thermal conductivity measurement

points, at intervals of 10 °C to 20 °C, are desirable around the transition temperature. A dead-weight compression system is required for solid state measurements.

Transition Temperature

Method: ASTM D3418, Transition Temperatures of Polymers by Thermal Analysis.

Description of Method: The cooling mode is to be used for these measurements. The transition temperatures are determined from the extrapolated onset, Tf, for crystalline transitions and from the glass transition temperature, Tg, as defined by the midpoint temperature, Tm, for amorphous materials.

Test Specification:

- *Specimen form*: pellets or plaques
- *Specimen pre-processing*: per resin suppliers instruction
- *Initial temperature*: mid-process temperature
- *Final temperature*: 50 °C
- *Cooling rate*: 20 °C/min
- *Equilibrium time*: 1 min
- *Sample pan material*: aluminum
- *Sample pan type*: standard or volatile
- *Purge gas*: 99.99 per cent pure N_2
- *Purge gas flow rate*: 30 cm3/s

General Guidelines

For semi-crystalline materials, the transition temperature depends on direction and cooling rate. A significant hysteresis may be observed between the melting and crystallization transitions; additionally, the faster the cooling rate, the lower the transition temperature. To approximate the actual molding conditions, the DSC cooling rate should be as fast as possible, at a minimum of 20 °C/min. The transition temperature is determined from the onset of the curve, as shown below for amorphous and crystalline polymers.

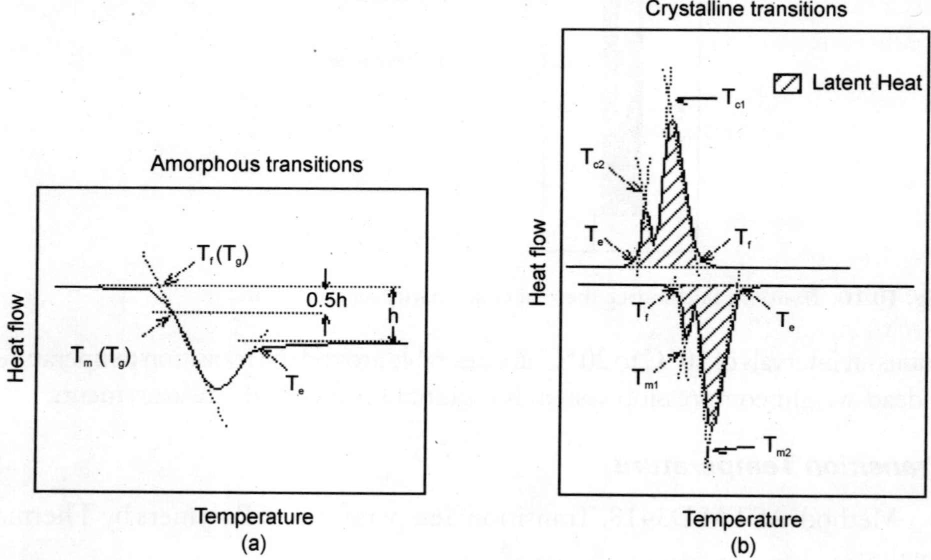

Fig. 10.11: Transition temperature of amorphous (left) and crystalline (right) polymers

A single transition temperature is easy to determine from the DSC cooling scan for homopolymers. Multiple transitions can be found in polymer blends, and it is difficult to determine which transition is used in the simulation. A classic case is the Xenoy resin, where the polymer flows below the crystallization temperature of the PBT component. An additional test for Vicat temperature (ASTM D1525) is required. The closest DSC cooling scan transition below the Vicat temperature is the appropriate one to use. It will be very helpful if the composition of the blend is known before the test.

Mechanical Properties

The most common instrument for performing mechanical property measurements is a Universal Testing Machine (UTM). Constant rate of elongation is the method of loading. Extensometers are to be used for all displacement measurements. Data are to be gathered only in the elastic region.

The warpage analysis requires data to be measured on specimens in the direction of flow and transverse to the flow. Some materials are essentially isotropic in nature, permitting a simpler isotropic elastic model to be used. Properties are measured on specimens cut in the flow direction; alternatively, ASTM dogbones may be used.

Fiber-filled materials, modified polymers, and some semi-crystalline resins will exhibit anisotropic behaviour, requiring the application of a transversely-isotropic model. Specimens are cut in the flow and transverse to flow directions. If the variation of properties is within the limits of accuracy of the test technique, the directional data should be averaged and reported as a single isotropic value.

Modulus of Elasticity

Method: ASTM D638, Tensile Properties of Plastics.

Description of Method: Elastic modulus, E, is defined as the ratio of stress, σ, to strain in the direction of load, εL, within the elastic range of the material. Specimens are placed in the self-aligning jaws of a tensile testing machine equipped with an extensometer (see Figure 10.12), and load is applied. Take care to ensure that all data used for these measurements are taken in the elastic region of the stress-strain curves.

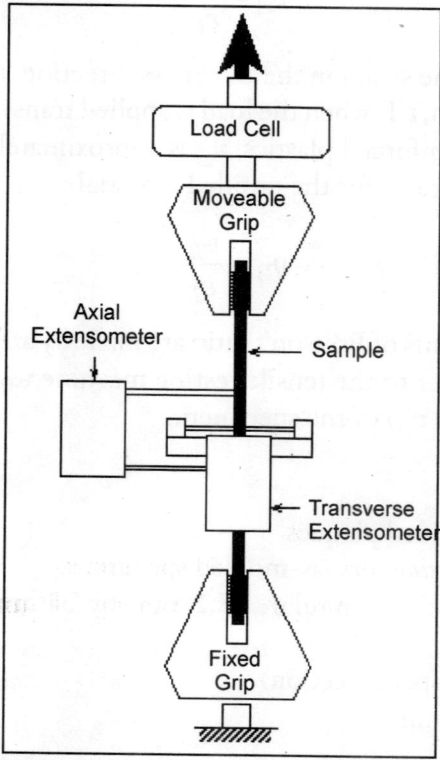

Fig. 10.12. Tensile test set-up

For the anisotropic model, elastic modulus in the flow direction, E_1, and in the direction transverse to flow, E_2, are to be measured. The simpler isotropic model requires a single elastic modulus, measured in the flow direction, E.

Test Specification:

- *Specimen form*: molded plaques
- *Specimen pre-processing*: dry-as-molded specimens
- *Specimen preparation*: high speed milling
- *Specimen geometry*: rectangular; 12.7 by 51 mm gage length, full thickness
- *Specimens tested*: 5 (isotropic) or 3 (per direction)
- *Strain rate*: 0.1 1/min

Poisson's Ratio

Method: ASTM E 132, Poisson's Ratio at Room Temperature

Description of Method: Poisson's ratio, v, is defined as the ratio of the lateral contraction strain to the longitudinal strain.

v_{12} is the ratio of the strain in the transverse direction, εT, to the strain in the longitudinal direction, εL, when the load is applied in the flow direction.

$$v_{12} = \frac{\varepsilon_T}{\varepsilon_L} \qquad \text{... (2.27)}$$

v_{23} is the ratio of the strain in the thickness direction, εTh, to the strain in the transverse direction, εT, when the load is applied transverse to the direction of flow. For weakly reinforced plastics, v_{23} is approximately equal to v matrix, which is the Poisson's ratio for the unfilled material.

$$v_{23} = \frac{\varepsilon_{Th}}{\varepsilon_T} \qquad \text{... (2.28)}$$

Direct measurements of Poisson's ratio are made by attaching an additional transverse extensometer to the tensile testing machine to measure the change of width (thickness for v_{23}) of the specimen.

Test Specification:

- *Specimen form*: molded plaques
- *Specimen pre-processing*: dry-as-molded specimens
- *Specimen geometry*: rectangular; 15.2 mm by 51 mm gage length, full thickness
- *Specimens tested*: 3 (per direction)
- *Strain rate*: 0.1 1/min

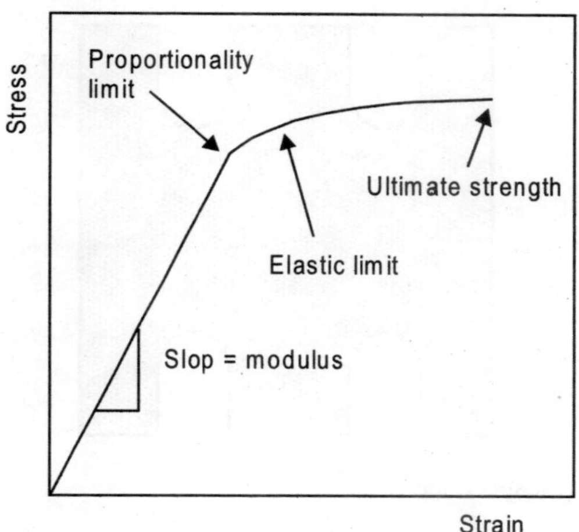

Fig. 10.13: Stress-strain curve

In-Plane Shear Modulus

Method: Rail Shear Test.

Description of Method: This technique uses a test fixture mounted in a basic lever loading frame (see Figure 10.15). Specimens are machined from the specimen and notched to fit the fixture, with round end profiles designed to minimize end effects and produce a homogeneous stress field. The specimen is mounted in the test fixture and bolted in position between the rails. An extensometer is placed on the specimen between the rails, with needle-point arms at 45° to the direction of loading. Load is applied to the specimen and shear strain is monitored by the extensometer. This is repeated in each of four positions (front right and left, back right and left) on the specimen. Shear modulus, G, is calculated from the data for each position as:

$$G = \frac{\tau}{\gamma} \qquad \text{... (2.29)}$$

where $\tau = P/2$, $\gamma = 2\varepsilon_{45°}$, P = load, A = area of one web, and $\varepsilon_{45°}$ = 45° strain. The results from each position are averaged.

Test Specification:

- *Specimen form*: molded plaques
- *Specimen pre-processing*: dry as molded
- *Specimen geometry*: 57.2 mm by 127 mm gage length, full thickness, as shown in Figure 10.2-14
- *Specimens tested*: 2
- *Crosshead speed*: 5 mm/min

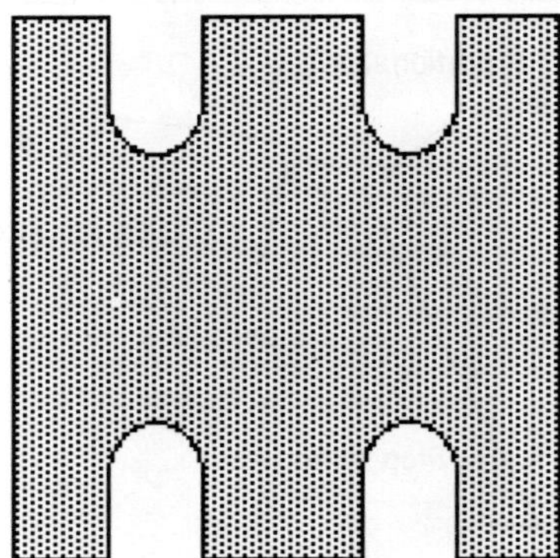

Fig. 10.14: Rail shear test specimen

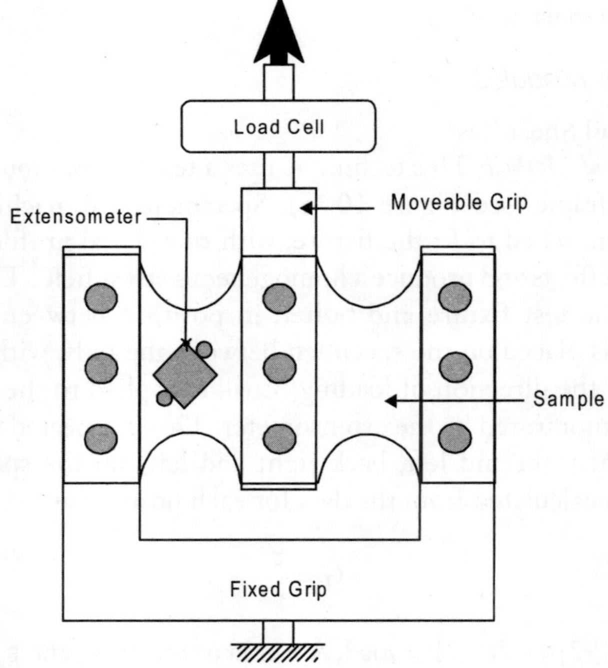

Fig. 10.15: Rail shear fixture

In many applications, it is found that the anisotropy in shear modulus is negligibly small, and the choice of the specimen has a less critical effect on the measurements.

Thermal Expansion Coefficients

Method: ASTM D 696, Coefficient of Linear Thermal Expansion of Plastics.
Description of Method: A quartz tube dilatometer (see Figure 10.16) is used.

A specimen is prepared and placed at the bottom of the outer dilatometer tube with the inner one resting on the specimen. The dial gauge, firmly attached to the outer tube, is placed in contact with the top of the inner tube so as to measure variations in the length of the specimen with changes in temperature. Temperature changes are brought about by immersing the outer tube in a liquid bath at the desired temperature. If the temperature of a specimen of initial length, L, is changed by ΔT, and the specimen undergoes a change in length, ΔL, the thermal expansion coefficient, α, is given by:

$$a = \frac{\Delta L}{L_0 \Delta T} \qquad \ldots (2.30)$$

Test Specification:

- ❖ *Specimen form*: molded plaques
- ❖ *Specimen pre-processing*: dry as molded
- ❖ *Specimen geometry*: rectangular; 8 mm by approximately 55 mm, full thickness
- ❖ *Specimens tested*: 2 (per direction)
- ❖ *Temperature range*: 0 to 60 °C

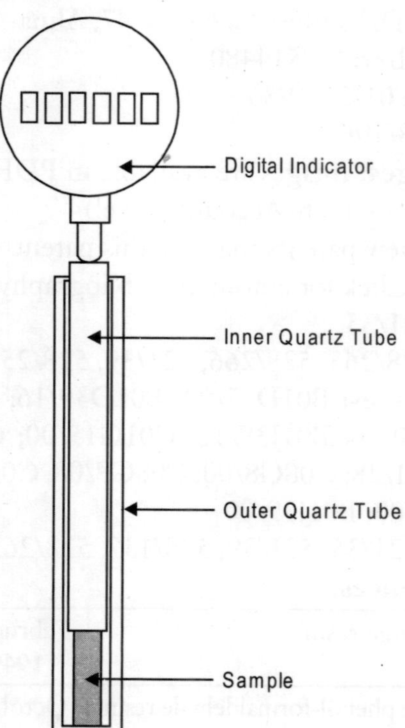

Fig. 10.16: Quartz tube dilatometer

10.9 Polymeric Material Adapted for Physico-Chemical Separation of Substances

Document Type and Number: United States Patent 4567207

Abstract: A polymeric material comprising a solid dispersion of a porous three-dimensional polymer with a pore diameter of 0.002 to 10 ìm and a permeability coefficient of from 2×10^{-7} to 2×10^{31} 2 cm?sec^{-1}. This material is produced by reacting formaldehyde with at least one monomer capable of forming, with formaldehyde, a polymer of a three-dimensional structure. The reaction is carried out in the presence of a polymerization catalyst in an aqueous medium at a pH of from 0.1 to 4, while maintaining the polymer concentration within the range of from 20 to 65 per cent by mass. Thereafter the solution with the polymer is maintained for a period sufficient to form a porous polymer in the form a solid dispersion.

Inventors: Ljubman, Nazar Y. (ulitsa Dzhandosova, 166, kv. 38, Alma-Ata, SU) Imangazieva, Gulsara K. (ulitsa Dzhandosova, 69, kv. 112, Alma-Ata, SU) Nugmanova, Lyalya T. (Mikroraion Sairan, 10, kv. 48, Alma-Ata, SU) Uskov, Alexandr I. (2 Mikroraion, 45, kv. 36, Alma-Ata, SU) Sydykova, Tokhtobubu C. (Mikroraion "Orbita-4", 6, kv. 41, Alma-Ata, SU) Kim, Zinaida I. (ulitas Dzhandosova, 69, kv. 47, Alma-Ata, SU)

Application Number: 06/514480

Publication Date: 01/28/1986

Filing Date: 07/18/1983

View Patent Images: Images are available in PDF form when logged in. To view PDFs, Login or Create Account (Free!)

Referenced by: View patents that cite this patent

Export Citation: Click for automatic bibliography generation

Primary Class: 521/35

Other Classes: 528/265, 528/266, 521/39, 528/254, 528/137

International Classes: B01D17/04; B01D39/16; B01D67/00; B01D71/72; B01J20/22; B01J20/26; B01J39/18; C01G15/00; C01G28/00; C01G39/00; C01G47/00; C02F1/28; C08G8/00; C08G8/08; C08G12/00; C08G12/02; C08G14/00; B01D71/00; B01J39/00

Field of Search: 521/35, 521/39, 528/137, 528/265, 528/254, 528/266

US Patent References:

2460516	Ion-exchange resin	February, 1949	Fuoces	528/137
2809178	Oil soluble phenol-formaldehyde resin	October, 1957	Turner *et al.*	528/137

Foreign References: JP5023390 March, 1975 521/35

Other References:

Chem. Abstracts, vol. 91, Entry 124326, Khose *et al.*
Chem. Abstracts, vol. 92, Entry 23252, Khose *et al.*
Chemical Abstracts, vol. 101, Entry 91975q.

Primary Examiner:

Michl, Paul R.

Assistant Examiner:

Kulkosky, Peter F.

Attorney, Agent or Firm:

Lilling & Greenspan

Claims:

What is claimed is:

1. A polymeric material adapted for physico-chemical separation of substances comprising a porous material of a three-dimensional structure in the state of a solid dispersion with a pore diameter of from 0.0025 to 10 ìm and a permeability coefficient of from 2×10^{-7} to 2×10^{-2} cm?sec^{-1}, and containing a polymer with an elementary unit of the formula: ##STR12## wherein ##STR13## wherein n=0.2.
2. A polymeric material adapted for an ion-exchange separation and mainly for recovery of heavy metal ions according to claim 1, wherein the elementary unit of the polymer has the formula: ##STR14## wherein n=0-2.
3. A polymeric material adapted for an anion-exchange separation of oxygen anions and for separation of molybdenum from thenium according to claim 1, wherein the elementary unit of the polymer has the formula: ##STR15## wherein n=0-2.
4. A polymeric material intended for filtration in strong-alkali media according to claim 1, wherein the elementary unit of the polymer has the formula: ##STR16## wherein n=0-2.
5. A method for producing a polymeric material adapted for physico-chemical separation of substances, comprising:
   - reacting formaldehyde with at least one monomer capable of forming, with formaldehyde, a three-dimensional polymer, in the presence of a polymerization catalyst in aqueous media at a pH of from 0.1 to 4, while maintaining the polymer concentration at 20 to 65 per cent by mass, for a time sufficient to form a solid dispersion at a porous three-dimensional polymer with a pore diameter of from 0.0025 to 10 ìm and a permeability coefficient of from 2×10^{-7} to 2×10^{-2} cm?sec^{-1};

- wherein said monomer is selected from the group consisting of phenol, polyhydric phenols, cresols, sulphophenolic compounds, and carbamide compounds;
- and wherein said carbamide compounds are selected from the group consisting of urea, thiourea, and melamine.

6. A method according to claim 5, wherein the temperature varies from within the range of from 20° to 90° C.
7. A method according to claim 5, wherein the molar ratio between formaldehyde and the monomer is 1.3-8.0:1.
8. A method according to claim 5, wherein as the monomer capable of forming a three-dimensional polymer use is made of a monomer of the formula: ##STR17## wherein R and R' are the same or different radicals selected from the group consisting of —OH, —CH_3, —SO_3H.
9. A polymeric composition comprising a solid dispersion of a porous three-dimensional structure having an elementary unit of the formula: ##STR18## wherein R is selected from the group consisting of: ##STR19## and n=0-2; said polymeric composition having a pore diameter of 0.002 to 10 m and a permeability coefficient of 2×10^{-2} cm^2sec^{-1} formed by a method comprising: reacting formaldehyde with at least one monomer capable of forming, with formaldehyde, a three-dimensional polymer in the presence of a polymerization catalyst in an aqueous medium maintained at a pH of from about 0.1 to 4 to form a resulting polymer whose concentration is maintained at 20 to 65 per cent by mass, residing for a period sufficient to form said solid dispersion.
10. The polymeric composition of claim 9, wherein the molar ratio between formaldehyde and said monomer is about 1.3-8.0:1.
11. The polymeric composition of claim 9, wherein said monomer has the formula: ##STR20## wherein R' and R" are the same or different radicals selected from the group consisting of —OH, —CH_3, —SO_3H.
12. The polymeric composition of claim 9, wherein said monomer is a carbamide.
13. The polymeric composition of claim 12, wherein said carbamide is selected from the group consisting of urea, thiourea, and melamine.
14. The polymeric composition of claim 9, wherein said reaction is carried out at a temperature of about 20°-90° C.
15. The polymeric composition of claim 9, wherein the polymer elementary unit has the formula: ##STR21##.

16. The polymeric composition of claim 9, wherein the elementary unit of the polymer has the formula: ##STR22##.
17. The polymeric composition of claim 9, wherein the elementary unit of the polymer has the formula: ##STR23##.
18. The polymeric composition of claim 9, wherein the elementary unit of the polymer has the formula: ##STR24##.
19. The polymeric composition of claim 9, wherein the elementary unit of the polymer has the formula: ##STR25##.
20. The polymeric composition of claim 9, wherein the elementary unit of the polymer has the formula: ##STR26## wherein n=0-2.
21. The composition of claim 9, wherein the pore diameter varies from about 0.002 to 10 microns.
22. The composition of claim 9, wherein the permeability coefficient varies from about 2×10^{-7} to 2×10^{-2} cm^2sec-1.

Description:

Field of the Invention: The present invention relates to high-molecular weight compounds and, more specifically, to a polymeric material adapted for a physico-chemical separation of substances.

The material according to the present invention can be used for separation of substances in various physico-chemical processes such as sorption, ion exchange, coalescence, back osmosis, hyperfiltration, ultrafiltration, microfiltration, purification of liquids and gases, dust-catching and purification of effluents, as well as in other processes necessitating recovery, separation or removal of substances.

Background of the Invention: At the present time, numerous industries in carrying out basic manufacturing steps, must perform processes of separation and recovery of substances. With the progress of technology increasing importance is given to the quality of the material which is used as an aid in processes of separation and recovery. The separating material quality should meet the requirements implicated in the solution of the following problems: (1) recovery of valuable substances; (2) regeneration of process water; (3) conditioning of process solutions and gases; (4) purification of effluents which is of prime importance for the environment protection.

It is known in the art to use naturally-occurring polymers as separation materials in the form of cotton and wool fabrics. At the present time there is a universal tendency to change cotton and wool filtering fabrics with filtration materials from synthetic polymeric fibres such as polypropylene, polytetrafluoroethylene and polyamide fibres.

The materials ensure only a mechanical separation of substances in accordance with the screening effect which does not enable differentiation of substances according to their type and properties.

The depth and selectivity of separation are substantially increased in the case where the filtration process is combined with back osmosis or ultrafiltration. Special membrane partitions are intentionally made for this purpose. The most extensively used are acetylcellulose membranes which are prepared by wet coagulation of a solution of acetylcellulose in formamide (cf. U.S. Pat. No. 3,666,508 Cl. 106-183).

A disadvantage of acetylcellulose filtering aids is their chemical instability; they feature low performance properties, limited service life and low resistance in acidic media.

To overcome the above-mentioned disadvantages and increase resistance to aggressive media, the company "Millipore Inc." his suggested production of membrane filters based on polytetrafluoroethylene or polyvinylchloride. However, these filtering materials at a high chemical resistance have hydrophobic characteristics. Therefore, they cannot be employed in processes of filtration of aqueous solutions.

The company "Akzona Inc." has developed a novel method for the preparation of microporous polymers based on polyolefins (cf. U.S. Pat. No. 4,247,498 Cl. 264-41). This method comprises dissolution of polyolefins at an elevated temperature in aromatic hydrocarbons, amines, alcohols or ketones, followed by cooling the heated solutions. The thus-prepared materials comprise, according to the company's contention, a new generation of porous materials with the brand name "Accurel" and a controlled pore diameter of 0.2 to 0.4 ìm.

At the present time the state of the art lacks a polymeric filtering material possessing required properties to satisfy modern requirements of substance purification, such as separation, recovery, removal, versatility in this particular area of application, and be simultaneously selective and efficient.

It is an object of the present invention to provide such a polymeric filtering material having a broad range of its filtering capacity, and which would be suitable for various processes of physico-chemical separation of substances and possess a high separation efficiency, stability in operation and simplicity in manufacture.

Summary of the Invention: This object is accomplished by the provision of a polymeric material adapted for physico-chemical separation of substances which, according to the present invention, comprises a porous polymer with a three-dimensional structure having a pore diameter of 0.002 to 10μm and permeability coefficient from 2×10^{-7} to 2×10^{-2} cm^{2}sec^{-1}.

This method for preparing this polymeric material, according to the present invention, comprises reacting formaldehyde with at least one monomer capable of forming, with formaldehyde, a polymer of a three-dimensional structure in the presence of a polymerization catalyst in an aqueous medium with its pH maintained within the range of from 0.1 to 4. The resulting polymer concentration is maintained within the range of from 20 to 65 per cent by mass, followed by residence for a period sufficient to form a solid dispersion with a pore diameter of 0.002 to 10 ìm and a permeability coefficient of from 2×10^{-7} to 2×10^{-2} cm^2sec^{-1}.

The polymeric filtering material according to the present invention has high separating power in various physico-chemical processes such as sorption, ion-exchange, coalescence, reverse osmosis, hyperfiltration and the like. This material also has high mechanical strength and osmotic stability. Being a solid body, it can be easily subjected to any kind of machining. From the material according to the present invention articles of substantially any shape can be produced.

The polymeric material according to the present invention comprises a polymer with the elementary unit of the formula: ##STR1## wherein ##STR2## wherein n=0-2.

Depending on the structure of the elementary unit of the polymer, as well its permeability and pore size, there are possible various embodiments of the material according to the present invention. A polymeric material wherein the elementary unit has the formula: ##STR3## wherein n=0-2, is useful for coalescent separation of emulsions.

A polymeric material wherein the elementary unit has the formula: ##STR4## wherein n=0-2, is useful mainly for recovery of arsenic.

A polymeric material wherein the elementary unit has the formula: ##STR5## wherein n=0-2, is useful for recovery of thallium.

A polymeric material wherein the elementary unit has the formula; ##STR6## wherein n=0-2, is useful for transfer of electrons and primarily for the removal of oxygen.

A polymeric material wherein the elementary unit has the formula: ##STR7## wherein n=0-2, is useful for cation exchange and primarily for water softening.

A polymeric material, wherein the elementary unit has the formula: ##STR8## wherein n=0-2, is useful for sorption of dissolved organic components.

A polymeric material wherein the elementary unit has the formula: ##STR9## wherein n=0-2, is useful for an ion-exchange separation and primarily for recovery of heavy metal ions.

A polymeric material wherein the elementary unit has the formula: ##STR10## wherein n=0-2, is useful for an anion-exchange separation of oxygen anions and for separation of molybdenum from rhenium.

A polymeric material wherein the elementary unit has the formula: ##STR11## wherein n=0-2 is useful for filtration in strong-alkali media.

An advantage of the material of the present invention is its effective filtering capacity in separation processes: sorption, ion exchange, coalescence, microfiltration, ultrafiltration, reverse osmosis and the like. It provides for quantitative recovery of dissolved components such as arsenic, antimony, molybdenum, thallium and the like. This material can be efficiently employed for carrying out filtration of liquids, purification of gases and dust catching.

Detailed Description of the Invention: As already mentioned hereinbefore, the polymeric material adapted for physico-chemical separation of substances according to the present invention comprises a solid dispersion of a porous polymer having a three-dimensional structure with a pore diameter of from 0.002 to 10 ìm and a coefficient of permeability ranging from 2×10^{-7} to 2×10^{-2} cm^2sec^{-1}.

One of the features characterizing the polymeric filtering material according to the present invention is the solid dispersion state of the polymer. By the term "solid dispersion" is meant the state of a pclymer, wherein properties of a suspension and those of a solid body are combined for the case where water is dispersed in the polymer.

Another feature characterizing the polymeric filtering material according to the present invention is its three-dimensional structure. A further feature of the material according to the invention is its porosity and permeability. The pore diameter is within the range of from 0.002 to 10 ìm and the permeability coefficient ranges from 2×10^{-7} to 2×10^{-2} cm^2sec^{-1}.

This polymeric material is prepared by the method which, according to the present invention, comprises reacting formaldehyde with at least one monomer capable of forming, with formaldehyde, a polymer having a three-dimensional structure. The reaction is carried out in the presence of a polymerization catalyst in an aqueous medium. During the reaction, the pH of the medium is maintained within the range of from 0.1 to 4 and the resulting polymer concentration is controlled within the limits of from 20 to 65 per cent by mass. When the polymer concentration within this range is achieved, the solution with the polymer is kept for a period sufficient to form a solid dispersion of the polymer.

As the starting monomer suitable for the reaction with formaldehyde use can be made of any monomer capable of forming, with formaldehyde, a three-dimensional structure. Such monomers can be exemplified by phenol, polyhydric phenols, cresols, sulphophenolic compounds, carbamide compounds and the like. The polymerization catalyst can such compounds as alkalis and acids.

The reaction is carried out at a temperature within the range of from 20° to 90° C. depending on the starting monomer and the type polymerization catalyst.

The molar ratio of formaldehyde to the monomer is varied within the range of from 1.3:1 to 8.0:1 respectively.

As mentioned hereinabove, during the reaction the concentration of the resulting polymer is maintained within the range of from 20 to 65 per cent by mass. The lower limit is determined from the fact that at a concentration of the polymer in the solution below 20 per cent the material has no solid dispersion condition. Thus, the polymer becomes dispersed and loses the above-specified properties. The upper limit of the polymer concentration is defined by the fact that at a concentration of the polymer above 65 per cent its porosity is decreased and, consequently, its permeability is reduced. Therefore, violation of the above-specified concentration range will result in failure to obtain a material with the desired properties. During the reaction the pH of the reaction mass is maintained within the range of from 0.1 to 4. At a pH below 0.1 the porosity is reduced and, consequently, permeability of the material is lowered. At a pH of above 4 the degree of adherence between particles is lowered and the material cannot be obtained in the state of a solid dispersion. If the polymerization reaction is carried out in special molds, it is possible to simultaneously obtain the material and an article therefrom. Articles may have different shape depending on their end use such as tubes, sheets, membranes, fibers, capillaries and the like.

The method according to the present invention is performed in the following manner. The abovementioned starting components—formaldehyde and monomer in a molar ratio of 1.3-8.0:1—are mixed in an aqueous medium and maintained until the formation of a product in the form of a solid dispersion, while maintaining the polymer concentration within the range of from 20 to 65 per cent and the pH at a value of from 0.1 to 4. The resulting product is porous, permeable and three-dimensional. This product is commercial and ready for use. As it is seen from the description, the method is simple to perform and can be readily implemented on a commercial scale, since it necessitates no special equipment and expensive hardly-available components. Another advantage of the present invention resides in the possibility of simultaneous production of the material and an article therefrom.

For a better understanding of the present invention, some specific examples are given hereinbelow which illustrate the polymeric filtering material and the process for producing same according to the present invention. Unless otherwise specified, in all of the examples the concentration of substances is expressed in percent by mass.

Example 1

Into a reaction vessel there are charged 670 g of phenol, 187 ml of a 35 per cent aqueous solution of formaldehyde, 190 g of paraform and 26 g of sodium hydroxide. The reaction mixture is maintained at a temperature of 60° C. for 90 minutes, whereafter there are added 996 ml of a 98 per cent acetic acid and hydrochloric acid to a pH=3.

The resulting solution with a concentration of the formed polymer of 43 per cent is maintained at a temperature of 70° C. for 25 hours. On expiration of this period the polymer is withdrawn from the reaction vessel. This polymer has a three-dimensional structure and comprises a solid dispersion with a pore diameter of 20 to 40 Å and a permeability coefficient of 2.2×10^{-7} cm^2sec^{-1}. This material is used in reverse osmosis processes as a separating membrane.

Example 2

Into a three-necked reactor provided with a stirrer, reflux condenser and a thermometer three are charged 323 g of phenol, 370 ml of a 35 per cent aqueous solution of formaldehyde and 13 g of sodium hydroxide. The mixture is stirred and maintained for 90 minutes at a temperature of 60° C., whereafter there are added 1,376 ml of 98 per cent acetic acid and hydrochloric acid to a pH=3. The resulting solution having a concentration of the formed polymer of 21 per cent is poured into the intertubular space of a stainless steel mold comprising a sleeve with a height of 280 mm and inside diameter of 120 mm, whereinto a rod with an outside diameter of 70 mm is coaxially inserted. The filled mold is sealed and placed into a thermal cabinet with a temperature of 80° C. After 25 hours of residence therein the mold is cooled and a yellow article with a height of 270 mm and thickness of filtering walls of 25 mm is withdrawn therefrom.

The resulting article has the following properties: pore diameter 8-10 µm, permeability coefficient 5.7×10^{-3} cm^2sec^{-1}. This article is tested in a process of coalescent separation of an emulsion. An aqueous emulsion containing 66 mg/l of kerosene is passed through the polymeric article at the rate of 590 l/hr. At the outlet from the article there is produced a readily and rapidly separable system consisting of two phases: an aqueous phase containing 0.7 mg/l of kerosene and an organic phase containing 0.18 g/l of water.

Example 3

Into a reaction vessel there are charged 454 ml of a 30.6 per cent aqueous solution of pyrocatechol, 144 ml of a 37 per cent aqueous solution of formaldehyde and hydrochloric acid to a pH=2. The solution is stirred at a temperature of 60° C. for 1 hour and 40 minutes. Then it is cooled to a temperature of 30.7° C. The resulting solution with the concentration of the polymer of 32 per cent is kept at a temperature of 30.7° C. for 46 hours and at a temperature of 82° C. for 30 hours. Thereafter, a three-dimensional polymer is withdrawn from the vessel. The product comprises a solid dispersion with a pore diameter of 4-7 μm and a coefficient of permeability of 8.6×10^{-4} cm^2sec^{-7}.

The thus-produced material is tested in a process of purification of solutions from electrolytic copper production. The starting solution having the following composition (g/l): copper—47.0, arsenic—11.8, nickel—18.7, antimony—0.8, slime—0.1, sulphuric acid—160, is passed through the material for one hour to give 24 liters of a filtrate having the following composition (g/l): copper—47.0, arsenic—10.9, nickel—18.7, antimony—0.1, slime—0.018, sulphuric acid—160. The filtering capacity is regenerated by purging with compressed air and elution of antimony is effected by treatment with 7N hydrochloric acid. During the elution there is obtained 1.0 liters of an eluate containing 16.3 g/l of antimony.

Example 4

Into a reaction vessel there are charged 454 ml of a 30.6 per cent aqueous solution of pyrocatechol, 144 ml of a 37 per cent aqueous solution of formaldehyde and hydrochloric acid to a pH=2. The solution is stirred at a temperature of 55° C. for 1 hour 50 minutes, and cooled to a temperature of 32.6° C. The resulting solution with the formed polymer concentration of 32 per cent is kept at a temperature of 32.6° C. for 46 hours and at a temperature of 82° C. for 30 hours. Thereafter, a polymer is withdrawn from the vessel. The polymer has a three-dimensional structure and comprises a solid dispersion with a pore diameter of 3-5 ìm and permeability coefficient of 1.2×10^{-4} cm^2sec^{-1}.

The thus-produced material is tested in the process of decontamination of sulphuric acid from arsenic. The starting solution having the following composition (g/l): arsenic—3.5, iron—0.98, sulphuric acid—380 is passed through the material at the volume rate of 200 unit volumes/hr. In doing so, a filtrate is obtained which has the following composition (g/l): arsenic—0.007, iron—0.83, sulphuric acid—366. The material is regenerated by treating it with water at a temperature of 70° C.

Example 5

A reaction mixture is prepared by intermixing 1.15 liters of a 44.0 per cent aqueous solution of resorcinol, 498 ml of a 35 per cent aqueous solution of formaldehyde and hydrochloric acid to a pH=4 and stirring is continued for an additional 3 hours. The resulting solution with the polymer concentration of 40 per cent is cast into 14 moulds made from polyethylene. The mold comprises two coaxially positioned tubes and the intertubular space is filled with a solution of the polymer obtained as above. The charged molds are placed into a water thermostat at a temperature of 20° C. After 24 hours the molds are transferred into a thermal cabinet at a temperature of 80° and kept therein for 48 hours. Afterwards, tubular articles having good mechanical strength are extracted from the molds. The articles have a length of 341 mm, filtering wall thickness of 4 mm, pore diameter of 0.03-0.06 μm and permeability coefficient of 8.6×10^{-6} cm^2sec^{-1}. The ultimate compression strength of the article is 250 kg/cm^2.

The thus-produced polymeric tubular articles are tested in the process of ultrafiltration. Subjected to the ultrafiltrational purification is a copper electrolyte having the following composition (g/l): sulphuric acid—153, copper—45, nickel—12, arsenic—8, antimony—0.8, dispersed slime inclusions—0.048 (particle size of 0.04-0.1 ìm), emulsifield organic substances—0.18.

The resulting filtrate contains (g/l): sulphuric acid—153, copper—45, nickel—12, arsenic—8, antimony—0.8, dispersed slime inclusions—none, emulsified organic substances—0.001. The yield of the filtrate is 98 per cent, the concentrate contains about 2.3 g/l of slime inclusions and 9 g/l of emulsified organic substances.

Operation of the ultrafiltration unit for 150 hours revealed no noticeable change of filtering properties of the tubular polymeric articles.

Example 6

Into a reaction vessel there are charged 452 ml of a 30 per cent aqueous solution of resorcinol, 138 ml of a 37 per cent aqueous solution of formaldehyde and hydrochloric acid to a pH=4 and the reaction mass is stirred for 2 to 6 hours. The resulting solution having a concentration of the formed polymer of 38 per cent is maintained at the temperature of 20° C. for 46 hours and then at the temperature of 82° C. for 24 hours. On expiration of this period a polymer is withdrawn from the reaction vessel. The product has a three-dimensional structure and comprises a solid dispersion with a pore diameter of 6-9 μm and permeability coefficient of 6.6×10^{-3} cm^2sec^{-1}.

This polymer is tested in the process of decontamination of waste liquiors, resulting from the production of lead and zinc, from thallium. The starting solution (pH=9) having the following composition (g/l): zinc—1.5, cadmium—0.2, thallium—0.02 is passed through the polymer at a rate of 400 unit volumes per hour. The filtrate contains (g/l): zinc—1.5, cadmium—0.2, thallium—traces. The polymer regeneration is effected by treatment with a 15 per cent sulphuric acid to give an eluate containing 11.8 g/l of thallium.

Example 7

Into a reaction vessel there are charged 200 ml of a 55 per cent aqueous solution of hydroquinone, 226 ml of a 37 per cent aqueous solution of formaldehyde and hydrochloric acid to a pH of 0.1 and stirring is effected for 90 minutes. The resulting solution with the formed polymer concentration of 45 per cent is maintained at a temperature of 50° C. for 24 hours and at the temperature of 83° C. for 48 hours. On expiration of this period a polymer product is withdrawn from the reaction vessel. The product has a three-dimensional structure and comprises a solid dispersion with a pore diameter of 5-7 μm and permeability coefficient of 1.9×10^{-3} cm^2sec^{-1}.

This material is tested in the process of oxygen removal from water.

Example 8

Into a three-necked reactor provided with a stirrer, reflux condenser and thermometer there are charged 324 g of m-cresol, 226 ml of a 37 per cent aqueous solution of formaldehyde, 98 g of paraform and 17 g of sodium hydroxide. The mixture is stirred and maintained at a temperature of 60° C. for 90 minutes, whereafter it is charged with 892 ml of a 98 per cent acetic acid and hydrochloric acid to a pH=1. The resulting solution with the formed polymer concentration of 31 per cent is kept at the temperature of 80° C. After 25 hours a three-dimensional polymer is discharged from the vessel. The product comprises a solid dispersion with a pore diameter of 3-5 μm and permeability coefficient of 1.1×10^{-4} cm^2sec^{-1}.

The thus-produced material is tested in the process of removal of dissolved organic components. The starting solution containing 5 mg/l of kerosene is passed through the material at a rate of 200 unit volumes per hour. In the effluent filtrate, kerosene is not detected by chromatography. The material is regenerated by treating with live steam.

Example 9

A reaction mixture is prepared by dissolving 160 g of urea in 220 ml of a 35 per cent aqueous solution of formaldehyde, followed by the addition of 10g of resorcinol. The mixture is stirred for 15 minutes at a pH=4. The resulting solution with the concentration of the formed polymer of 59 per cent is cast into molds. The molds are made of coaxially positioned polyethylene tubes and the solution is poured into the intertubular space. The filled moulds are kept at room temperature. After 24 hours they are placed into a thermal cabinet at a temperature of 80° C. and kept therein for 48 hours. Thereafter articles are extracted from the molds. The articles have a tubular shape with a high mechanical strength and a length of 170 mm, filtering wall thickness of 4 mm, pore diameter of 0.003-0.008μm and permeability coefficient of 3.6×10^{-7} cm^2sec^{-1}.

The thus-produced polymeric tubular articles are used in the process of reverse osmosis.

Example 10

A reaction mixture is produced by combining 1.3 liters of a 37 per cent aqueous solution of urea, 72 ml of a 17 per cent phosphoric acid and 811 ml of a 35 per cent aqueous solution of formaldehyde. The mixture is stirred for 2 minutes at a pH=3.5. The resulting solution with the concentration of the formed polymer of 34 per cent is poured into the intertubular space of a mold made of stainless steel and comprising a sleeve with a height of 280 mm and inside diameter of 120 mm, whereinto a rod with an outside diameter of 70 mm is inserted. The filled mold is kept at room temperature for 2 hours and at a temperature of 80° C. for 24 hours. After cooling the article is withdrawn in the form of a matted white thick-walled tube. The thus-produced article comprises a polymer in a solid dispersion state with a pore diameter of 0.7-0.9 μm, permeability coefficient of 7.3×10^{-5} cm^2sec^{-1} and breaking compression strength of 0.74 kgf/cm^2. It is chemically resistant in strong-alkali media.

The article is used as a filtering cartridge. Its catching power relative to the solid phase of an argillaceous suspension with a particle size of 1 μm is substantially equal to 100 per cent at a complete regenerability by back purging with compressed air.

Example 11

A reaction mixture is prepared by combining 1.0 liter of a 43 per cent aqueous solution of urea, 232 ml of a 19 per cent aqueous solution of resorcinol, 55 ml of 15 per cent phosphoric acid and 875 ml of a 35 per cent aqueous solution of formaldehyde. The mixture is stirred for 2 minutes at a pH=4.0. The resulting solution with a concentration of the formed polymer of 34 per cent is poured into the intertubular space of a stainless steel mold comprising a sleeve with a height of 280 mm and an inside diameter of 120 mm, whereinto a rod is inserted with an outside diameter of 70 mm. The filled mold is kept at room temperature for 2 hours and at a temperature of 80° C. for 24 hours. After cooling the article is removed. Its material has a three-dimensional structure and comprises a solid dispersion with a pore diameter of 0.9-1.8 μm, permeability coefficient of 1.2×10^{-4} cm^2sec^{-1} and breaking compression strength of 8.7 kgf/cm^2.

The article is used as a filtering cartridge.

Example 12

Into a three-necked reactor provided with a stirrer, reflux condenser and thermometer there are charged 396 g of melamine, 739 ml of a 35 per cent aqueous solution of formaldehyde and 53 ml of a 26 per cent aqueous solution of ammonia. The mixture is stirred at a temperature of 60° C. for 120 minutes, then charged with 351 ml of water along with 351 ml of a 98 per cent acetic acid. The resulting solution with a pH=4 and concentration of the formed polymer of 33 per cent is poured into the intertubular space of a mold made from two coaxially positioned stainless-steel tubes. The filled mould is placed into a heating cabinet at a temperature of 70° C. After residence for 25 hours the article is withdrawn. It has a white colour, height of 275 mm and filtering wall thickness of 25 mm. The material of the article has a three-dimensional structure and comprises a solid dispersion with a pore diameter of 0.4-0.6 μm and permeability coefficient of 3.6×10^{-5} cm^2sec^{-1}.

The thus-produced article is tested in a process of separation of molybdenum and rhenium. The starting solution with a pH=1 and having the following composition (g/l): rhenium—0.18, molybdenum—0.42 is filtered through the article at the rate of 200 unit volumes per hour. The solution effluent from the article does not substantially contain molybdenum, while the concentration of rhenium remains substantially unchanged, i.e. in this manner an absolute separation of molybdenum from rhenium is achieved. The filtering element is regenerated by way of treatment with a 10 per cent solution of ammonia to give an eluate containing 8.8 g/l of molybdenum.

Example 13

Into a three-necked reactor provided with a stirrer, reflux condenser and thermometer there are charged 500 g of melamine, 945 ml of a 35 per cent aqueous solution of formaldehyde and 61 ml of a 26 per cent aqueous solution of ammonia. The mixture is stirred at a temperature of 60° C. for 120 minutes, followed by the addition of 430 ml of a solution having the following composition (g/l): hydrochloric acid—37, resorcinol—117. The resulting solution with a pH=4 and the concentration of the formed polymer of 52 per cent is cast into the intertubular space of a mold comprising two coaxially positioned stainless-steel tubes. The filled mold is placed into a heating cabinet at a temperature of 70° C. After 25 hours a white article is withdrawn which has a height of 275 mm and a filtering wall thickness of 25 mm. The material of the article is of a three-dimensional structure and comprises a solid dispension with a pore diameter of 0.8-2.2 μm and permeability coefficient of 3.7×10^{-4} cm^2sec^{-1}.

The thus-produced article is tested in the process of removing acids from aqueous media. The starting solution containing 100 mg/l of hydrochloric acid is passed through the article at a rate of 200 unit volumes per hour. The concentration of hydrochloric acid in 200 liters of the filtrate is reduced down to 0.3 mg/l. The filtering member is regenerated by way of treatment with a 10 per cent aqueous solution of ammonia.

Example 14

Into a three-necked reactor provided with a stirrer, reflux condenser and thermometer there are charged 456 g of thiourea, 140 g of resorcinol and 1.1 liter of a 30 per cent aqueous solution of formaldehyde. The mixture is stirred at a temperature of 60° C. for 120 minutes, whereafter there are added 540 ml of a solution having the following composition (g/l): sulphuric acid 28, urea 170. The resulting solution has a pH=0.1 and the concentration of the formed polymer of 20 per cent is poured into the intertubular space of a mold made of coaxially positioned polyethylene tubes. The filled mold is placed into a heating cabinet at a temperature of 70° C. and kept therein for 25 hours. The resulting article has an orange colour, height of 275 mm and filtering wall thickness of 25 mm. The material of the article has a three-dimensional structure and comprises a solid dispersion with a pore diameter of 5-10 μm and permeability coefficient of $1.8\times10^{-2}cm^2sec^{-1}$.

The thus-produced article is tested in recovering heavy metals from process solutions. The starting solution containing 80 g/l of bismuth is passed through the article at a rate of 200 unit volumes per hour. The concentration of bismuth in 420 liters of the filtrate is lowered to 0.08 mg/l. The filtering element is regenerated by way of its treatment with a solution of the following composition (g/l): thiourea—100, sulphuric acid—56 to give an eluate containing 3.6 g/l bismuth.

Example 15

Into a three-necked reactor provided with a stirrer, dropping funnel and cooler there are placed 611 g of phenol. The reactor is heated to a temperature of 90°C. on a water bath under stirring and charged, from the dropping funnel, with 340 ml of a 86 per cent sulphuric acid for 1 hour. The mixture is stirred for 1 hour at 90° C., then charged with 1.2 liters of a 35 per cent aqueous solution of formaldehyde, 450 ml of 98 per cent acetic acid and 520g of paraform. The resulting solution has a pH=0.1 and the concentration of the formed polymer of 31 per cent is poured into the intertubular space of a mold comprising coaxially mounted stainless-steel tubes. The filled mould is placed into a boiling water bath. After 25 hours of residence a dark-cherry article is withdrawn. The material in the article has a three-dimensional structure and comprises a solid dispersion with a pore diameter of 0.3-0.5 mm and permeability coefficient of $2.2\times10^{-5}cm^2sec^{-1}$.

The thus-produced article is tested in the process of water softening. The starting solution containing 100 mg/l of calcium is passed through the article at the rate of 200 unit volumes per hour. The concentration of calcium in 500 liters of the filtrate is reduced to 1.2 mg/l. The filtering material is regenerated by treatment with a 5 per cent solution of NaCl and can then be used again.

11

Polymers Strength Enhancement and Degradation

11.1 Polymer Degradation

Polymer degradation is a change in the properties - tensile strength, colour, shape, etc - of a polymer or polymer based product under the influence of one or more environmental factors such as heat, light or chemicals. These changes are usually undesirable, such as changes during use, cracking and depolymerisation of products or, more rarely, desirable, as in biodegradation or deliberately lowering the molecular weight of a polymer for recycling.

In a finished product such a change is to be prevented or delayed. However degradation can be useful for recycling/reusing the polymer waste to prevent or reduce environmental pollution. Degradation can also be induced deliberately to assist structure determination.

Polymeric molecules are very large (on the molecular scale), and their unique and useful properties are mainly a result of their size. Any loss in chain length lowers tensile strength and is a primary cause of premature cracking.

Commodity Polymers

Today there are primarily six commodity polymers in use, namely polyethylene, polypropylene, polyvinyl chloride, polyethylene terephthalate or PET, polystyrene and polycarbonate. These make up nearly 98 per cent of all polymers and plastics encountered in daily life. Each of these polymers has its own characteristic modes of degradation and resistances to heat, light and chemicals. Polyethylene and polypropylene are sensitive to oxidation and UV radiation, while PVC may discolour at high temperatures due to loss of hydrogen chloride gas, and become very brittle. PET is sensitive to hydrolysis and attack by strong acids, while polycarbonate depolymerizes rapidly when exposed to strong alkalis.

For example, polyethylene usually degrades by *random scission* - that is by a random breakage of the linkages (bonds) that hold the atoms of the polymer together. When this polymer is heated above 450 Celsius it becomes a complex mixture of molecules of various sizes which resemble gasoline. Other polymers - like polyalphamethylstyrene - undergo 'unspecific' chain scission with breakage occurring only at the ends; they literally unzip or depolymerize to become the constituent monomers.

Examples

Many polymers, especially step-growth polymers, are degraded by specific chemicals such as strong acids and strong alkalis. They are made by condensation polymerization, so degradation is a reversal of the synthesis reaction. Other degradation routes involve interaction with strong oxidising agents and interaction with UV radiation.

Hydrolysis: Nylon is sensitive to degradation by acids, a process known as hydrolysis, and nylon mouldings will crack when attacked by strong acids. A fuel pipe fractured when a small drip of 40 per cent sulphuric acid from a nearby lead-acid battery fell onto a nylon 6,6 moulded connector in the diesel line. The crack grew with time until it penetrated the interior, so initiating a slow leak of diesel. The crack continued to grow until final separation occurred, and diesel fuel poured into the road. Diesel is especially hazardous when it is present on road surfaces because it forms an extremely slippery surface which cannot be seen easily by road users (just like black ice).

The leak caused several accidents to other cars, one of which caused serious injuries to the driver. The fracture surface of the connector showed the progressive growth of the crack from the initial acid attack (Ch) to the final cusp (C) of polymer. The problem is known as stress corrosion cracking, and in this case was caused by hydrolysis of the polymer. It was the reverse reaction of the synthesis of the polymer:

$$n\ \mathrm{HO{-}C(=O){-}R{-}C(=O){-}OH} + n\,\mathrm{H2N{-}R{-}NH_2} \longrightarrow \left[\mathrm{{-}C(=O){-}R{-}C(=O){-}N(H){-}R{-}N(H){-}}\right]_n$$

The owner of the vehicle on which the fuel pipe leak should have spotted the leak well before the final accident, and the injured driver was awarded compensation by the insurers.

Ozonolysis: Cracks can be formed in many different elastomers by ozone attack. Tiny traces of the gas in the air will attack double bonds in rubber chains, with Natural rubber, Styrene-butadiene rubber and NBR being most sensitive to degradation. Ozone cracks form in products under tension, but the critical strain is very small. The cracks are always oriented at right angles to the strain axis, so will form around the circumference in a rubber tube bent over. Such cracks are very dangerous when they occur in fuel pipes because the cracks will grow from the outside exposed surfaces into the bore of the pipe, so fuel leakage and fire may follow. The problem of ozone cracking can be prevented by adding anti-ozonants to the rubber before vulcanization. Ozone cracks were commonly seen in automobile tire sidewalls, but are now seen rarely thanks to the use of these additives. On the other hand, the problem does recur in unprotected products such as rubber tubing and seals.

Oxidation: IR spectrum showing carbonyl absorption due to oxidative degradation of polypropylene crutch moulding

Polymers are susceptible to attack by atmospheric oxygen, especially at elevated temperatures encountered during processing to shape. Many process methods such as extrusion and injection moulding involve pumping molten polymer into tools, and the high temperatures needed for melting may result in oxidation unless precautions are taken. For example, a forearm crutch suddenly snapped and the user was severely injured in the resulting fall. The crutch had fractured across a polypropylene insert within the aluminium tube of the device, and infra-red spectroscopy of the material showed that it had oxidised, possible as a result of poor moulding.

Oxidation is usually relatively easy to detect owing to the strong absorption by the carbonyl group in the spectrum of polyolefins. Polypropylene has a relatively simple spectrum with few peaks at the carbonyl position (like polyethylene). Oxidation tends to start at tertiary carbon atoms because free radicals here at more stable, so last longer and are attacked by oxygen. The carbonyl group can be further oxidised to break the chain, so weakening the material by lowering the molecular weight, and cracks start to grow in the regions affected.

Chlorine-induced Cracking: Another highly reactive gas is chlorine, which will attack susceptible polymers such as acetal resin and polybutylene pipework. There have been many examples of such pipes and acetal fittings failing in properties in the USA as a result of chlorine-induced cracking. Essentially the gas attacks sensitive parts of the chain molecules (especially secondary, tertiary or allylic carbon atoms), oxidising the chains and ultimately causing chain cleavage. The root cause is traces of chlorine in the water supply, added for its anti-bacterial action, attack occurring even at parts per million traces of the dissolved gas. The chlorine attacks weak parts of a product, and in the case of an acetal resin junction in a water supply system, it was the thread roots which were attacked first, causing a brittle crack to grow. The discolouration on the fracture surface was caused by deposition of carbonates from the hard water supply, so the joint had been in a critical state for many months. When it finally failed, it did so at the worst possible time, at the weekend when no-one was around to sort the problem. The leak flooded computer labs below, and caused substantial damage. The problems in the USA also occurred to polybutylene pipework, and led to the material being removed from that market, although it is still used elsewhere in the world.

Stabilisers

Hindered-amine light stabilisers (HALS) stabilise against weathering by scavenging free radicals that are produced by photo-oxidation of the polymer matrix. UV-absorbers stabilises against weathering by absorbing ultraviolet light and converting it into heat. Antioxidants stabilizes the polymer by terminating the chain reaction due to the adsorption of UV light from sunlight. The chain reaction initiated by photo-oxidation leads to cessation of crosslinking of the polymers and degradation the property of polymers.

Almost 50 per cent of failures of engineering plastics result from environmental degradation. The increasing utilisation of plastics in more exacting applications and the pressure for increased life, without uneconomic over design, are imposing a requirement for improved characterisation of the performance of polymeric materials. The major challenge is to predict long term behaviour from short term laboratory or field exposures.

The most common mode of environmental degradation is caused by exposure of engineering polymers to chemicals, resulting in environmental induced stress cracking. The steps in the degradation process involve stress enhanced absorption and concentration of the chemical molecules at susceptible microstructural sites. Localised plasticisation then ensues, leading to crazing and subsequent crack development.

Failure may often be associated with exposure to secondary fluids, such as cleaning agents or lubricants, rather than the primary design environment, and with residual moulding stresses. Improved awareness of environment induced stress cracking can reduce the incidence of a failure but more data are required characterising the cracking resistance of plastics to a range of common fluids. More generally, there is a need to provide a methodology for life prediction involving testing and predictive models.

Accelerated Testing

It is essential to balance the need for accelerated testing with the inherent time dependence of the chemical and physical processes involved. The timescale for chemical ingress into the plastic or for leaching of mobile additives is important, and misleading results in ranking the aggressiveness of different chemicals, or the resistance of different materials to a specific chemical, are conceivable if chemical diffusivity is not considered.

Temperature Testing

The adoption of higher temperatures is commonly used for accelerated testing of the resistance of plastics to environment induced cracking, but there is an increasing trend towards simulating practical environments more realistically and deducing long term behaviour from short term measurements.

Natural Weathering

Natural weathering of plastics is also a significant cause of degradation. The interaction of UV radiation, oxygen, atmospheric pollutants, alternate wetting and drying cycles, and temperature variations can result in complex photomechanical reaction processes leading to a degradation of physical, mechanical and chemical properties. Failure may take the form of fracture as a consequence of impact, design or residual stresses, but more commonly loss of transparency, gloss or colour render the material unfit for purpose.

In conducting meaningful accelerated durability tests for weathering, whether for plastics, organic coatings or adhesives, the range of variables and their time and spatial variation pose intrinsic difficulties in devising standard tests of general applicability. When coupled with the complexity of the reaction processes in polymers, establishment of predictive models or damage functions is a formidable challenge. There is a prevalent view that there is no correlation between natural and artificial weathering. This view is expressed even in some of the standards. However, the term correlation is often used without clear definition.

Work at (National Physical Laboratory, UK) NPL has gone into evaluating the effectiveness of accelerated laboratory testing in predicting long term performance of plastics in natural weathering conditions, using a range of indices of degradation including microhardness, carbonyl index, colour and gloss.

Predicting Degradation Rates

The quantitative prediction of long term performance in natural conditions depends on the ability to predict the rate of degradation in natural weathering. It is insufficient to demonstrate that 2 months in a weatherometer with a specified weathering cycle is equivalent to 2 years, say, in natural conditions, for which data may be known as a reference point, unless there is a framework for predicting the extent of degradation at 15 years for example. The problem is illustrated in Figure 11.1 by the weathering performance of UPVC window profiles, using microhardness as an index of degradation. For this system and for many others, the rate of degradation changes with time in a complex way due to the development of a modified surface layer. The intrinsic dependence of the development of the degraded layer on exposure conditions means that it can be inherently difficult to establish rate equations which can be transferred meaningfully from artificial weathering tests to natural exposure.

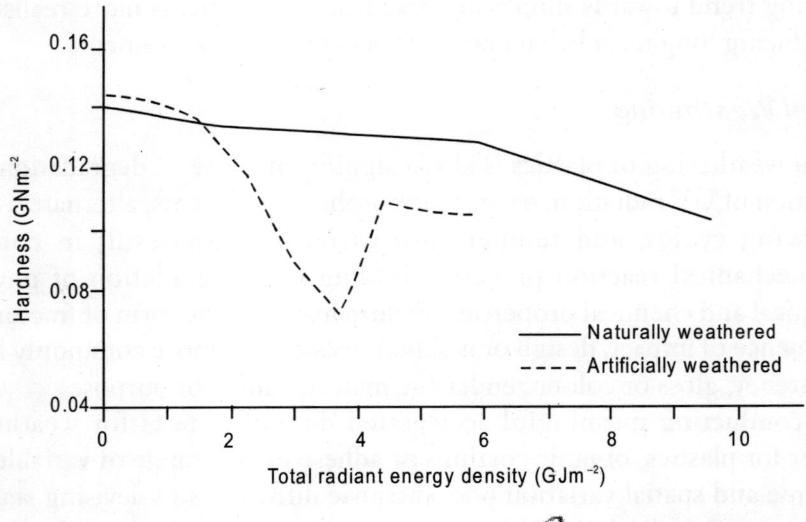

Fig. 11.1: Microhardness comparison for naturally and artificially weathered UPVC window profiles. 10GJm^{-2} is equivalent to 0.78 years of laboratory testing and 5.4 years of natural exposure in NE England. The accelerating factors were the increased temperature (65°C black standard) and continuous UV exposure for the laboratory tests.

Ranking of Materials

For ranking of materials it is necessary simply to demonstrate that the relative performance of the different materials, in terms of the extent of degradation, is consistent with that observed in natural conditions. However, work at NPL has shown that UPVC window-grade materials exhibiting a relatively high propensity to yellowing and to production of carbonyl degradation products can be more resistant to mechanical degradation and erosion of the surface. Hence, in ranking materials it is important to focus on the property of most practical relevance for the specific application, or to measure a range of properties.

Summary

The objective in laboratory testing is to establish a combination of artificial weathering parameters which can simulate the impact on material properties of natural weathering. It should not be expected that a specific set of artificial weathering parameters can be established that simulate temperature conditions, for example, for all materials, but for classes of material this would appear feasible. In this context, all artificial weathering tests would involve testing reference standards of the class of material in parallel. This would provide a basis for quality control on the test conditions and, if appropriate, natural exposure data for reference samples could be obtained, for relative ranking. In the latter context, there is a considerable virtue in collaborative programmes on an international basis so that a world wide 'degradation map' for classes of material could be established.

Considerable progress in artificial weathering testing in the laboratory is being made through adoption of sources of UV radiation capable of matching the solar spectrum, such as xenon arc and fluorescent tubes. Depth profiling of the material properties at NPL, using techniques such as nanoindentation for modulus and hardness measurement and FTIR with photoacoustic cell for chemical analysis, is revealing more details of the properties of the degraded surface layer and enabling quantification of the rate of material degradation as a function of time and proximity to the surface. Hence, the laboratory tools available now are increasingly more effective. There is considerable research to be done in relation to the adoption of dark periods in the exposure cycle and the role of pollutants and of stress. The complexity of weathering suggests that cooperative effort on an international basis will provide the most effective return on individual investment.

11.2 Weather Testing of Polymers

Weather testing of polymers is the controlled polymer degradation and polymer coating degradation under lab or natural conditions.

Just like errosion of rocks, natural phenomena can cause degratation in polymer systems. The elements of most concern to polymers are Ultraviolet radiation, moisture and humidity, high temperatures and temperature fluctuations. Polymers are used in every day life, so it is important for scientists and polymer produces to understand durability and expected lifespan of polymer products. Paint, a common polymer coating, is used to change the colour, change the reflectance (gloss), as well as forming a protective coating. The structure of paint consists of pigments in a matrix of resin. A typical example is painted steel roofing and walling products, which are constantly exposed to harmful weathering conditions.

Typical Result of Polymer Surface after Weathering

Figure 11.1 shows typical weathering results of a sample of painted steel; the paint on the steel is an example of a common polymer system. The sample on the left had been placed in an outdoor exposure rack, and weathered for a total of 6 years. It can be seen that the sample has a chalky appearance and has undergone a colour change in comparison to the unweathered sample on the right.

Fig. 11.1: Weathered and unweathered samples of painted steel. Note the colour change

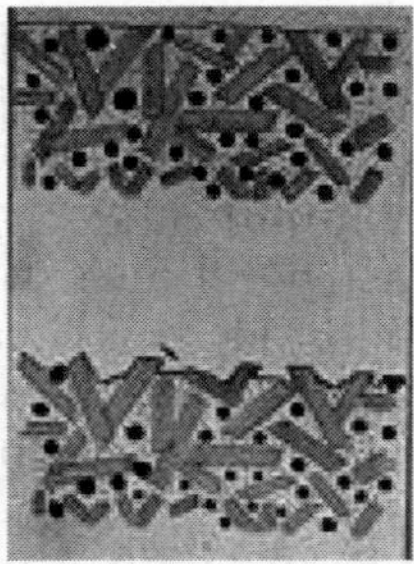

Fig. 11.2: Diagram showing paint pigments in a resin matrix. It can be seen that weathering only removes the smaller pigments, resulting in the colour change

Colour is determined by light-reflecting chemical particles, pigments, in the paint. These particles can have very different physical sizes, as shown in the diagram in figure 11.1.2. In this example, the black pigments are the small black dots; red pigments are larger spheres, while the yellow pigments are acicular. This combination of pigments produces the original brown colour. The upper diagram has had no weathering, and the surface is still smooth and undamaged. The lower diagram shows the painted surface after weathering has occurred. The surface has eroded with significant loss of the black and red pigments from the surface layer. The pitted surface scatters light, therefore reducing the gloss and creating the chalky affect. The larger acicular yellow pigments are more difficult to remove, resulting in a colour change towards a more yellowy appearance. Weather testing was paramount is discovering this mechanism. Pigment composition has recently been modified to help minimise this effect.

Types of Weather Testing

There are 3 main testing techniques; Natural Weathering, Accelerated Natural Weathering and Artificial Weathering. Because natural weathering can be a slow process, each of the techniques is a tradeoff between realistic weathering results and the duration of testing before results are collated.

Natural Weathering

Natural Weathering involves placing samples on inclined racks orientated at the sun. In Northern hemisphere these racks are at an angle of 45 degrees in a southerly direction. In Southern hemisphere these racks are at an angle of 45 degrees in a northerly direction. This angle ensures exposure to the full spectrum of solar radiation, from infrared to Ultra violet. Sites used for this type of testing are usually in tropical areas as high temperature, UV intensity and humidity are needed for maximum degradation. Florida, for example is the world standard as it possesses all three characteristics. Despite the harsh conditions, testing takes several years before significant results are achieved.

Accelerated Natural Weathering

To speed up the weathering process while still using natural weather conditions, accelerated natural testing can be applied. One method called Emma uses mirrors to amplify available UV radiation.

This is usually used in dry condition\s, for example the Arizona desert, which has 180 kiloLangley per year. Sometimes this can be used with watersprays, which is referred Emmaqua to simulate a more humid climate. Sometimes the water contains up to 5 per cent sodium chloride, putting the sample in a permanent salt fog. It is typical for this to accelerate weathering by a factor 5, in comparison to weathering in Florida.

An Atrac in Townsville Australia, uses follow the sun technology in which the samples are rotated so that they always face the sun. In 17 months this produced the equivalent of 2 years of weathering.

Artificial Weathering

The weather testing process can be greatly accelerated through the use of specially designed weathering chambers. While this speeds up the time needed to get results, the conditions are not always representative of real world conditions.

Prior to the early1920's, chemists doubted the existence of molecules having molecular weights greater than a few thousand. This limiting view was challenged by Hermann Staudinger, a German chemist with experience in studying natural compounds such as rubber and cellulose. In contrast to the prevailing rationalization of these substances as aggregates of small molecules, Staudinger proposed they were made up of **macromolecules** composed of 10,000 or more atoms. He formulated a **polymeric** structure for rubber, based on a repeating isoprene unit (referred to as a monomer). For his contributions to chemistry, Staudinger received the 1953 Nobel Prize. The terms **polymer** and **monomer** were derived from the Greek roots poly (many), mono (one) and meros (part).

Recognition that polymeric macromolecules make up many important natural materials was followed by the creation of synthetic analogs having a variety of properties. Indeed, applications of these materials as fibers, flexible films, adhesives, resistant paints and tough but light solids have transformed modern society. Some important examples of these substances are discussed in the following sections.

Writing Formulas for Polymeric Macromolecules

The repeating structural unit of most simple polymers not only reflects the monomer(s) from which the polymers are constructed, but also provides a concise means for drawing structures to represent these macromolecules. For polyethylene, arguably the simplest polymer, this is demonstrated by the following equation. Here ethylene (ethene) is the monomer, and the corresponding linear polymer is called high-density polyethylene (HDPE). HDPE is composed of macromolecules in which n ranges from 10,000 to 100,000 (molecular weight $2{*}10^5$ to $3{*}10^6$).

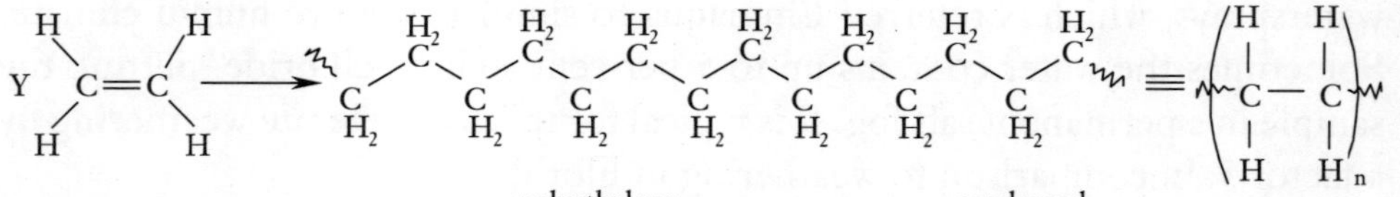

ethylene

polyethylene
a macromolecule

condensed
formula for HDPE

If Y and Z represent moles of monomer and polymer respectively, Z is approximately 10^{-5} Y. This polymer is called polyethylene rather than polymethylene, $(-CH_2-)_n$, because ethylene is a stable compound (methylene is not), and it also serves as the synthetic precursor of the polymer. The two open bonds remaining at the ends of the long chain of carbons (coloured magenta) are normally not specified, because the atoms or groups found there depend on the chemical process used for polymerization. The synthetic methods used to prepare this and other polymers will be described later in this chapter.

Unlike simpler pure compounds, most polymers are not composed of identical molecules. The HDPE molecules, for example, are all long carbon chains, but the lengths may vary by thousands of monomer units. Because of this, polymer molecular weights are usually given as averages. Two experimentally determined values are common: M_n, the number average molecular weight, is calculated from the mole fraction distribution of different sized molecules in a sample, and M_w, the weight average molecular weight, is calculated from the weight fraction distribution of different sized molecules. These are defined below. Since larger molecules in a sample weigh more than smaller molecules, the weight average M_w is necessarily skewed to higher values, and is always greater than M_n. As the weight dispersion of molecules in a sample narrows, M_w approaches M_n, and in the unlikely case that all the polymer molecules have identical weights (a pure monodisperse sample), the ratio M_w / M_n becomes unity.

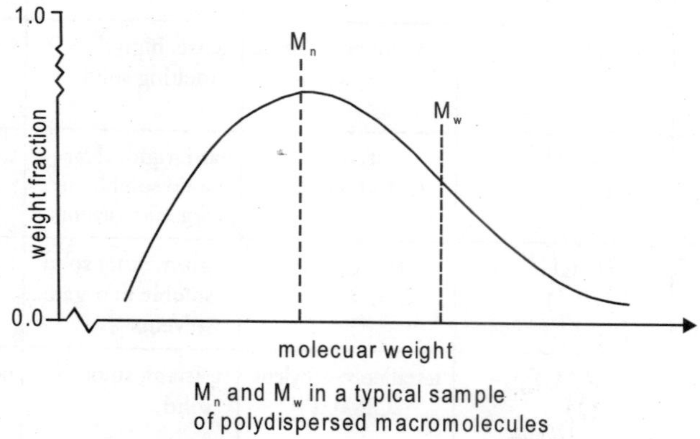

M_n and M_w in a typical sample of polydispersed macromolecules

$M_n = Sn_iM_i$ n_i = mole fraction of chains with molecular weight M_i
$M_w = \Sigma w_iM_i$ w_i = mole fraction of chains with molecular weight M_i

$$\text{note } w_i = \frac{n_i M_i}{M_n}$$

Many polymeric materials having chain-like structures similar to polyethylene are known. Polymers formed by a straightforward linking together of monomer units, with no loss or gain of material, are called addition polymers or chain-growth polymers. A listing of some important addition polymers and their monomer precursors is presented in the Table 11.1.

Properties of Macromolecules

A comparison of the properties of polyethylene (both LDPE & HDPE) with the natural polymers rubber and cellulose is instructive. As noted above, synthetic HDPE macromolecules have masses ranging from 10^5 to 10^6 amu

Table 11.1: Some Common Addition Polymers

Name(s)	Formula	Monomer	Properties	Uses
Polyethylene low density (LDPE)	$-(CH_2\text{-}CH_2)_n-$	ethylene $CH_2{=}CH_2$	soft, waxy solid	film wrap, plastic bags
Polyethylene high density (HDPE)	$-(CH_2\text{-}CH_2)_n-$	ethylene $CH_2{=}CH_2$	rigid, translucent solid	electrical insulation bottles, toys
Polypropylene (PP) different grades	$-[CH_2\text{-}CH(CH_3)]_n-$	propylene $CH_2{=}CHCH_3$	atactic: soft, elastic solid isotactic: hard, strong solid	similar to LDPE carpet, upholstery
Poly (vinyl chloride) (PVC)	$-(CH_2\text{-}CHCl)_n-$	vinyl chloride $CH_2{=}CHCl$	strong rigid solid	pipes, siding, flooring
Poly (vinylidene chloride) (Saran A)	$-(CH_2\text{-}CCl_2)_n-$	vinylidene chloride $CH_2{=}CCl_2$	dense, high-melting solid	seat covers, films
Polystyrene (PS)	$-[CH_2\text{-}CH(C_6H_5)]_n-$	styrene $CH_2{=}CHC_6H_5$	hard, rigid, clear solid soluble in organic solvents	toys, cabinets packaging (foamed)
Polyacrylonitrile (PAN, Orlon, Acrilan)	$-(CH_2\text{-}CHCN)_n-$	acrylonitrile $CH_2{=}CHCN$	high-melting solid soluble in organic solvents	rugs, blankets clothing
Polytetrafluoroethy lene (PTFE, Teflon)	$-(CF_2\text{-}CF_2)_n-$	tetrafluoro-ethylene $CF_2{=}CF_2$	resistant, smooth solid	non-stick surfaces electrical insulation
Poly (methyl methacrylate) (PMMA, Lucite, Plexiglas)	$-[CH_2\text{-}C(CH_3)CO_2CH_3]_n-$	methyl methacrylate $CH_2{=}C(CH_3)CO_2CH_3$	hard, transparent solid	lighting covers, signs skylights
Poly (vinyl acetate) (PVAc)	$-(CH_2\text{-}CHOCOCH_3)_n-$	vinyl acetate $CH_2{=}CHOCOCH_3$	soft, sticky solid	latex paints, adhesives
cis-Polyisoprene natural rubber	$-[CH_2\text{-}CH{=}C(CH_3)\text{-}CH_2]_n-$	isoprene $CH_2{=}CH\text{-}C(CH_3){=}CH_2$	soft, sticky solid	requires vulcanization for practical use
Polychloroprene (cis + trans) (Neoprene)	$-[CH_2\text{-}CH{=}CCl\text{-}CH_2]_n-$	chloroprene $CH_2{=}CH\text{-}CCl{=}CH_2$	tough, rubbery solid	synthetic rubber oil resistant

(LDPE molecules are more than a hundred times smaller). Rubber and cellulose molecules have similar mass ranges, but fewer monomer units because of the monomer's larger size. The physical properties of these three polymeric substances differ from each other, and of course from their monomers.

❖ HDPE is a rigid translucent solid which softens on heating above 100° C, and can be fashioned into various forms including films. It is not as easily stretched and deformed as is LDPE. HDPE is insoluble in water and most organic solvents, although some swelling may occur on immersion in the latter. HDPE is an excellent electrical insulator.

❖ LDPE is a soft translucent solid which deforms badly above 75° C. Films made from LDPE stretch easily and are commonly used for wrapping. LDPE is insoluble in water, but softens and swells on exposure to hydrocarbon solvents. Both LDPE and HDPE become brittle at very low temperatures (below -80° C). Ethylene, the common monomer for these polymers, is a low boiling (-104° C) gas.

❖ Natural (latex) rubber is an opaque, soft, easily deformable solid that becomes sticky when heated (above. 60° C), and brittle when cooled below -50° C. It swells to more than double its size in nonpolar organic solvents like toluene, eventually dissolving, but is impermeable to water. The C5H8 monomer isoprene is a volatile liquid (b.p. 34° C).

❖ Pure cellulose, in the form of cotton, is a soft flexible fiber, essentially unchanged by variations in temperature ranging from -70 to 80° C. Cotton absorbs water readily, but is unaffected by immersion in toluene or most other organic solvents. Cellulose fibers may be bent and twisted, but do not stretch much before breaking. The monomer of cellulose is the C6H12O6 aldohexose D-glucose. Glucose is a water soluble solid melting below 150° C.

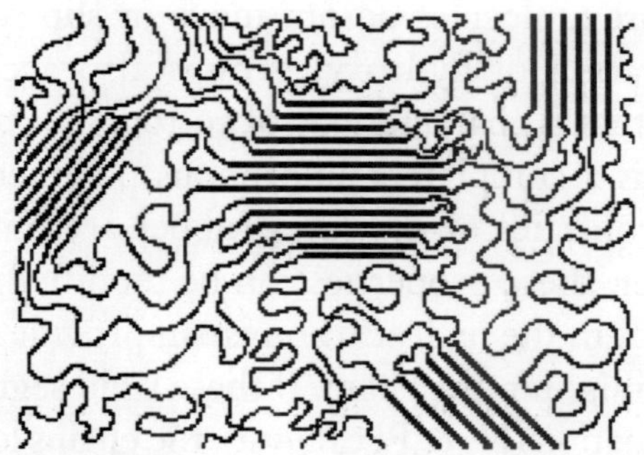

To account for the differences noted here we need to consider the nature of the aggregate macromolecular structure, or **morphology**, of each substance. Because polymer molecules are so large, they generally pack together in a non-uniform fashion, with ordered or crystalline-like regions mixed together with disordered or amorphous domains. In some cases the entire solid may be amorphous, composed entirely of coiled and tangled macromolecular chains. Crystallinity occurs when linear polymer chains are structurally oriented in a uniform three-dimensional matrix. In the diagram on the right, crystalline domains are coloured blue.

Increased crystallinity is associated with an increase in rigidity, tensile strength and opacity (due to light scattering). Amorphous polymers are usually less rigid, weaker and more easily deformed. They are often transparent.

Three factors that influence the degree of crystallinity are:

i) Chain length
ii) Chain branching
iii) Interchain bonding

The importance of the first two factors is nicely illustrated by the differences between LDPE and HDPE. As noted earlier, HDPE is composed of very long unbranched hydrocarbon chains. These pack together easily in crystalline domains that alternate with amorphous segments, and the resulting material, while relatively strong and stiff, retains a degree of flexibility. In contrast, LDPE is composed of smaller and more highly branched chains which do not easily adopt crystalline structures. This material is therefore softer, weaker, less dense and more easily deformed than HDPE. As a rule, mechanical properties such as ductility, tensile strength, and hardness rise and eventually level off with increasing chain length.

The nature of cellulose supports the above analysis and demonstrates the importance of the third factor (iii). To begin with, cellulose chains easily adopt a stable rod-like conformation. These molecules align themselves side by side into fibres that are stabilized by inter-chain hydrogen bonding between the three hydroxyl groups on each monomer unit. Consequently, crystallinity is high and the cellulose molecules do not move or slip relative to each other. The high concentration of hydroxyl groups also accounts for the facile absorption of water that is characteristic of cotton.

Natural rubber is a completely amorphous polymer. Unfortunately, the potentially useful properties of raw latex rubber are limited by temperature dependence; however, these properties can be modified by chemical change. The cis-double bonds in the hydrocarbon chain provide planar segments that stiffen, but do not straighten the chain. If these rigid segments are completely removed by hydrogenation (H_2 & Pt catalyst), the chains lose all constrainment, and the product is a low melting paraffin-like semisolid of little value. If instead, the chains of rubber molecules are slightly cross-linked by sulfur atoms, a process called **vulcanization** which was discovered by Charles Goodyear in 1839, the desireable elastomeric properties of rubber are substantially improved. At 2 to 3 per cent crosslinking a useful soft rubber, that no longer suffers stickyness and brittleness problems on heating and cooling, is obtained. At 25 to 35 per cent crosslinking a rigid hard rubber product is formed. The following illustration shows a cross-linked section of amorphous rubber. By clicking on the diagram it will change to a display of the corresponding stretched section.

The more highly-ordered chains in the stretched conformation are entropically unstable and return to their original coiled state when allowed to relax (click a second time).

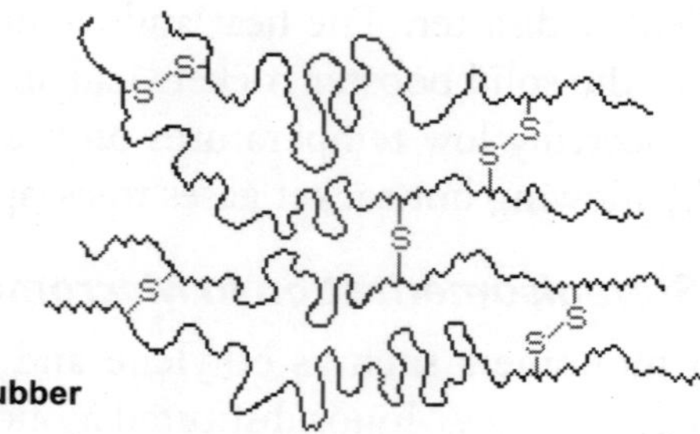

On heating or cooling most polymers undergo thermal transitions that provide insight into their morphology. These are defined as the **melt transition, T_m**, and the **glass transition, T_g**.

T_m is the temperature at which crystalline domains lose their structure, or melt. As crystallinity increases, so does T_m.

T_g is the temperature below which amorphous domains lose the structural mobility of the polymer chains and become rigid glasses.

T_g often depends on the history of the sample, particularly previous heat treatment, mechanical manipulation and annealing. It is sometimes interpreted as the temperature above which significant portions of polymer chains are able to slide past each other in response to an applied force. The introduction of relatively large and stiff substituents (such as benzene rings) will interfere with this chain movement, thus ncreasing T_g (note polystyrene below). The introduction of small molecular compounds called plasticisers into the polymer matrix increases the interchain spacing, allowing chain movement at lower temperatures. with a resulting decrease in T_g. The outgassing of plasticisers used to modify interior plastic components of automobiles produces the "new-car smell" to which we are accustomed.

T_m and T_g values for some common addition polymers are listed below. Note that cellulose has neither a T_m nor a T_g.

Polymer	*LDPE*	*HDPE*	*PP*	*PVC*	*PS*	*PAN*	*PTFE*	*PMMA*	*Rubber*
T_m (°C)	110	130	175	180	175	>200	330	180	30
T_g (°C)	-110	-110	-20	80	90	95	-110	105	-70

Rubber is a member of an important group of polymers called **elastomers**. Elastomers are amorphous polymers that have the ability to stretch and then return to their original shape at temperatures above T_g. This property is important in applications such as gaskets and O-rings, so the development of synthetic elastomers that can function under harsh or demanding conditions remains a practical goal. At temperatures below T_g elastomers become rigid glassy solids and lose all elasticity. A tragic example of this caused the space shuttle Challenger disaster. The heat and chemical resistant O-rings used to seal sections of the solid booster rockets had an unfortunately high T_g near 0 °C. The unexpectedly low temperatures on the morning of the launch were below this T_g, allowing hot rocket gases to escape the seals.

Regio and Stereoisomerization in Macromolecules

Symmetrical monomers such as ethylene and tetrafluoroethylene can join together in only one way. Monosubstituted monomers, on the other hand, may join together in two organized ways, described in the following diagram, or in a third random manner. Most monomers of this kind, including propylene, vinyl chloride, styrene, acrylonitrile and acrylic esters, prefer to join in a head-to-tail fashion, with some randomness occurring from time to time. The reasons for this regioselectivity will be discussed in the synthetic methods section.

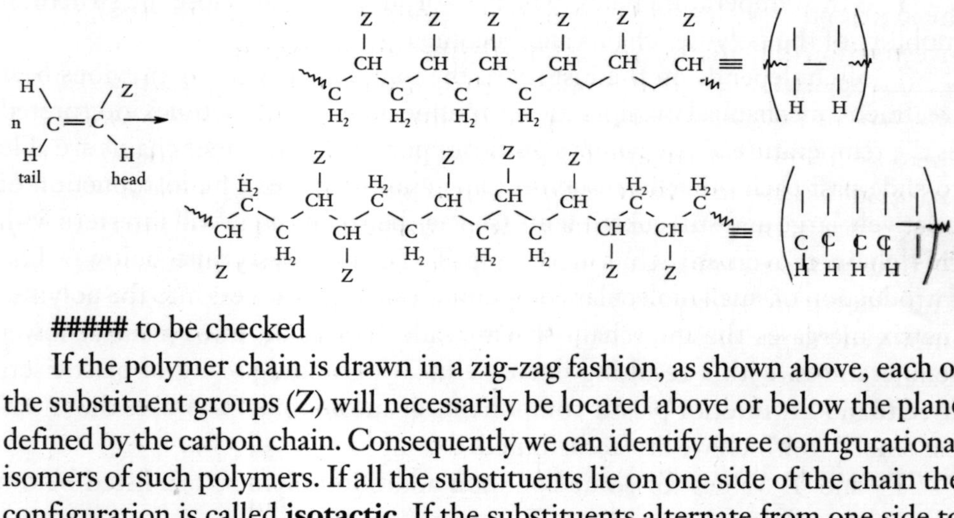

tail head

to be checked

If the polymer chain is drawn in a zig-zag fashion, as shown above, each of the substituent groups (Z) will necessarily be located above or below the plane defined by the carbon chain. Consequently we can identify three configurational isomers of such polymers. If all the substituents lie on one side of the chain the configuration is called **isotactic**. If the substituents alternate from one side to another in a regular manner the configuration is termed **syndiotactic**. Finally, a random arangement of substituent groups is referred to as **atactic**. Examples of these configurations are shown here.

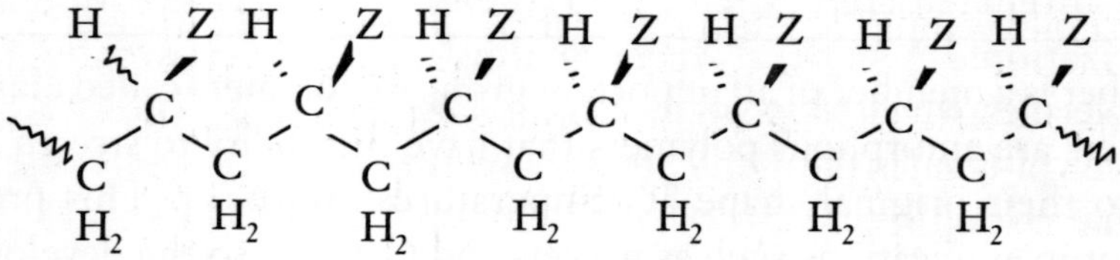

isotactic

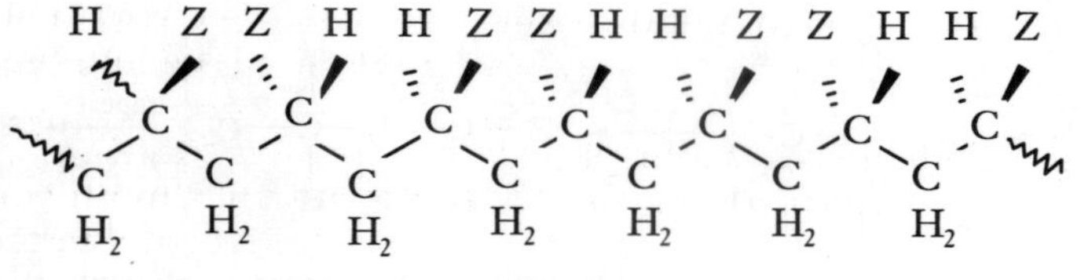

syndiotactic

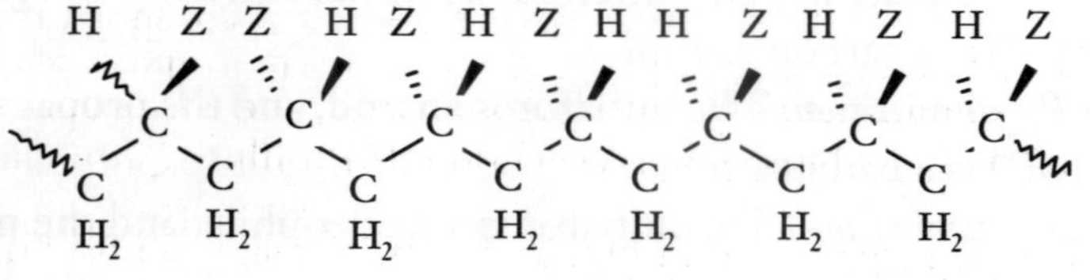

atactic

Many common and useful polymers, such as polystyrene, polyacrylonitrile and poly (vinyl chloride) are atactic as normally prepared. Customized catalysts that effect stereoregular polymerization of polypropylene and some other monomers have been developed, and the improved properties associated with the increased crystallinity of these products has made this an important field of investigation. The following values of T_g have been reported.

Polymer	T_g *atactic*	T_g *isotactic*	T_g *syndiotactic*
PP	–20 °C	0 °C	–8 °C
PMMA	100 °C	130 °C	120 °C

The properties of a given polymer will vary considerably with its tacticity. Thus, atactic polypropylene is useless as a solid construction material, and is employed mainly as a component of adhesives or as a soft matrix for composite materials. In contrast, isotactic polypropylene is a high-melting solid (ca. 170 °C) which can be molded or machined into structural components.

Synthesis of Addition Polymers

All the monomers from which addition polymers are made are alkenes or functionally substituted alkenes. The most common and thermodynamically favoured chemical transformations of alkenes are addition reactions. Many of these addition reactions are known to proceed in a stepwise fashion by way of reactive intermediates, and this is the mechanism followed by most polymerizations. A general diagram illustrating this assembly of linear macromolecules, which supports the name **chain growth polymers**, is presented here. Since a pi-bond in the monomer is converted to a sigma-bond in the polymer, the polymerization reaction is usually exothermic by 8 to 20 kcal/mol. Indeed, cases of explosively uncontrolled polymerizations have been reported.

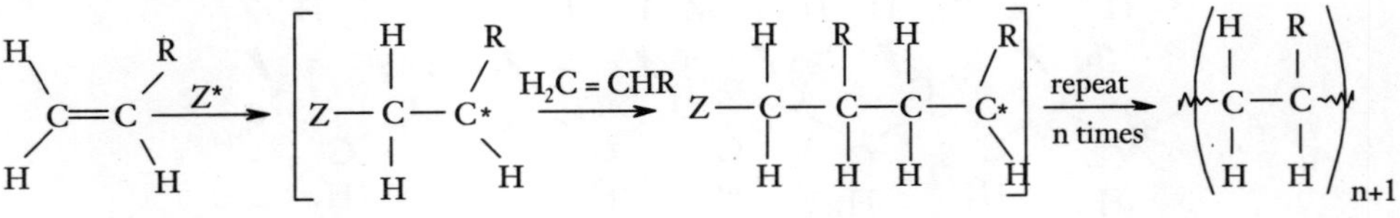

$$\mathrm{H_2C{=}CHR} \xrightarrow{Z^*} \left[\mathrm{Z{-}CH_2{-}C^*HR}\right] \xrightarrow{\mathrm{H_2C=CHR}} \left[\mathrm{Z{-}CH_2{-}CHR{-}CH_2{-}C^*HR}\right] \xrightarrow[\text{n times}]{\text{repeat}} \left(\mathrm{{-}CH_2{-}CHR{-}}\right)_{n+1}$$

Z* is an initiating species

[* may be a radical, a cation or an anion]

7◆ *Radical Polymerization*: The initiator is a radical, and the propagating site of reactivity (*) is a carbon radical.

◆ *Cationic Polymerization*: The initiator is an acid, and the propagating site of reactivity (*) is a carbocation.

◆ *Anionic Polymerization*: The initiator is a nucleophile, and the propagating site of reactivity (*) is a carbanion.

◆ *Coordination Catalytic Polymerization*: The initiator is a transition metal complex, and the propagating site of reactivity (*) is a terminal catalytic complex.

Radical Chain-Growth Polymerization

Virtually all of the monomers described above are subject to radical polymerization. Since this can be initiated by traces of oxygen or other minor impurities, pure samples of these compounds are often "stabilized" by small amounts of radical inhibitors to avoid unwanted reaction. When radical polymerization is desired, it must be started by using a **radical initiator**, such as a peroxide or certain azo compounds. The formulas of some common initiators, and equations showing the formation of radical species from these initiators are presented below.

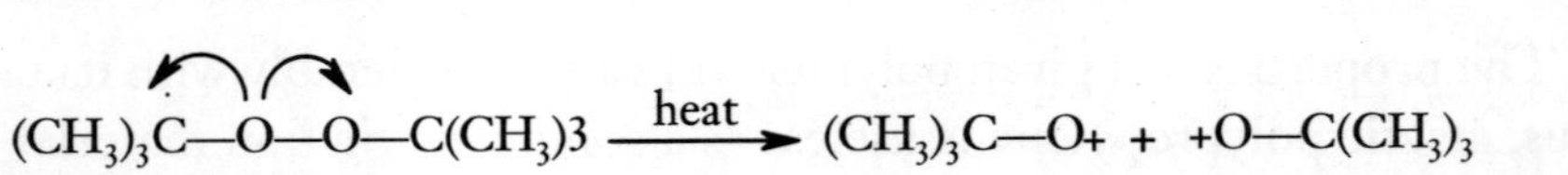

$$\mathrm{(CH_3)_3C{-}O{-}O{-}C(CH_3)3} \xrightarrow{\text{heat}} \mathrm{(CH_3)_3C{-}O\bullet + \bullet O{-}C(CH_3)_3}$$

tert-butyl peroxide

$$\mathrm{C_6H_5{-}C(=O){-}O{-}O{-}C(=O){-}C_6H_5} \xrightarrow{\text{heat}} \mathrm{C_6H_5{-}C(=O)O\bullet + \bullet OC(=O){-}C_6H_5}$$

benzoyl peroxide

$$\mathrm{H_3C{-}C(CN)(CH_3){-}N{=}N{-}C(CN)(CH_3){-}CH_3} \xrightarrow{\text{heat}} \mathrm{2\ H_3C{-}C\bullet(CN)(CH_3) + N_2}$$

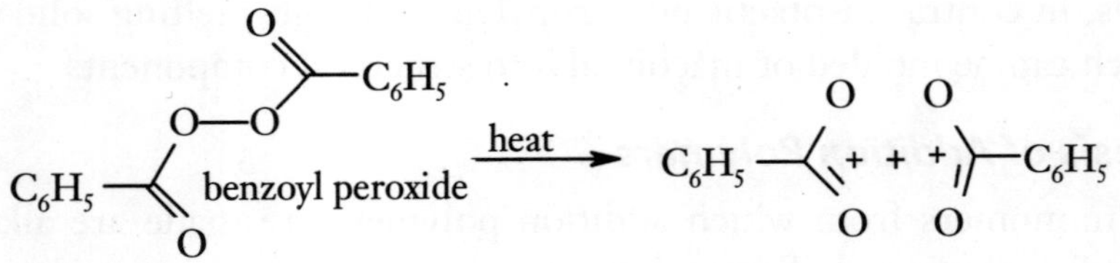

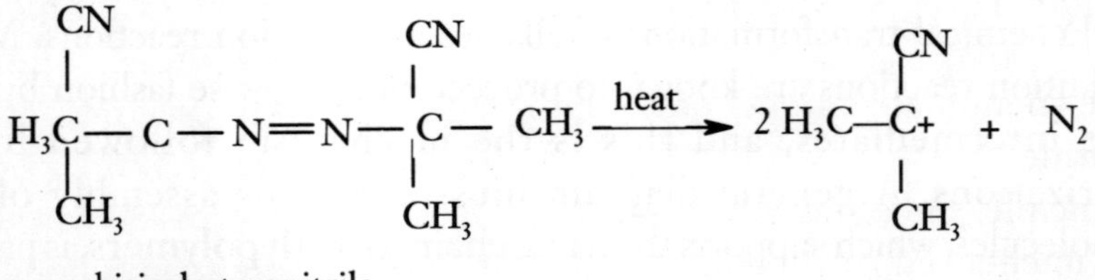

azobisisobutyronitrile

Subsequent Reactions

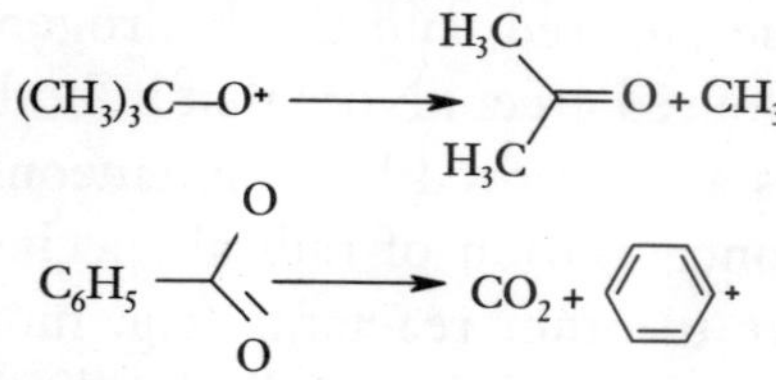

$$(CH_3)_3C{-}O\bullet \longrightarrow (H_3C)_2C{=}O + CH_3\bullet$$

$$C_6H_5{-}C(=O)O\bullet \longrightarrow CO_2 + C_6H_5\bullet$$

By using small amounts of initiators, a wide variety of monomers can be polymerized. One example of this radical polymerization is the conversion of styrene to polystyrene, shown in the following diagram. The first two equations illustrate the **initiation** process, and the last two equations are examples of **chain propagation**. Each monomer unit adds to the growing chain in a manner that generates the most stable radical. Since carbon radicals are stabilized by substituents of many kinds, the preference for head-to-tail regioselectivity in most addition polymerizations is understandable. Because radicals are tolerant of many functional groups and solvents (including water), radical polymerizations are widely used in the chemical industry.

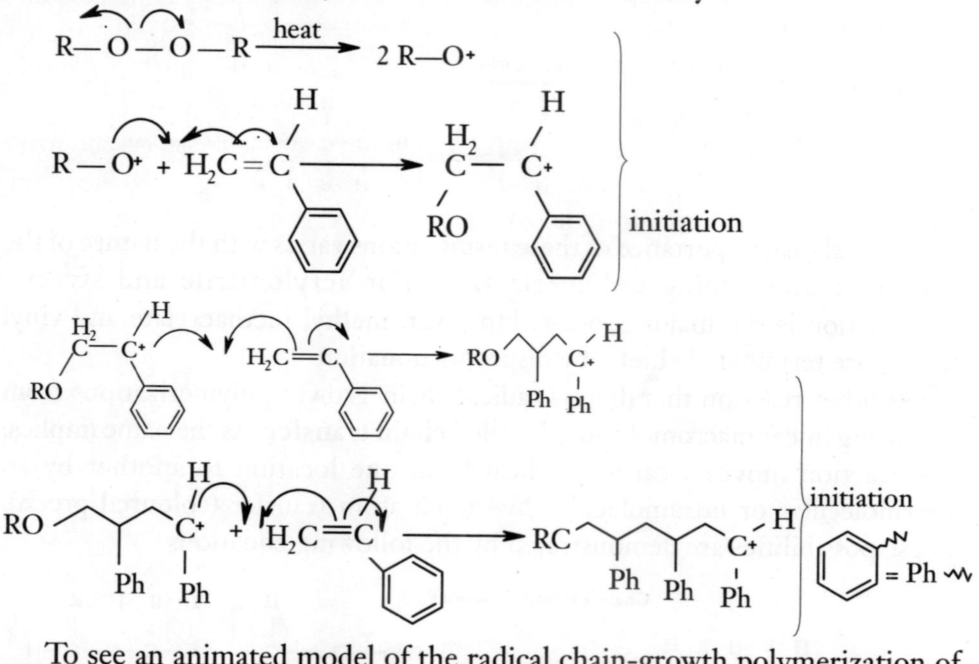

$$R{-}O{-}O{-}R \xrightarrow{\text{heat}} 2\,R{-}O\bullet$$

$$R{-}O\bullet + H_2C{=}CH(C_6H_5) \longrightarrow RO{-}CH_2{-}\dot{C}H(C_6H_5)$$

initiation

$$RO{-}CH_2{-}\dot{C}H(C_6H_5) + H_2C{=}CH(C_6H_5) \longrightarrow RO{-}CH_2{-}CH(Ph){-}CH_2{-}\dot{C}H(Ph)$$

$$RO{-}CH_2{-}CH(Ph){-}CH_2{-}\dot{C}H(Ph) + H_2C{=}CH(C_6H_5) \longrightarrow RO{-}CH_2{-}CH(Ph){-}CH_2{-}CH(Ph){-}CH_2{-}\dot{C}H(Ph)$$

initiation

C_6H_5– = Ph

To see an animated model of the radical chain-growth polymerization of vinyl chloride

In principle, once started a radical polymerization might be expected to continue unchecked, producing a few extremely long chain polymers. In practice, larger numbers of moderately sized chains are formed, indicating that chain-terminating reactions must be taking place. The most common termination processes are **Radical Combination** and **Disproportionation.**

These reactions are illustrated by the following equations. The growing polymer chains are coloured blue and red, and the hydrogen atom transferred in disproportionation is coloured green. Note that in both types of termination two reactive radical sites are removed by simultaneous convertion to stable product(s). Since the concentration of radical species in a polymerization reaction is small relative to other reactants (e.g. monomers, solvents and terminated chains), the rate at which these radical-radical termination reactions occurs is very small, and most growing chains achieve moderate length before termination.

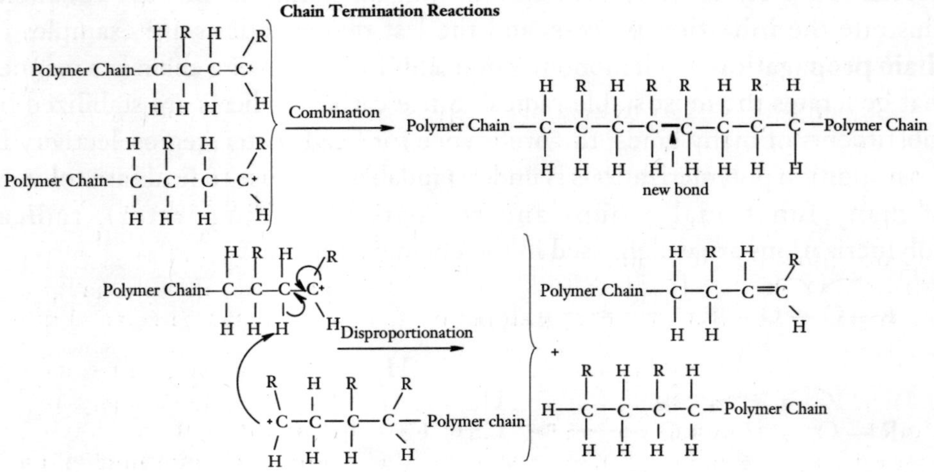

Chain Termination Reactions

Combination

new bond

Disproportionation

The relative importance of these terminations varies with the nature of the monomer undergoing polymerization. For acrylonitrile and styrene combination is the major process. However, methyl methacrylate and vinyl acetate are terminated chiefly by disproportionation.

Another reaction that diverts radical chain-growth polymerizations from producing linear macromolecules is called **chain transfer**. As the name implies, this reaction moves a carbon radical from one location to another by an intermolecular or intramolecular hydrogen atom transfer (coloured green). These possibilities are demonstrated by the following equations

Chain Transfer Reactions

Intermoleculer Hydrogen Transfer

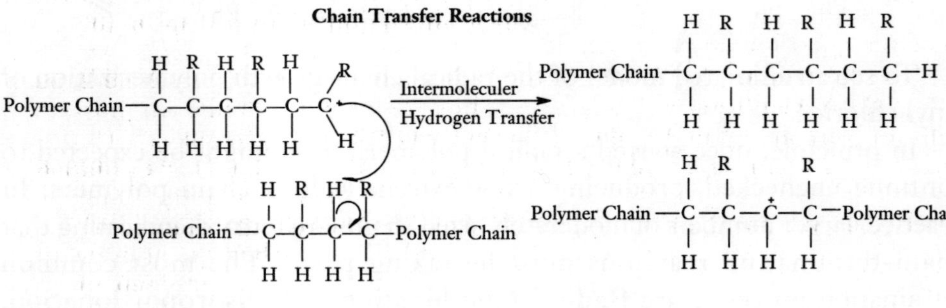

Polymer Chain — C(H)(H) — C(R)(H) — CH_2 — CHR — CH_2 — $\overset{+}{C}$ → (Intramolecular Hydrogen Transfer) → Polymer Chain — C(H)(H) — $\overset{+}{C}$(R) — C(H)(H) — C(R)(H) — C(H)(H) — C(R)(H) — H

Chain transfer reactions are especially prevalent in the high pressure radical polymerization of ethylene, which is the method used to make LDPE (low density polyethylene). The 1°-radical at the end of a growing chain is converted to a more stable 2°-radical by hydrogen atom transfer. Further polymerization at the new radical site generates a side chain radical, and this may in turn lead to creation of other side chains by chain transfer reactions. As a result, the morphology of LDPE is an amorphous network of highly branched macromolecules.

Cationic Chain-Growth Polymerization

Polymerization of isobutylene (2-methylpropene) by traces of strong acids is an example of cationic polymerization. The polyisobutylene product is a soft rubbery solid, T_g = -70° C, which is used for inner tubes. This process is similar to radical polymerization, as demonstrated by the following equations.

$H_2C=C(CH_3)_2$ → (BF_3, H_2O (trace), CH_2Cl_2, –78°C) → $H-CH_2-C^{\oplus}(CH_3)_2$ + $H_2C=C(CH_3)_2$ → $H-CH_2-C(CH_3)_2-CH_2-C^{\oplus}(CH_3)_2$ + $H_2C=C(CH_3)_2$ →

$H-[CH_2-C(CH_3)_2]_2-CH_2-C^{\oplus}(CH_3)_2$ → (repeat) → Polymer Chain $-[CH_2-C(CH_3)_2]_n-CH_2-C(CH_3)=CH_2$

Polyisobutylene (Butyl Rubber) [one possible term group is shown]

Monomers bearing cation stabilizing groups, such as alkyl, phenyl or vinyl can be polymerized by cationic processes. These are normally initiated at low temperature in methylene chloride solution. At low temperatures, chain transfer reactions are rare in such polymerizations, so the resulting polymers are cleanly linear (unbranched).

Anionic Chain-Growth Polymerization

Treatment of a cold THF solution of styrene with 0.001 equivalents of n-butyllithium causes an immediate polymerization. Chain growth may be terminated by water or carbon dioxide, and chain transfer seldom occurs. Only monomers having anion stabilizing substituents, such as phenyl, cyano or carbonyl are good substrates for this polymerization technique. Many of the resulting polymers are largely isotactic in configuration, and have high degrees of crystallinity.

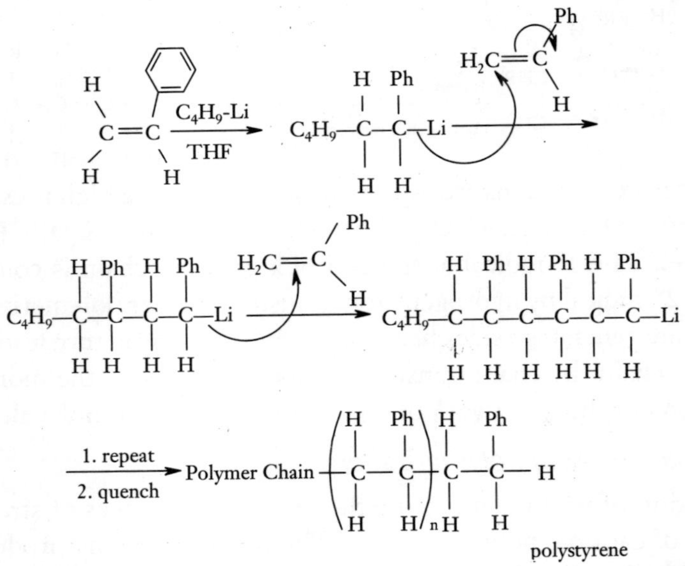

$$\mathrm{H_2C{=}CH(C_6H_5)} \xrightarrow[\text{THF}]{\mathrm{C_4H_9\text{-}Li}} \mathrm{C_4H_9{-}CH_2{-}CH(Ph){-}Li} \xrightarrow{\mathrm{H_2C{=}CH(Ph)}}$$

$$\mathrm{C_4H_9{-}CH_2{-}CH(Ph){-}CH_2{-}CH(Ph){-}Li} \xrightarrow{\mathrm{H_2C{=}CH(Ph)}} \mathrm{C_4H_9{-}CH_2{-}CH(Ph){-}CH_2{-}CH(Ph){-}CH_2{-}CH(Ph){-}Li}$$

$$\xrightarrow[\text{2. quench}]{\text{1. repeat}} \text{Polymer Chain}{-}\left(\mathrm{CH_2{-}CH(Ph)}\right)_n{-}\mathrm{CH_2{-}CH(Ph){-}H}$$

polystyrene

Species that have been used to initiate anionic polymerization include alkali metals, alkali amides, alkyl lithiums and various electron sources. A practical application of anionic polymerization occurs in the use of superglue. This material is methyl 2-cyanoacrylate, $CH_2=C(CN)CO_2CH_3$. When exposed to water, amines or other nucleophiles, a rapid polymerization of this monomer takes place.

Ziegler-Natta Catalytic Polymerization

An efficient and stereospecific catalytic polymerization procedure was developed by Karl Ziegler (Germany) and Giulio Natta (Italy) in the 1950's. Their findings permitted, for the first time, the synthesis of unbranched, high molecular weight polyethylene (HDPE), laboratory synthesis of natural rubber from isoprene, and configurational control of polymers from terminal alkenes like propene (e.g. pure isotactic and syndiotactic polymers). In the case of ethylene, rapid polymerization occurred at atmospheric pressure and moderate to low temperature, giving a stronger (more crystalline) product (HDPE) than that from radical polymerization (LDPE). For this important discovery these chemists received the 1963 Nobel Prize in chemistry.

Ziegler-Natta catalysts are prepared by reacting certain transition metal halides with organometallic reagents such as alkyl aluminum, lithium and zinc reagents. The catalyst formed by reaction of triethylaluminum with titanium tetrachloride has been widely studied, but other metals (e.g. V & Zr) have also proven effective. The following diagram presents one mechanism for this useful reaction. Others have been suggested, with changes to accomodate the heterogeneity or homogeneity of the catalyst. Polymerization of propylene through action of the titanium catalyst gives an isotactic product; whereas, a vanadium based catalyst gives a syndiotactic product.

A Mechanism for Ziegler-Natta Catalysis

'R–M

M = Al, Li, Mg, Zn...

L is an unspecified ligand

$H_2C{=}CHR$ → rearrange → $H_2C{=}CHR$ → rearrange → repeat many times → Final Product

Copolymers

The synthesis of macromolecules composed of more than one monomeric repeating unit has been explored as a means of controlling the properties of the resulting material. In this respect, it is useful to distinguish several ways in which different monomeric units might be incorporated in a polymeric molecule. The following examples refer to a two component system, in which one monomer is designated A and the other B.

Statistical Copolymers	Also called random copolymers. Here the monomeric units are distributed randomly, and sometimes unevenly, in the polymer chain: **~ABBAAABAABBBABAABA~**.
Alternating Copolymers	Here the monomeric units are distributed in a regular alternating fashion, with nearly equimolar amounts of each in the chain: **~ABABABABABABABAB~**.
Block Copolymers	Instead of a mixed distribution of monomeric units, a long sequence or block of one monomer is joined to a block of the second monomer: **~AAAAA-BBBBBBB~AAAAAAA~BBB~**.
Graft Copolymers	As the name suggests, side chains of a given monomer are attached to the main chain of the second monomer: **~AAAAAAA(BBBBBBB~)AAAAAAA(BBBB~)AAA~**.

Addition Copolymerization

Most direct copolymerizations of equimolar mixtures of different monomers give statistical copolymers. The copolymerization of styrene with methyl methacrylate, for example, proceeds differently depending on the mechanism. Radical polymerization gives a statistical copolymer. However, the product of cationic polymerization is largely polystyrene, and anionic polymerization favours formation of poly (methyl methacrylate). In cases where the relative reactivities are different, the copolymer composition can sometimes be controlled by continuous introduction of a biased mixture of monomers into the reaction.

Formation of alternating copolymers is favoured when the monomers have different polar substituents (e.g. one electron withdrawing and the other electron donating), and both have similar reactivities toward radicals. For example, styrene and acrylonitrile copolymerize in a largely alternating fashion.

Some Useful Copolymers

Monomer A	*Monomer B*	*Copolymer*	*Uses*
$H_2C{=}CHCl$	$H_2C{=}CCl_2$	Saran	films & fibers
$H_2C{=}CHC_6H_5$	$H_2C{=}C\text{-}CH{=}CH_2$	SBR	
styrene butadiene rubber	tires		
$H_2C{=}CHCN$	$H_2C{=}C\text{-}CH{=}CH_2$	Nitrile Rubber	adhesives
hoses			
$H_2C{=}C(CH_3)_2$	$H_2C{=}C\text{-}CH{=}CH_2$	Butyl Rubber	inner tubes
$F_2C{=}CF(CF_3)$	$H_2C{=}CHF$	Viton	gaskets

A terpolymer of acrylonitrile, butadiene and styrene, called ABS rubber, is used for high-impact containers, pipes and gaskets.

Block Copolymerization

Several different techniques for preparing block copolymers have been developed, many of which use condensation reactions (next section). At this point, our discussion will be limited to an application of anionic polymerization. In the anionic polymerization of styrene described above, a reactive site remains at the end of the chain until it is quenched. The unquenched polymer has been termed a **living polymer**, and if additional styrene or a different suitable monomer is added a block polymer will form. This is illustrated for methyl methacrylate in the following diagram.

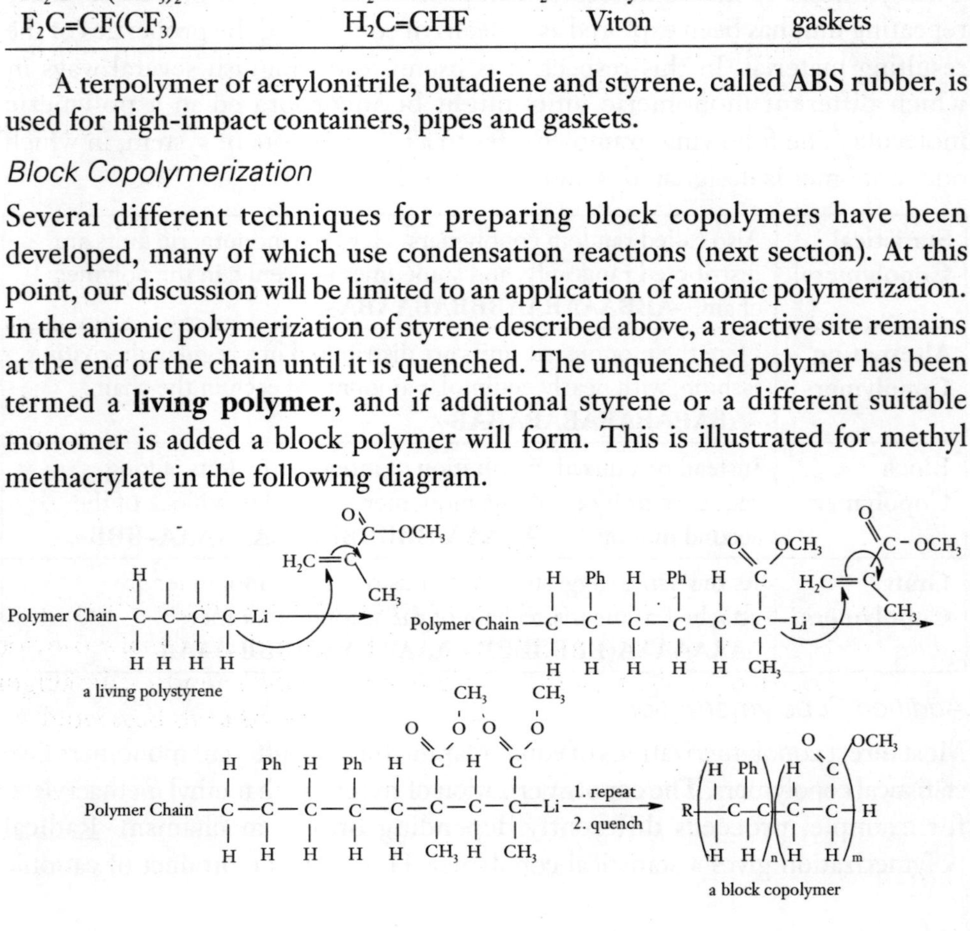

Polymer Chain–C–C–C–C–Li + $H_2C{=}C(CH_3)C(=O)OCH_3$ → Polymer Chain–C–C–C–C–C(CH$_3$)(C(=O)OCH$_3$)–Li + $H_2C{=}C(CH_3)C(=O)OCH_3$ →

a living polystyrene

Polymer Chain–C–C–C–C–C–C–C–C–Li (1. repeat, 2. quench) → R–[CH$_2$–CHPh]$_n$–[CH$_2$–C(CH$_3$)(CO$_2$CH$_3$)]$_m$–H

a block copolymer

Condensation Polymers

A large number of important and useful polymeric materials are not formed by chain-growth processes involving reactive species such as radicals, but proceed instead by conventional functional group transformations of polyfunctional reactants. These polymerizations often (but not always) occur with loss of a small byproduct, such as water, and generally (but not always) combine two different components in an alternating structure. The polyester Dacron and the polyamide Nylon 66, shown here, are two examples of synthetic condensation polymers, also known as **step-growth** polymers. In contrast to chain-growth polymers, most of which grow by carbon-carbon bond formation, step-growth polymers generally grow by carbon-heteroatom bond formation (C-O & C-N in Dacron & Nylon respectively). Although polymers of this kind might be considered to be alternating copolymers, the repeating monomeric unit is usually defined as a combined moiety. Examples of naturally occurring condensation polymers are cellulose, the polypeptide chains of proteins, and poly (â-hydroxybutyric acid), a polyester synthesized in large quantity by certain soil and water bacteria.

Examples of Condensation Polymers

Polyester

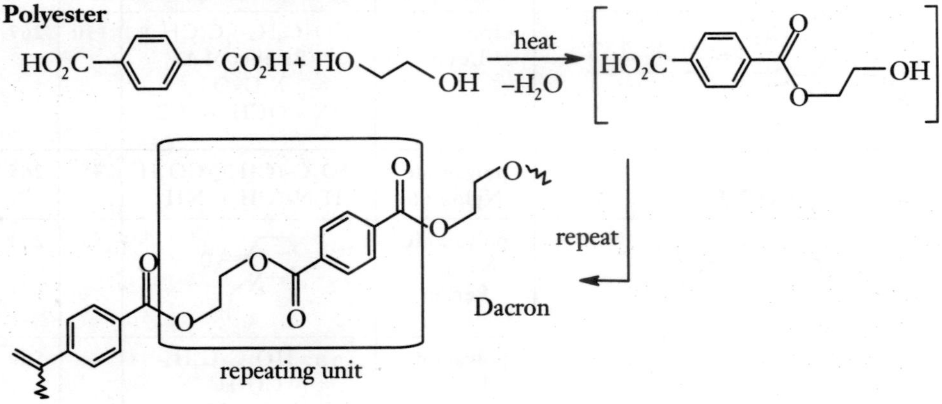

HO_2C–C₆H₄–CO_2H + HO–CH₂CH₂–OH $\xrightarrow[-H_2O]{\text{heat}}$ [HO_2C–C₆H₄–C(=O)–O–CH₂CH₂–OH]

repeat

Dacron

repeating unit

Polymide

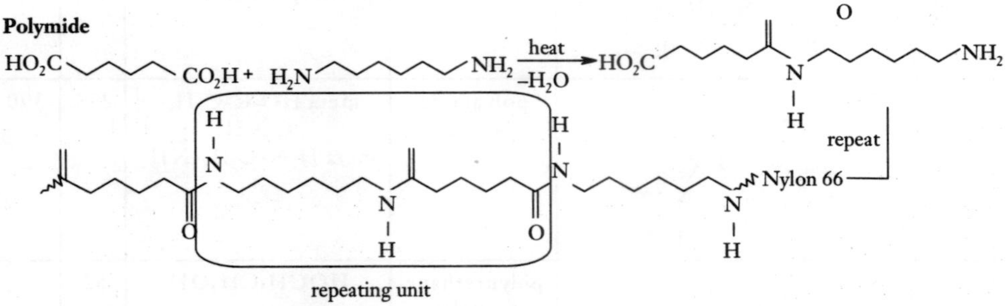

HO_2C–(CH₂)₄–CO_2H + H_2N–(CH₂)₆–NH_2 $\xrightarrow[-H_2O]{\text{heat}}$ HO_2C–(CH₂)₄–C(=O)–N(H)–(CH₂)₆–NH_2

repeat

Nylon 66

repeating unit

Characteristics of Condensation Polymers

Condensation polymers form more slowly than addition polymers, often requiring heat, and they are generally lower in molecular weight. The terminal functional groups on a chain remain active, so that groups of shorter chains combine into longer chains in the late stages of polymerization. The presence of polar functional groups on the chains often enhances chain-chain attractions, particularly if these involve hydrogen bonding, and thereby crystallinity and tensile strength. The following examples of condensation polymers are illustrative.

Some Condensation Polymers

Formula	*Type*	*Components*	T_g °C	T_m °C
$\sim[CO(CH_2)_4CO\text{-}OCH_2CH_2O]_n\sim$	polyester	$HO_2C\text{-}(CH_2)_4\text{-}CO_2H$ $HO\text{-}CH_2CH_2\text{-}OH$	< 0	50
	polyester Dacron Mylar	para $HO_2C\text{-}C_6H_4\text{-}CO_2H$ $HO\text{-}CH_2CH_2\text{-}OH$	70	265
	polyester	meta $HO_2C\text{-}C_6H_4\text{-}CO_2H$ $HO\text{-}CH_2CH_2\text{-}OH$	50	240
	polycarbonate Lexan	$(HO\text{-}C_6H_4\text{-})_2C(CH_3)_2$ (Bisphenol A) $X_2C{=}O$ ($X = OCH_3$ or Cl)	150	267
$\sim[CO(CH_2)_4CO\text{-}NH(CH_2)_6NH]_n\sim$	polyamide Nylon 66	$HO_2C\text{-}(CH_2)_4\text{-}CO_2H$ $H_2N\text{-}(CH_2)_6\text{-}NH_2$	45	265
$\sim[CO(CH_2)_5NH]_n\sim$	polyamide Nylon 6 Perlon		53	223
	polyamide Kevlar	para $HO_2C\text{-}C_6H_4\text{-}CO_2H$ para $H_2N\text{-}C_6H_4\text{-}NH_2$	—	500
	polyamide Nomex	meta $HO_2C\text{-}C_6H_4\text{-}CO_2H$ meta $H_2N\text{-}C_6H_4\text{-}NH_2$	273	390
	polyurethane Spandex	$HOCH_2CH_2OH$	52	—

Note that for commercial synthesis the carboxylic acid components may actually be employed in the form of derivatives such as simple esters. Also, the polymerization reactions for Nylon 6 and Spandex do not proceed by elimination of water or other small molecules. Nevertheless, the polymer clearly forms by a step-growth process.

The difference in T_g and T_m between the first polyester (completely aliphatic) and the two nylon polyamides (5th & 6th entries) shows the effect of intra-chain hydrogen bonding on crystallinity. The replacement of flexible alkylidene links with rigid benzene rings also stiffens the polymer chain, leading to increased crystalline character, as demonstrated for polyesters (entries 1, 2 &3) and polyamides (entries 5, 6, 7 & 8). The high T_g and T_m values for the amorphous polymer Lexan are consistent with its brilliant transparency and glass-like rigidity. Kevlar and Nomex are extremely tough and resistant materials, which find use in bullet-proof vests and fire resistant clothing.

Interchain Hydrogen Bonding Enhances Crystallinity

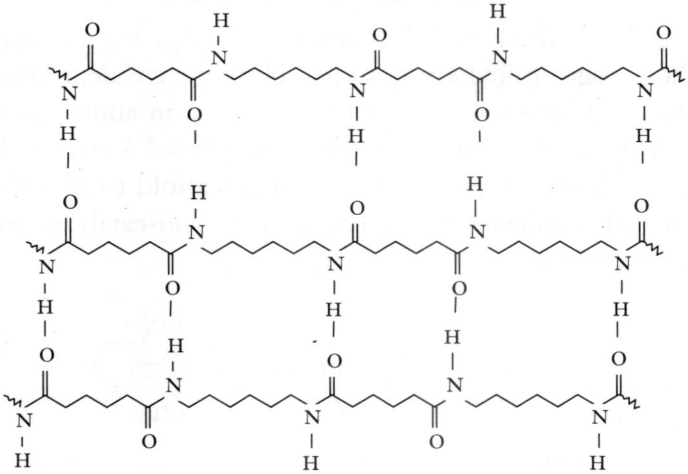

Many polymers, both addition and condensation, are used as fibers The chief methods of spinning synthetic polymers into fibers are from melts or viscous solutions. Polyesters, polyamides and polyolefins are usually spun from melts, provided the T_m is not too high. Polyacrylates suffer thermal degradation and are therefore spun from solution in a volatile solvent. **Cold-drawin** is an important physical treatment that improves the strength and appearance of these polymer fibers. At temperatures above T_g, a thicker than desired fiber can be forcibly stretched to many times its length; and in so doing the polymer chains become untangled, and tend to align in a parallel fashion. This cold-drawing procedure organizes randomly oriented crystalline domains, and also aligns amorphous domains so they become more crystalline. In these cases, the physically oriented morphology is stabilized and retained in the final product. This contrasts with elastomeric polymers, for which the stretched or aligned morphology is unstable relative to the amorphous random coil morphology. By clicking on the following diagram, a cartoon of these changes will toggle from one extreme to the other. This cold-drawing treatment may also be used to treat polymer films (e.g. Mylar & Saran) as well as fibers.

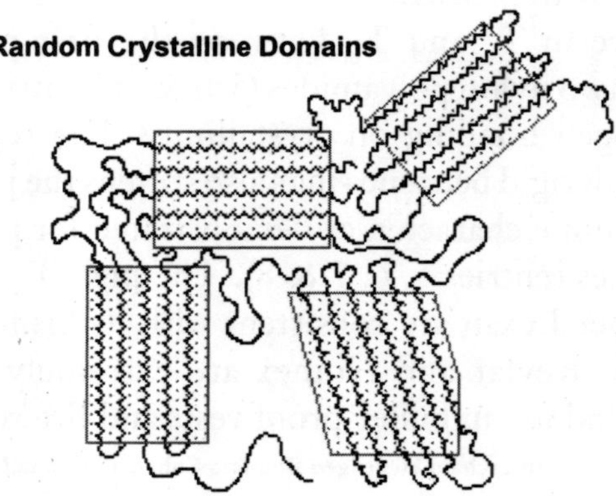

Step-growth polymerization is also used for preparing a class of adhesives and amorphous solids called epoxy resins. Here the covalent bonding occurs by an S_N2 reaction between a nucleophile, usually an amine, and a terminal epoxide. In the following example, the same bisphenol A intermediate used as a monomer for Lexan serves as a difunctional scaffold to which the epoxide rings are attached. Bisphenol A is prepared by the acid-catalyzed condensation of acetone with phenol.

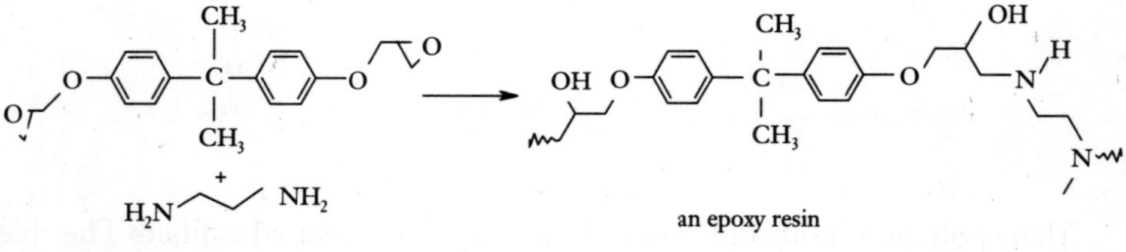

an epoxy resin

Thermosetting vs. Thermoplastic Polymers

Most of the polymers described above are classified as **thermoplastic.** This reflects the fact that above T_g they may be shaped or pressed into molds, spun or cast from melts or dissolved in suitable solvents for later fashioning. Because of their high melting point and poor solubility in most solvents, Kevlar and Nomex proved to be a challenge, but this was eventually solved. Another group of polymers, characterized by a high degree of cross-linking, resist deformation and solution once their final morphology is achieved. Such polymers are usually prepared in molds that yield the desired object. Because these polymers, once formed, cannot be reshaped by heating, they are called **thermosets**. Partial formulas for four of these will be shown below by clicking the appropriate button. The initial display is of Bakelite, one of the first completely synthetic plastics to see commercial use (circa 1910).

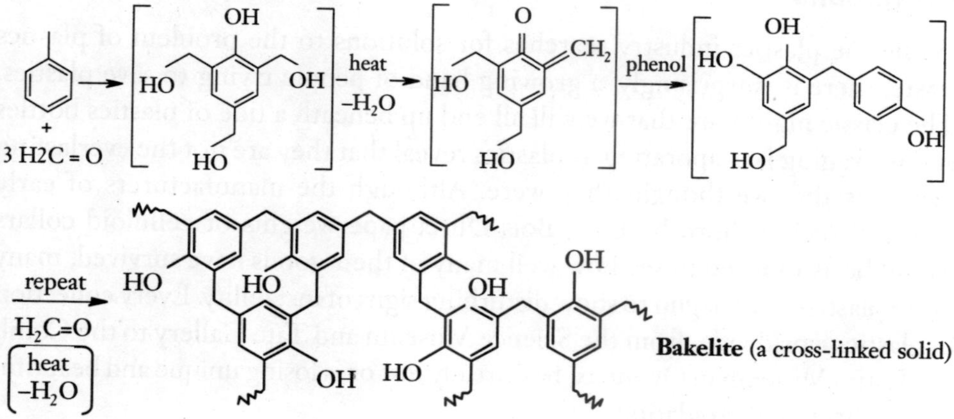

Bakelite (a cross-linked solid)

A natural resinous polymer called lignin has a cross-linked structure similar to bakelite. Lignin is the amorphous matrix in which the cellulose fibers of wood are oriented. Wood is a natural composite material, nature's equivalent of fiberglass and carbon fiber composites. A partial structure for lignan is shown here.

Partial Constitution of Lignan

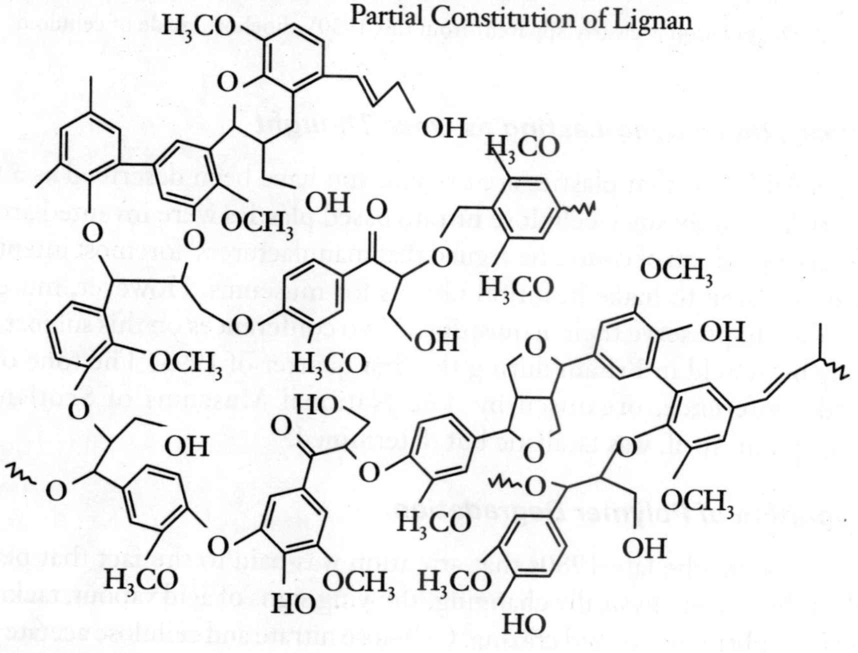

Problems

The following problems focus on concepts and facts associated with the trearment of polymer chemistry in this text.

11.3 Degradation of Polymers

Background

While the plastics industry searches for solutions to the problem of plastics waste, there is, surprisingly, a growing band of people trying to save plastics. The classic nightmare that we will all end up beneath a tide of plastics bottles and packaging is evaporating as plastics reveal that they are not the everlasting materials that we thought they were. Although the manufacturers of early 'plastics' such as horn buttons, Bois Durci paperweights or celluloid collars would be astonished to see how well many of their goods have survived, many other plastics have begun to show disturbing signs of instability. Every collection of plastics worldwide, from the Science Museum and Tate Gallery to the Comb & Plastics Museum in Oyonnax, has already lost or is losing unique and beautiful pieces through degradation.

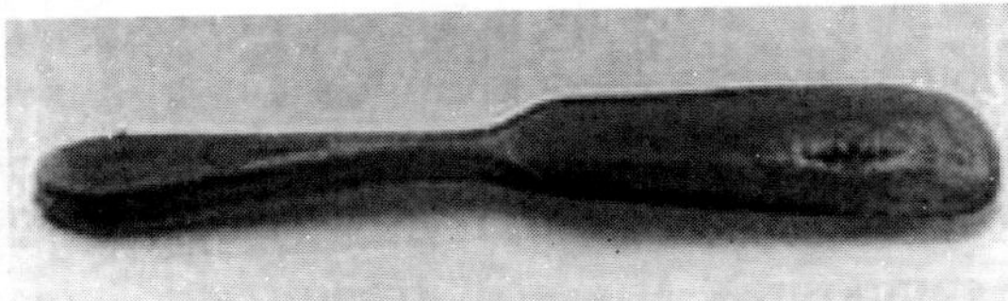

Fig. 11.1: Degradation is clearly apparent from this 1930's shoehorn made of celluloid

Plastics, Not as Long-Lasting as Once Thought

The crucial fact is that plastics are organic and have been described as a time bomb ticking away since cellulose nitrate based plastics were invented around 130 years ago. It can of course be argued that manufacturers' foremost intentions have never been to make beautiful objects for museums. However, museums have a duty to preserve their acquisitions. Two conferences on this subject have already been held in Britain during the first quarter of 1995. The tone of the second conference, organised by The National Museums of Scotland in Edinburgh in April, was fatalistic but determined.

Recognition of Polymer Degradation

It was not until the late 1980s that attention was paid to the fact that plastics artefacts had been physically changing, showing signs of acid vapour, tackiness, warping, embrittlement and crazing. Cellulose nitrate and cellulose acetate were particularly affected. By 1991 John Morgan of the Plastics Historical Society had collected enough data to write Conservation of Plastics -An Introduction, a joint PHS/Conservation Unit publication, and the Conservation Unit launched a survey to identify objects at risk with the aim of setting up a research programme. The survey included everything from radios and cables to textiles and sculptures.

Deterioration of Acrylic Paintings and Pieces of Art

By 1992 acrylic based paintings worth millions of pounds by leading artists of the 1960s including David Hockney and Jackson Pollock had begun to suffer discolouration, cracking and greyness due to the absorption of dust and atmospheric pollutants. These paints seemed particularly vulnerable. At room temperature they are relatively soft and attract dirt which becomes embedded. However, to date no method has been found of cleaning them.

Impact on the Photographic Film Industry

The photographic film industry was also badly hit when irreplaceable archive nitrate stock started to decompose. Today, the National Film Archive transfers cellulose nitrate and cellulose triacetate onto more stable polyester at the rate of a million metres a year.

Preserving Plastics Pieces in Museums

The PHS/CU survey unearthed some interesting facts. For example, 40 per cent of museums surveyed contain plastics objects manufactured and collected since 1980, and modem plastics are also showing symptoms of decay. Polyurethane foam appears to be one of the worst victims, and many early video and audio tapes on magnetic media are already unplayable. The curator who has to supervise a collection of high-tech, mixed material products such as space suits is confronted with a conservation dilemma: which material deserves priority treatment when each separate plastic has different requirements?

Factors Affecting Polymer Degradation

The degradation of plastics can be said to begin as soon as the polymer is synthesised, and is increased by residual stresses left by moulding processes. This can be followed by exposure to light (especially UV), humidity, oxygen, heat, bacteria and stress. Plastics can also be contaminated by other materials, including other plastics. A polystyrene camera body, for example, can be attacked by plasticiser migrating from a PVC strap. Ideally the conservator needs information about the history of an object before prescribing treatment, but even before this, the plastics 'doctor' must jump another hurdle. Specific conservation action cannot be taken until the polymer has been identified, and this is a technical area full of pitfalls. A 1920s black brooch could be made of at least six different plastics materials, or simply painted as was common practice in the 19th century. Even a patent number, one of the few 'hallmarks' found on plastics and an obvious aid to identification, may refer to a fixing mechanism and not to the moulding.

Preserving Plastics

Derek Pullen, a conservator at the Tate Gallery, explains, 'Plastics are giant molecules held together by forces which can be broken by attacking energy forces such as light. All the conservator can do is to keep mouldings in a very stable, low energy environment (the burial chambers of the Pyramids were ideal)'.

Types of Polymer Degradation

There are two main types of plastics degradation being researched at present: physical and chemical, and both are closely inter-connected. Physical degradation can involve environmental stress cracking and plasticiser migration and loss. Chemical reactions include oxidation and hydrolysis, and are a problem particularly affecting the cellulose esters (cellulose nitrate and cellulose acetate), which emit acidic degradation products. If not removed, these catalyse further reactions and eventually cause serious crazing and total destruction of the object. If degrading cellulose esters are not isolated, the acidic fumes will infect similar objects stored close by and initiate degradation there.

Solutions to Polymer Degradation

As the recognition of polymer degradation improves, conservation guidelines are beginning to emerge. High-tech solutions which could help in theory are prohibitively expensive, but tailor made scavengers such as activated charcoal or Ageless help to create a low oxygen environment. Ageless is a reactive powdered iron and is normally used to prolong the shelf-life of dry foods by absorbing oxygen. Epoxidised soyabean oil (ESBO), has also been tested with encouraging results as an acid absorbing coating on degrading cellulose nitrate.

Conclusion

Compared to traditional materials with long established technologies such as metals and glass, the complex chemical nature of plastics is providing conservators with possibly their most formidable challenge yet.

11.4 Design and Fabrication of Polyester-Fiber and Matrix Composites for Totally Absorbable Biomaterials

Several semicrystalline polyesters exhibiting blood compatibility are widely used as resorbable material for implantable devices and surgical articles (see Figure 11.1). These polyesters biodegrade hydrolytically, and the resulting biologically active or nontoxic carboxylic acids are processed through normal metabolic pathways. Cross-linked analogs of semicrystalline polyesters can be used as matrix resins in bioabsorbable composites that possess physical properties uniquely suited to rigid fixation applications such as bone plates. (Figures not yet available on-line.)

For example, replacing high-modulus metal bone plates with a material whose modulus more closely resembles that of bone could promote healing of fractures. Current rigid fixation devices shield the fracture site and surrounding bone area; loads are transferred through the plate, often leading to stress-protection osteopenia. The shielded bone resorbs, becomes porous, and remains susceptible to refracture after surgical removal of the metal bone plate. Instead of acting as a shield, a plate made from absorbable polymers would gradually transfer loads into the fracture area as it degrades, enhancing bone regeneration. In addition, hydroxyapatite or other minerals and nutrients can be incorporated into the composite and, once released, can often stimulate bone growth. Of course, totally resorbable composites eliminate the need for surgical removal.

The degradation behaviour of cross-linked, amorphous, absorbable matrices is also advantageous for rigid fixation applications. Semicrystalline polymers display heterogeneous degradation due to distinct amorphous and crystalline regions.[5] The differing rates of degradation yield a product that decreases in physical strength at a faster rate than it decreases in mass. Heterogeneous degradation produces a composite matrix that will prematurely lose its physical strength, whereas degradation of wholly amorphous, cross-linked polyesters should show a more linear decrease in physical strength with loss of mass, thus retaining properties over time.

Monomers that typically yield semicrystalline polymers can be used to produce cross-linkable, amorphous prepolymers through careful selection of comonomer composition and control over molecular weight. Totally absorbable composites may be fabricated by using biodegradable fibers for directional reinforcement in a degradable matrix. The matrix resin used in the present work, poly (D, L-lactide-*co*-glycolic acid) fumarate, was chosen for its design flexibility and the manner in which it facilitates composite fabrication via free-radical curing techniques. Polyglycolic acid surgical mesh, a knitted fabric with high tensile strength, provided the directional reinforcement for the composite. The physical properties exhibited by this combination of materials justify continued research into the development of totally absorbable composites for surgical applications.

Experimental

Matrix Resin Synthesi: Unsaturated poly (D, L-lactide-*co*-glycolic acid) fumarate oligomer was synthesized as reported previously. To a 500-ml, three-necked reaction flask equipped with a mechanical stirrer and West condenser were added 44.05 g (0.50 mole) 2-butene-1,4-diol as initiator; 144.02 g (1.0 mole) racemic D, L-lactide; 76.05 g (1.0 mole) glycolic acid; and 1.02 g stannous octoate. Under nitrogen purge, the reaction temperature was slowly raised to 120°C and held for 24 hours, at which time a 2-torr vacuum was applied, maintained for 8 hours, and slowly increased to 0.1 torr over an additional 8 hours. The hydroxy-terminated oligomer that resulted was reacted under nitrogen purge with 116.12 g (1.0 mole) fumaric acid at 160°C for 40 hours, and at 180°C for an additional 10 hours. At this time the reaction was continued for 48 hours at 165°C, as the vacuum was slowly increased from 10 to 4 torr. The resulting oligomer was dissolved in chloroform, vacuum filtered, precipitated into cold methanol, stored under methanol at -5°C for 16 hours, collected by filtration, and vacuum dried for 72 hours at 60°C.

Curing Techniques: Oligomer (175 g) was dissolved in 35 g tetrahydrofuran (THF) to make a stock resin that was prepromoted with 1.05 g cobalt naphthenate and stored under dry nitrogen until use. For the curing studies, 2-butanone peroxide was used in varying concentrations as a free-radical initiator. This curing system was chosen because of its known efficiency; its biological activity has not yet been addressed.

Samples were cured by adding approximately 2 g of prepromoted resin to an aluminum dish; 2-butanone peroxide was subsequently added in varying amounts from 1 to 10 wt per cent relative to polymer, and cured for 48 hours at 60°C. All cured samples were then stored in a desiccator until evaluation by differential scanning calorimetry (DSC).

Composite Fabrication: Polyglycolic acid surgical mesh was cut into 1.5 x 3-in. samples and saturated with initiated polyester prepolymer. In some cases, the mesh—prior to saturation with resin—was immersed for 2 hours in a mixture of 3 g of *bis* (dimethylamino)-methylvinylsilane diluted in 100 g of absolute ethanol, and then dried overnight at room temperature. Laminated plates consisting of four layers of mesh were fabricated using standard vacuum bagging techniques. These plates were cured for 5 hours at room temperature, removed from the vacuum bag, and cured for an additional 40 hours at 60°C. The reinforcement-to-matrix weight ratio for all composites was approximately 40:60, reflecting minimum air voids.

Characterization: Tensile strengths for the composite samples were measured on an Instron Model-1130 universal test machine using a 500-kg load cell at 40 per cent range, with chart and crosshead speeds of 5 cm/min. Fourier transform infrared spectroscopy (FTIR) was performed on a Perkin Elmer Model 1600 spectrophotometer. The molecular weight of the matrix prepolymer was obtained using a Waters Associates gel permeation chromatograph (GPC) equipped with a Rheodyne injector, a Model 6000A solvent-delivery module, four Ultrastyragel columns with nominal pore sizes of 100, 500, 10^3, and 10^4 Å connected in series, and a 410 differential refractometer.

THF—freshly distilled from calcium hydride—served as the mobile phase and was delivered at a flow rate of 1.0 ml/min. Adalab Chromatochart software with GPC enhancement was employed to determine molecular weight as compared with polystyrene standards (Polysciences Corp.). Scanning electron microscopy (SEM) was performed with an Electro Scan Model E-20 and DSC was carried out using a Mettler Model 30 calorimeter with a heating rate of 10°C/min. The glass-transition temperature (T_g) was taken as the midpoint of a straight line drawn between the inflection points of the peak's onset and end point.

Discussion

Chain extension reactions by esterification of the hydroxy-terminated oligomers with the Kreb's cycle intermediate fumaric acid have proven effective for incorporating unsaturated moieties into the backbone of a bioabsorbable polyester, and for providing control over the polymer's final molecular weight. The presence of hydroxyl end groups after the initial synthetic step has been confirmed by FTIR. Based on initial reaction conditions, the original hydroxy-terminated oligomer has a theoretical molecular weight of 546 g/mole—below the range of the polymer standards used in our GPC calibration yet within the range of molecular sizes separable by our particular combination of gel columns. In fact, based on the sharp individual peaks in Figure 11.3a, the product is most likely a series of dimers, trimers, tetramers, etc. Reaction of this hydroxy-terminated oligomer with excess fumaric acid gave an acid-terminated polymer with a final molecular weight of approximately 9200 g/mole (see Figure 11.3b).

Free-radical curing of unsaturated polymer with 2- butanone peroxide yielded matrices for which the glass-transition temperatures increased with increasing peroxide concentrations. The breadth of the polymer's glass transition from onset to end point also increased with increasing peroxide concentration.

Curing was monitored using FTIR by observing a decrease in the absorbance due to the olefinic groups. Although the unsaturations in the resin are all 1,2-disubstituted, and thus low kinetic chain lengths are expected, sufficient network properties may still be obtained with high conversion. The nearly complete reaction of double bonds was confirmed by observing the essentially quantitative disappearance, after complete curing, of the C=C peak in the FTIR spectrum of the matrix polymer. Cross-linking was further confirmed by insolubility of the matrix in THF after curing.

Completely absorbable composites were fabricated by reinforcing the resin with polyglycolic acid surgical mesh. The authors have demonstrated that increasing interfacial adhesion between fiber and matrix is necessary for maximizing the properties of the bioabsorbable composite.

Bis-(dimethylamino)-methylvinylsilane was evaluated as a coupling agent for reducing the interfacial surface-energy difference between fiber and matrix. Dimethylamino groups of the coupling agent are expected to form a strong dipole-dipole interaction with carbonyl groups of the polyglycolic acid mesh in the initial pretreatment of the fiber. The vinyl group of the silane will then cross-link—via free radicals—into the matrix resin, forming a covalent bond within the network. Improved wetting of the fiber surface with the matrix resin should yield an increased physical strength for the composite. An approximately 10 per cent overall increase in the tensile strength—from 84 to 92 MPa—was observed following pretreatment of the surgical mesh fibers with the selected coupling agent. SEM revealed a visible gap between the fiber and matrix and clear regions of fiber pullout in untreated samples. In the composites formed with silane-treated mesh, SEM indicated improved fiber wetting through more complete encapsulation of the fiber with matrix resin and less occurrence of fiber pullout.

DSC thermograms for the composite show a glass transition for the amorphous, cross-linked matrix at 55°C and a crystalline melt transition for the fiber at 225°C. Thus, the composite is pliable at temperatures above 55°C but maintains rigidity at biological temperature. This means that the composite can be custom formed for use in specific surgical devices through the application of heat—a feature that could prove quite useful.

Degradation of the composite was studied in vitro by immersion in a buffered saline at 37°C.

Conclusion

Poly (D, L-lactide-*co*-glycolic acid) fumarate, when cured by free-radical initiation, demonstrates properties suitable for use as a matrix resin in totally absorbable composites. The incorporation of polyglycolic acid surgical mesh provides directional strength for the material. A critical design consideration for resorbable composites is the use of a silane coupling agent to reduce the interfacial energy between the fiber and matrix. Thermal evaluation shows that the composite can be easily contoured at a temperature slightly above biological temperature. Because of their unique properties, completely absorbable composites warrant further investigation as candidates for biomaterials.

Acknowledgments

This paper is based on research supported in part by the National Science Foundation (grant #RII-8902064), the State of Mississippi, and the University of Southern Mississippi. The authors gratefully acknowledge Robert Pope of the University of Southern Mississippi for providing the SEM analysis and T.J. Nash of Davis and Geck for generously supplying polyglucolic acid surgical mesh.

11.7 Torlon® Polyamide-Imide

Torlon® polyamide-imide (PAI) is a high strength plastic with the highest strength and stiffness of any thermoplastic up to 275°C (525°F). It has outstanding resistance to wear, creep, and chemicals—including strong acids and most organics—and is ideally suited for severe service environments.

Key Features

- Exceptional long-term strength and stiffness up to 275°C
- Outstanding wear resistance
- Superior toughness from cryogenic up to 275°C
- Resistant to strong acids and most organics
- Inherent flame resistance
- Low CLTE

Typical Applications of Torlon High Strength Plastic

- Aircraft
  - Hardware and Fasteners
  - Mechanical Components
- Auto/truck transmission components
- Coatings, composites, additives
- Electric motor components
- Semicon fabrication, testing
- Oil and gas processing equipment
- Bearing retainers
- Gears
- Bushings
- Seals
- Thrust washers

Featured Application

Thrust Washers

- ❖ Wear resistance in lubricated and non-lubricated environments
- ❖ Impact and compressive strength
- ❖ Dimensional stability
- ❖ Non-seizing vs. metal

11.8 Radiation Sterilization

Polymer Materials Selection for Radiation-Sterilized Products

> Choosing the right polymer for a radiation-sterilized device requires an understanding of radiation effects, manufacturing processes, and the product's intended use. *Karl J. Hemmerich*

Gamma and electron-beam irradiation are among the most popular and well established processes for sterilizing polymer-based medical devices. It has been long known, however, that these techniques can lead to significant alterations in the materials being treated. High-energy radiation produces ionization and excitation in polymer molecules. These energy-rich species undergo dissociation, abstraction, and addition reactions in a sequence leading to chemical stability. The stabilization process—which occurs during, immediately after, or even days, weeks, or months after irradiation—often results in physical and chemical cross-linking or. chain scission. Resultant physical changes can include embrittlement, discolouration, odor generation, stiffening, softening, enhancement or reduction of chemical resistance, and an increase or decrease in melt temperature. This article discusses how and why irradiated polymeric materials change, presents data on the radiation stability of various polymers, and offers some general guidelines for material selection.

Ionizing radiation is a unique and powerful means of modifying polymers, particularly since the changes occur when materials are in a solid state, as opposed to chemical or thermal reactions carried out in hot or melted polymers. While solid-state modification may have significant advantages, any changes in material characteristics or performance brought about by radiation are governed by structure/property relationships that are perfectly analogous to those generated by other chemical and thermal processes. These include polymerization, grafting, cross-linking, changes in saturation, chain scission (degradation), oxidation, cyclization, isomerization, amorphization, and crystallization.

Many important physical or chemical properties of polymers can be modified with radiation. Among these are molecular weight, chain length, entanglement, polydispersity, branching, pendant functionality, and chain termination. Understanding how and to what extent these characteristics can be altered as a function of the level of radiation exposure (dose) is crucial to predicting the performance and utility of irradiated plastics.

The influence of radiation on the properties and performance of a polymer differs according to whether the material degrades or cross-links, and this in turn depends on specific sensitivities or susceptibilities inherent in the polymer backbone. All materials have been found to break down at very high radiation doses; however, the range of doses under which a given plastic will maintain its desirable properties depends greatly on the chemical structure of the polymer. Indeed, below the destructive level of exposure, radiation treatment can impart many benefits and enhance properties of commercial value. By gaining sufficient knowledge about these beneficial radiation-induced effects, device manufacturers can make thoughtful choices regarding polymers used in sterile medical products and ensure that critical elements of material and product performance are not compromised.

Radiation Effects

Because the effects of ionizing radiation depend greatly on polymer chemical structure, the dose necessary to produce similar significant effects in two different materials can vary from values as low as 4 kGy in polytetraflouroethylene to 4000 kGy or more in styrene or polyimide. Even the lowest doses, however, can engender significant alterations. For example, in a typical polyethylene with an average molecular weight of about 100,000 and readily soluble in a solvent such as Decalin, only one cross-link per 14,000 monomer (ethylene) units can cause gelation (insolubility).

Radiation effects on the properties of a polymer can also be difficult to predict, especially in the presence of certain additives that can help to prevent radiation damage to plastics. These compounds are frequently termed "antirads," and generally are substances that also act as antioxidants. They function either as reactants, combining readily with radiation-generated free radicals in the polymer, or as primary energy absorbers, preventing the radiation's interaction with the polymer itself.

Radiation normally affects polymers in two basic manners, both resulting from excitation or ionization of atoms. The two mechanisms are chain scission, a random rupturing of bonds, which reduces the molecular weight (i.e., strength) of the polymer, and cross-linking of polymer molecules, which results in the formation of large three-dimensional molecular networks.

Most often, both of these mechanisms occur as polymeric materials are subjected to ionizing radiation, but frequently one mechanism predominates within a specific polymer. As a result of chain scission, very-low-molecular-weight fragments, gas evolution, and unsaturated bonds may appear. Cross-linking generally results in an initial increase in tensile strength, while impact strength decreases and the polymer becomes more brittle with increased dose.

For polymers with carbon-carbon chains (backbones), it has been observed that cross-linking generally will occur if the carbons have one or more hydrogen atoms attached, whereas scission occurs at tetra-substituted carbons. Polymers containing aromatic molecules generally are much more resistant to radiation degradation than are aliphatic polymers; this is true whether or not the aromatic group is directly in the chain backbone or not. Thus, both polystyrenes, with a pendant aromatic group, and polyimides, with an aromatic group directly in the polymer backbone, are relatively resistant to high doses (>4000 kGy).

From a product use standpoint, the loss of mechanical properties is the most important characteristic effected by irradiation of polymers. These properties include tensile strength, elastic modulus, impact strength, shear strength, and elongation. Embrittlement may occur even as irradiated polymers decrease in hardness. Crystallinity and, hence, density characteristics may also change as chain scission continues.

When a polymer is subjected to irradiation by ionizing radiation—such as gamma rays, x-rays, or accelerated electrons—various effects can be expected from the ionizations that occur. The ratio of resultant recombination, cross-linking, and chain scission will vary from polymer to polymer and to some degree from part to part based on the chemical composition and morphology of the polymer, the total radiation dose absorbed, and the rate at which the dose was deposited. The ratio is also significantly affected by the residual stress processed into the part, the environment present during irradiation (especially the presence or absence of oxygen), and the postirradiation storage environment (temperature and oxygen).

Effects of Oxygen and Dose Rate

The environmental conditions under which radiation processing is conducted can significantly affect the properties of the polymer material. For example, the presence of oxygen or air during irradiation produces free radicals that are often rapidly converted to peroxidic radicals. The fate of these radicals depends on the nature of the irradiated polymer, the presence of additives, and other parameters such as temperature, total dose, dose rate, and sample size. A variation in processing conditions in the presence of polymer additives can result in gas evolution and formation of other degradation products from these small molecules, with the possibility of producing irritants or other undesirable compounds.

A significant difference in irradiation processing exists between electron-beam and gamma sterilization related to dose rate and, ultimately, to the oxidation degradation of material at or near the surface. Dose rate refers to how fast energy is absorbed and depends on many factors including the source, strength, and size of the radiation field; its distance from the source; and the type of radiation. For electron and gamma sources of the same strength, the dose rate of the electron source is many times greater than that of the gamma source. This is because the electron beam is unidirectional and is concentrated in a much smaller region, and because the interaction of electrons with other electrons is much stronger than with photons.

Although the difference may be minute, it can be said in an absolute sense that all polymeric materials undergo less material embrittlement in electron-beam than in gamma-ray sterilization as a result of reduced oxidative chain scission. For many products, this is not critical, but is important to keep in mind when selecting a radiation sterilization process for oxidation-sensitive materials such as polypropylene, nylon, or Teflon or for products containing thin profiles, such as films and fibers.

Additional explanation will help clarify this phenomenon. As a polymer is irradiated, radicals are formed in a concentration proportional to the local dose. However, the associated stabilizing chemical reactions that follow are proportional to the local concentrations of reactants. Because the concentration of reactants differs by location (i.e., higher oxygen near exterior surfaces), the resultant radio-chemical stabilizing reactions are thus heterogeneous.

In both gamma and electron-beam irradiation systems, available oxygen is quickly consumed within the polymer. However, in the case of electron-beam processing, the time of energy application is so short that before more oxygen can permeate into the material from its external surfaces, the application of radiant energy has been terminated, the direct formation of additional radicals ceases, and the stabilizing chemistry pursues alternate, less-degrading routes (recombination or cross-linking). In simple terms, the chemistry changes as a result of a starved chemical reaction—much as a fire goes out when the oxygen is consumed in a closed container. In gamma irradiation, the application of ionizing energy continues over a much longer period of time, allowing reactants, such as oxygen, to permeate back into depleted areas of the material, resulting in a greater degree of chain scission.

Colour and Odor

A common undesirable effect resulting from the irradiation of some polymers is discolouration (usually yellowing) from the development of specific chromophores or colour centers in the polymer. Colour development, which occurs at widely differing doses in various polymers, may diminish or increase with storage time after irradiation. Often, discolouration appears prior to any measurable loss in physical properties. Such as is the case, for example, with PVC, in which radiation-induced yellowing from conjugated double bonds develops at a dose much lower than is necessary to cause any reduction in the material's physical properties.

Another problematic effect in some polymers that results from specific radio-stabilizing chemistries is odor. The polymers that most commonly exhibit post-irradiation odor are polyethylene, PVC (rancid oil odor from oxidized soybean and linseed oils in the plasticizer), and polyurethane.

If the reaction chemistries causing the odors are understood they can often be mitigated through the use of antioxidants, different processing temperatures, or selection of a higher-molecular-weight polymer. Odor reduction can also be accomplished through the use of gas-permeable packaging (for example, Tyvek, paper) and elevated temperature conditioning.

Material Selection

In addition to personal experience, materials databases, supplier information, and literature searches can be used to help avoid potential problems with polymers that are less radiation resistant than required for a particular product's design and function. Material selection should be done with diligence, especially when dealing with thin sections such as films, coatings, and fibers as well as with materials having low radiation resistance such as acetal, polypropylene, or Teflon (these three polymers can be remembered as those that are "APT" to fail). It must be remembered that the physical properties of irradiated polymers are subject to significant variations due to residual or functional stress, section thickness, molecular weight, morphology, moisture, and storage environment (oxygen/temperature), and must be tested in the specific application under consideration.

The selection of materials for radiation sterilization should start with the following basic rules:

- Most medical plastics are durable in radiation.
- Use the highest-molecular-weight material (with the narrowest molecular-weight distribution) that is possible for the application.
- Aromatic materials are more radiation resistant than are aliphatic materials.
- Amorphous materials are more radiation resistant than are semicrystalline materials.
- Higher levels of antioxidants improve radiation resistance.
- Low-density materials are more radiation resistant than high-density materials.
- Materials with small pendant (side) groups are more radiation resistant.
- For semicrystalline materials, the lower the crystallinity, the greater the radiation resistance.
- Materials with low oxygen permeability are more radiation resistant.
- Avoid materials that are "APT" to fail: acetal, polypropylene (unstabilized), or Teflon (PTFE).

Information derived from government, industrial, and scientific studies and publications concerning radiation effects on polymer properties after exposure to various doses, which graphically display the dose at which a number of common thermoplastics and thermosets experience a 25 per cent loss in elongation. (Loss of elongation is a commonly used measure of the effect of irradiation.) These figures provide a visual means of making an initial estimate of a polymer's ability to withstand a typical sterilization dose (10-50 kGy) or a higher dose used in a more specialized radiation process. A more qualitative summary of the radiation stability of selected polymeric materials is provided.

Conclusion

When used to enhance polymer properties or to sterilize polymer-based medical products, ionizing radiation interacts with polymers via two primary mechanisms: chain scission to reduce molecular weight and cross-linking to generate large polymer networks. Both mechanisms occur in all polymers during irradiation, but one generally dominates. Polymers vary in sensitivity to radiation from PTFE and polyacetal, with damage occurring at doses as low as 4 kGy, to polystyrenes, polyimides, and LCP, which can tolerate doses as high as 105 kGy without significant damage.

Polymers containing aromatic groups have much greater resistance to radiation damage than those with aliphatic structure. Most thermoplastics, essentially all thermosets, and most elastomers can withstand at least one radiation sterilization (<50 kGy) without significant damage. Because of oxidative effects, polymeric materials used in adhesives, fibers, films, and encapsulates show slightly lower radiation tolerance compared with bulk polymers; however, the use of antioxidant additives can significantly offset the effects of radiation.

With a basic understanding of the effects of radiation on polymers, reference data available from polymer manufacturers and other sources, and a thorough understanding of the product's manufacturing process and intended use, wise choices of radiation-tolerant materials can be made for the majority of medical product applications.

11.9 Bio-Plastics

Introduction

Industrial revolution has resulted in many materials which are man-made and not resembling to natural ones. The term 'xenobiotic' (stranger to life) is derived from the Greek word 'xenon' - a strange and bios - life. For the environmental chemist, xenobiotic implies foreign to the biosphere. 'Anthropogenic' is a specific term for "man-made". These materials may be products, intermediates or wastes of our industrial and other activities. And at whatever stage they enter into nature, they can not become a part of cycle of matter. This is for the simple reason that microorganisms do not find any resemblance between these materials and their natural food habits. However simultaneously it is also true that all foreign to nature man-made substances need not be non-degradable.

Those compounds which are not degraded in nature following their release into environment even when the conditions appear to be adequate for microbial growth, are termed as 'recalcitrant'. Recalcitrant compounds persist in all natural environments regardless of whether they are inherently biodegradable or not. Without being bias to any industry, it is well accepted fact that xenobiotic compounds, toxic chemicals, hazardous wastes come from the petrochemical industry, pesticide industry, chemical industry, mining, metal processing etc. The list of pollutants which pose environmental and health hazard and are tough for biodegradation is long one and includes solvents, wood preservative chemicals, plasticizers, refrigerants, coaltar wastes, pesticides, biphenyls, polychlorinated and polybrominated biphenyls, synthetic fibers, plastics, polyvinyl chloride, polystyrene, detergents like alkyl-benzene sulphonates, oils etc.

Aiming for Biodegradable and Ecofriendly Products

Physical and chemical methods of pollution control were always in the forefront because they were easy to understand, easy to control and were reproducible. Biodegradation, the real mechanism of nature of balancing the material, was always found to be incompletely understood, unpredictable and uncontrollable if we have to adapt it in the form of biological treatment methods. A better option then is to modify our materials, processes and products in such a way that we can rely upon the biodegradation in nature and recalcitrance, bioaccumulation problems are overcome. We are slowly changing our philosophy and are not merely targeting for clean-up or removal of pollutant but are aiming for prevention of pollution or facilitating biodegradation. Manufacturing processes are rapidly changing and biodegradable products are fast replacing man-made, difficult to degrade products.

After understanding the nature, its role, its superiority and by knowing the fact that biodegradation is the ultimate fate of any material that enters into environment, it was inevitable for us to change our psychology - industrial ecology - industrial ecosystem - industrial metabolism - going close to nature. Objectives then are to improve upon the method of production, searching for alternative raw materials, recycling, conversion to suitable forms of certain wastes, so that we do not add any material, waste in nature which nature cannot take care of. "Biodegradation mechanisms" which occur in the soil, aquatic environment though slow are important for us as they do not involve any cost of treatment when it occurs naturally, and are safer and bring about complete degradation and not mere conversion. Hence incorporating biodegradability or aiming for biodegradability is an obvious approach while carrying out production of different items. The agricultural products, food products, commodities hereafter will carry a label of 'ecofriendly' in most of the countries. The trend is to spread to every material and product. The consumables, coal, oil, petrol, synthetic fibre, plastic which are polluting/non-biodegradable will be replaced slowly by the one which is non-polluting/biodegradable.

The Problem of Plastic

Plastic is one of the few new chemical materials which pose environmental problem. Polyethylene, polyvinyl chloride, polystyrene is largely used in the manufacture of plastics. Synthetic polymers are easily molded into complex shapes, have high chemical resistance, and are more or less elastic. Some can be formed into fibers or thin transparent films. These properties have made them popular in many durable or disposable goods and for packaging materials. These materials have molecular weight ranging from several thousands to 1,50,000. Excessive molecular size seems to be mainly responsible for the resistance of these chemicals to biodegradation and their persistence in soil environment for a long time. Plastic in the environment is regarded to be more an aesthetic nuisance than a hazard, since the material is biologically quite inert. The plastic industry in the US alone is $ 50 billion per year and is obviously a tempting market for biotechnological enterprises. Biotechnological processes are being developed as an alternative to existing route or to get new biodegradable biopolymers. 20 per cent of solid municipal wastes in US is plastic. Non-degradable plastics accumulate at the rate of 25 million tonnes per year.

According to an estimate more than 100 million tonnes of plastic is produced every year all over the world. In India it is only 2 million tonnes. In India use of plastic is 2 kg per person per year while in European countries it is 60 kg per person per year while that in US it is 80 kg per person per year.

Monomer Polymer

$H_2C = CH_2$ Polyethylene - CH_2 - CH_2 - [CH_2 - CH_2 -] n

$H_2C = CH$ Polyvinyl chloride - CH_2 - CH - CH_2 - CH -
| | | n
Cl Cl Cl

$H_2C = CH$ Polystyrene - CH_2 - CH - H_2C - CH -
| | | n

The Solutions for Plastic

Plastics may be either (a) photodegradable or (b) semi-biodegradable or (c) 100 per cent biodegradable. Photodegradable plastics have light sensitive groups incorporated directly into backbone of the polymer as additives. This produces non-degradable smaller fragments, which cause loss of material integrity. Example of semibiodegradable plastic can be blends of starch and polyethylene. PHB is an example of 100 per cent biodegradable plastic. Approximately a dozen of inherently biodegradable plastics are now in the market, with range of properties suitable for various consumer products.

Estimate of current global market for biodegradable plastic is 1.3 billion kg per year. Estimated global market for biodegradable polymers will reach 1.4 billion tons by the year 2000.

In response to increasing public concern over environmental hazard caused by plastic, many countries are conducting various solid waste management programmes including plastic waste reduction by development of biodegradable plastic material. There is an intense research for making the biodegradable plastic. Some biodegradable plastic materials under development are - (1) PHAs

(2) Polylactides (3) Aliphatic polyesters (4) Polysaccharides (5) Co-polymers and/or blends of above.

Different approaches are:

I. modification of existing material,
II. chemical co-polymerisation of known biodegradable material, and
IIIuse of biopolymers for making plastics.

(I) Partially biodegradable shopping bags are already manufactured from thin matrix of conventional polythene filled in with starch. After the bag has been thrown away, microorganisms eat away starch, leaving polythene film structure which soon disintegrates. Fertec (Ferruzzi, Ricerca e Technologia) of Italy and Warner Lambert of the US are developing fully biodegradable starch-based plastics. Starch forms upto 50 per cent by weight of the Fertec material and remaining is the synthetic polymer. Material shows partial biodegradability in Warburg test. Warner Lambert's starch-based plastic is called Novon and contains 80 per cent starch. Additives like plasticizers are used to make the material tougher and to improve processing. Novon is biodegradable in accelerated landfill, controlled compost, aerobic and anaerobic aquatic environments. Warner's starch-based plastic can be used for capsules for drugs, disposable single-use items like cups and food trays. The company has 25000 tonnes per annum production capacity from 1992. At present, biodegradable plastic represents just a tiny market compared with the conventional petrochemical material whose production amounts to >100 million tonnes per year. Bioplastics will comparatively prove cheaper when oil prices will continue to hike up.

(II) 'Bioceta' is the new biodegradable plastic which is cellulose diacetate-based product. It has been developed by Rhone Poulenc's Belgium subsidiary, Tubiz plastics. Bioceta uses additives which both plasticize and accelerate degradation by micro-organisms.

The Sekisui chemical company has developed a new biodegradable plastic by co-polymerisation of two different biodegradable chemicals based on aliphatic polyester derivatives. The plastic has both good properties and complete biodegradability. The biodegradability and properties of new plastic can be adjusted to a larger extent by controlling the conditions of polymerization. It is thermoplastic and can be recycled. The plastic can be used effectively for agriculture, for goods packaging, but high-value added plastic can also be produced using the co-polymer. The plastic was developed in co-operation with the Government Industrial Research Institute, Osaka. Plastic is based on polyester which is decomposed by means of enzymes such as lipase to H2O + CO2 causing no secondary contamination as in other degradable plastics. According to firm's claim, this plastic is stronger than polyethylene, has a higher melting point (over 900C) than the ordinary polyester resin and have fastest biodegradability. A film of 100 micron thickness of this plastic is totally decomposed in soil in just two months.

Biodegradable plastic film comparable in strength to the general purpose polyethylene has been developed in Japan by the Agency of Industrial Science and Technology's (AIST) Fermentation Research Institute. The film has tensile strength of 200 kg/cm2 and is produced from a mixture of polycaprolactone (PCL) which is completely biodegradable and special compatible polyolefin. The mixture has a PCL content of 50-80 per cent by weight and is formed with higher proportion of biodegradable PCL towards the surface, thus promoting high biodegradability while retaining the strength of special polyolefin. If buried in soil for a year, the film is degraded by micro-organisms into a powder with particles ranging from 1-10 microns. It releases no harmful gases when incinerated and has calorific value of 8000 k.cal per kg which is 80 per cent of ordinary films.

Co-polymers of succinic acid, glycerol and polyethylene glycol are found to be 100 per cent biodegradable in 90 days in soil. They have glass transition temperature (Tg), 210C.

(III) *Bioplastics*: Biopolymers obtained from growth of micro-organisms or from plants which are genetically-engineered to produce such polymers are likely to replace currently used plastics at least in some of the fields. Poly b - hydroxy butyrate and polylactic acid are the kind of polymers which are used as materials of bioplastics.

Biopol

PHB material was first described in 1926 by Lemoigne. PHB is a very common and widespread storage material in many micro-organisms. PHB has been found to be a very basic polymer of variety of chemically similar polymers, the polyhydroxyalkonates.

Poly b - hydroxy butyrate (PHB) accumulates as energy reserve material in many micro-organisms like Alcaligenes, Azotobacter, Bacillus, Nocardia, Pseudomonas, Rhizobium etc. PHB has physical properties comparable with polypropylene (PP). Poly b - hydroxy butyrate (PHB) consists repeat units of CH(CH3)-CH2-CO-O. The difference is that PP shows insignificant degradation while PHB shows complete degradation. PHB sinks while PP floats. Therefore, degradation is easy at sediment. Alcaligenes eutrophus and Azotobacter beijerinckii can accumulate upto 70 per cent of their dry weight of PHB. These micro-organisms can produce the polymer in environment of N and P limitation. Minimum 40-50 per cent of the dry weight of this polymer is required for making the process commercially viable. Extraction of PHB is done by using solvents like halogenated hydrocarbons and purification is done.

Moulding and extrusion of dried cells directly is possible when PHB contents are high. A lot of work is done on engineering polymeric properties of PHB. However PHB is suitable for specialized areas like biomedical use and speciality coatings.

Table 11.2: PHB Accumulation in Microorganisms

Organisms with PHB i.e. (PHA) Polyhydroxy Alkonates Accumulation per cent of dry weight of cell	
Alcaligenes eutrophus	96
Azospirillum	75
Azotobacter	73
Baggiatoa	57
Leptothrix	67
Methylocystis	70
Pseudomonas	67
Rhizobium	57
Rhodobacter	80

General structure of PHA and some representative members - Polyhydroxyalkonates (PHAs) are polyesters of various hydroxyalkonates that are synthesized and intracellularly accumulated by numerous micro-organisms as energy reserve material. More than 100 different monomer units have been identified as the constituents of PHAs. This creates the possibility of producing biodegradable polymers with wide range of properties. PHB, Poly (3-hydroxybutyrate) is the best characterized PHA. PHB has lowest molecular weight and is most common in nature. Their molecular weight can be upto 2 million (i.e. 20000 monomers per polymer molecule). The monomer units of PHA are all in D-(-) configuration owing to the stereospecificity of the biosynthetic enzymes.

R O
| ||
[- O - CH - (CH2) n - C -] 100 - 30 000

n = 1 R = hydrogen Poly (hydroxy propionate)

R = methyl Poly (3-hydroxybutyrate)

R = ethyl Poly (3-hydroxyvalerate)

R = propyl Poly (3-hydroxyhexanoate)

R = pentyl Poly (3-hydoxyoctanoate)

R = nonyl Poly (3-hydroxy doedcanoate)

n = 2 R = hydrogen Poly (4-hydroxybutyrate)

R = methyl Poly (4-hydroxyvalerate)

n = 3 R = hydrogen Poly (5-hydroxyvalerate)

R = methyl Poly (5-hydroxyhexanoate)

n = 4 R = hexyl Poly (6-hydroxydodecanoate)

There are three enzymes present in A.eutrophus for PHA biosynthesis. These are - PHA synthase, b ketothiolase and reductase.

Natural producers also have PHA depolymerase that degrades the polymer and uses the breakdown metabolites for cell growth. Metabolism of PHB occurs as -

PHB D (-) hydroxybutyric acid acetoacetate or acetoacetyl CoA.

Properties of PHB

1. PHB is water insoluble and relatively resistant to hydrolytic degradation. This differentiates PHB from most other currently available biodegradable plastics, which are either water soluble or moisture sensitive.
2. PHB shows good oxygen permeability.
3. PHB has good ultra-violet resistance but has poor resistance to acids and bases.
4. PHB is soluble in chloroform and other chlorinated hydrocarbons.
5. PHB is biocompatible and hence is suitable for medical applications.
6. PHB has melting point 1750C., and glass transition temperature 150C.
7. PHB has tensile strength 40 MPa which is close to that of polypropylene.
8. PHB sinks in water while polypropylene floats. But sinking of PHB facilitates its anaerobic biodegradation in sediments.
9. PHB is nontoxic.

Parameter Polypropylene (pp) PHB	
Melting point Tm [°C]	171-186 171-182
Glass transition temperature Tg [°C]	-15 5-10
Crystallinity [%]	65-70 65-80
Density [g cm^{-3}]	0.905 - 0.94 1.23 - 1.25
Mol. Weight Mw	(x10-5) 2.2 - 7 1 - 8
Mol. Weight distribution	5 - 12 2.2 - 3
Flexural modulus [GPa]	1.7 3.5 - 4
Tensile strength [MPa]	39 40
Extension to break [%]	400 6 - 8
UV resistance poor good	

Solvent resistance good poor

Oxygen permeability[cm^{3}m-2atm-1d-1] 1700 45

Biodegradability - good

US annual production [Mio. T.] 1.8 not determined

Other due to low density floats in aquatic system due to more density goes to sediment in aquatic system

Bioplastics are making commercial and scientific progress continuously. W.R. Grace was the American company which carried out the work on PHB as early as 1960s. The work was then curtailed because at that time the techniques available for extraction were not able to provide a product thermally stable in processing. The development of PHB was begun by ICI in 1975-6 as a response to increase in oil prices. ICI started marketing BIOPOL in 1982. ICI, the UK chemical group, has opened a plant at Billingham in North-east of England to make 300 tonnes of 'Biopol' a year, which it says is the first fully biodegradable commercial plastic. The company plans to raise its annual production of this "Nature's plastic" to 5000 tonnes very soon. At present, Biopol costs about pounds 10 per kg., 20 times more than conventional plastic. Costs can be reduced to some extent by scaling up of the production. Even at its current price, ICI has plenty of buyers for limited amounts of Biopol hat they produce.

Industrial Production of PHA

A subsidiary of ICI Ltd. in Great Britain produces 'Biopol' from Alcaligenes eutrophus (H16). Either PHB homopolymer or copolymers of PHB and B-hydroxyvalerate can be produced by these cells depending on the substrate or substrate mixture used for growth and production.

For industrial applications it is desirable to control the incorporation of different repeating units into the polymer in order to produce polyesters with specific material characteristics because their physical and chemical properties depend strongly on copolymer composition. "Tailor-made" copolymers can be made for this purpose by the use of controlled conditions. If a defined mixture of nutrients for certain type of microorganisms is supplied for growth, a defined and reproducible copolymer is formed. Biochemically there are 2 different ways of achieving polymer formation in microbes

Substrate Substrate PHA producing substrate

Cell growth PHA formation Cell growth PHA formation

Parallel process Serial process

Parallel process has been demonstrated in R. rubrum and P. oleovorans. During cell growth, however part of the polymer forming potential is lost because of its utilization of substrate to maintain the cell's metabolism.

In serial process of industrial polymer production micro-organisms are first grown on C source to obtain large biomass, then the medium is depleted of an essential nutrient and polymer forming substrate is added. This is converted directly to polymers and essentially only little growth occurs. This approach is used for large scale PHA production by A. eutrophus.

List of Limiting Components Leading to PHA Formation—

- Ammonia Alcaligenes eutrophus, also others
- Carbon Spirillum spp., Hypomicrobium spp.
- Iron, Mg Pseudomonas spp.
- Mn, O2 Azospirillum, Rhodobacter spp.
- PO4 Rhodospirillum, Rhodobacter spp.
- Potassium sulfate Bacillus, Rhodospirillum, Rhodobacter etc.

Biopol of ICI is produced by using Alcaligenes eutrophus. Polymer separation and purification is accomplished by using a proprietary aqueous wash process followed by drying. PHB is too stiff and brittle for most applications, so the ICI adds a small amount of simple organic acid to the sugar feed stock to make the plastic stronger and more flexible. In technical terms, bacteria then produce co-polymers composed of PHB with varying amounts of hydroxyvalerate (HV). In this way ICI can produce a range of thermoplastic polymers which can be processed with conventional techniques to make bottles, mouldings, fibres and films. High grade Biopol is being made for medical applications, including woven patches for use inside the body to protect tissues from scarring after surgery. After the wound is healed, enzymes in the blood dissolve away the patch. ICI's biodegradable plastic (thermoplastic Biopol) has found its North American applications as a blow-molded bottle and injection-molded caps for hair care products. The Berlin Packaging corporation, Chicago will produce bottles and caps. Biopol currently sells for \$ 8-10 per lb but is expected to sell for \$ 4 per lb. The ICI company has currently a capacity of 600,000 lb per year in Billingham, UK. Annual production of 11 billion lbs is predicted by late 1990s. Biotehnologische Forschungsgesellschaft mbH in Austria produces PHB on a technical scale from cells of Alcaligenes latus with sucrose as a carbon source.

'Biopol' has been used in Germany since 1990 to make the bottles of Wella's Sanara Shampoo. Biopol's US launch came in 1995, in the form of bottles for Brocanto International's Evanesce shampoo and material is being tested in the UK for cosmetic containers.

Japan also has shown interest in Biopol. Biopol has been introduced there in 1991 as a container for Ishizawa Kenkyujo's Earthic Alga shampoos and conditioners. Now three more hair care companies of Japan have started using

Biopol containers and shortly Kai will use it for disposable razor with Biopol handle. Rubbish bags, disposable nappies, paper plates, cups coated with thin plastic film can be made with Biopol and will get degraded when thrown in landfill. Biopol shampoo bottle disappears in two years in typical dung.

With development in yields, productivity, use of newer strains PHB concentration and productivity of 100 g.l-1 and 2.5 g.l-1 h-1 can be reached.

Chemi Linz AG began to develop a process for PHB production by fermentation in 1980s. Their polymer group Petrochemia Danubia (PCD) carries out the fermentation process through btF, a biotechnological research unit. The PCD process is different in that it uses Alcaligenes lactus as the producing organism and sucrose as the substrate. PHB formation is growth associated and nutrient limitation is not used to induce polymer accumulation. Solvent extraction is used to extract the product. It is a PHB homopolymer which is made.

PCD-Polymere, an Austrian polyolefin producer developed a production process for PHB and processing technology for injection moulding and blow moulding. An Austrian biotechnology research company has developed a process for PCD-Polymere.

The fermentation process is based on a unique bacterial strain Alkaligenes latus. The strain is isolated from soil in California and Australia. It produces PHB in large amounts (80% of cell dry weight) during unlimited growth. Proprietary fermentation process developed is scaled upto 10 m3 of fermentation volume. In a fed batch mode more than 1 ton of PHB can be produced in less than a week. These fermentations are carried out in common stirred tank reactors, but other reactors like air lift or bubble column may also be used to get similar good results. The fermentation is carried out in mineral salts medium, sucrose or glucose as source of carbon. If precursors like propionic acid are added then co-polymers of PHB/HV can be produced as well as co-polymers of 3HB/4HB (3-hydroxybutyrate-co-4hydroxybutyrate) if 1,4 butanediol is added as precursor.

After fermentation cells are harvested, washed with tap water and a concentrated suspension of 200 g/l is prepared. This suspension is directly used for the extraction process. The cell suspension is treated with solvent (methylene chloride). After this extraction step solvent is separated by centrifugation. The dissolved PHB is precipitated in water, recovered as white powder and dried. After one extraction step and one precipitation step PHB of 99 per cent purity is obtained. This powder is directly used for compounding and further processing. Biomass can be recovered after the extraction process. It was tested as soil enhancer.

PCD-Polymere films and fibres are under development. Good results were obtained for injection moulding and blow moulding parts. PCD also helps others to develop further the technology

The cost of producing PHB can be substantially reduced if methanol is used as substrate. Methanol is one of the cheapest noble substrates available, and has several advantages as fermentation substrate (purity, solubility, availability etc.) and is non-food substrate. Methanol can be easily obtained from natural gas and may also be obtained from biomass if necessary.

Methylobacterium extorquens is the isolate (gram negative, motile, pink pigmented bacterium can be used. Average PHB contents are found to be 25-30 per cent of dry weight. It was also capable of producing PHV with the ratio of PHV to PHB of 0.2. Good process control was essential in the development of high cell-density fed-batch fermentation process for PHB production from methanol. Biomass level of 120 g/l and PHB levels of 60 g/l could be reached.

Evaluatiuon of Different Extraction and Cell Rupture Methods for PHA Isolation from Rhodospirillum Rubrum:

❖ Hot chloroform extraction Lysozyme sonication French press Hypochlorite
❖ PHA content (%dry wt.) 13.6 13.6 13.6 13.6
❖ Isolated PHA (%dry wt) 12.5 9.7 7.7 12.1
❖ Recovery % 92 71 58 89
❖ Mw (x 10-6) 1.5 0.92 0.65 0.94
❖ Mn (x 10-6) 1.1 0.36 0.46 0.37
❖ Mw/Mn (Polydispersity) 1.4 2.6 1.4 2.5

Market for Biodegradable Material :

❖ During the early 1990s, annual production of PHAs was just several hundred tonnes. In 1996 Zeneca, the main producer has stopped producing PHBV and sold the assets to Monsanto and Astra.
❖ Currently, the market for biodegradable material is largest in Scandinavia and other European countries, where companies are willing to pay up to pounds 4 per kg for PHAs. The estimate given in table below is based on the development of a composting infrastructure, which is now becoming highly developed in many parts of Europe, particularly in Germany, Holland and Belgium.
❖ Western European market estimates for biodegradable plastics (tonnes per annum)
❖ Application Conventional resins Biodegradable resins
❖ Waste disposal bags for composting - 30 000
❖ Disposable fast food utensils 100 000 50 000
❖ Hygiene films 110 000 20 000
❖ Paper coatings 420 000 40 000

- ❖ Agricultural sector 65 000 30 000
- ❖ Total 695 000 170 000

Source: BASF (1993)

Production of PHA by genetically Engineered Plants

Biopol is made in industrial fermentor by bacteria that converts sugars (refined from corn or beet) into polymer. But US scientists recently announced an important step towards making biodegradable plastic directly in plants. Genetically engineered Arabidopsis thaliana a type of cress was used for the purpose. Costs can be reduced tenfold if the plastics are produced by plants. With transgenic plants producing PHA price comes to US 20 cents per kg. which is close to that of starch. Potatoes can be genetically engineered to make and store plastics instead of starch in their tubers.

The idea of producing PHAs in transgenic plants was first described by MIT researchers in a 1989 patent application. By the mid-1990s several research groups had successfully produced PHB in various plants, ranging from Arabidopsis thaliana, an experimental research plant to commercial oilseed crops and even cotton. Full scale commercial development of transgenic PHA crops is still 5-7 years away. The cost of plant-derived PHAs will depend upon several factors, such as plant crop used, content of PHAs, price of crop, location, scale, ease of extraction. If yields of PHAs are 20-50 per cent of the crop then price of PHAs can be competitive with synthetic plastics. 1 m hectares (ca 1010m2) of farmland would produce ca 375000t of plastic. This quantity of plastic is only 5 per cent of the US demand of plastic packaging market.

Chris Somerville of the US Department of Energy's plant research laboratory at Michigan state University led a team collaborating with the James Madison University in Virginia to modify the cress. The investigators inserted two foreign genes into cress, taken from bacteria Alcaligenes eutrophus which makes PHB naturally. Agrobacterium tumefacins is used as 'Trojan horse' bacterium to transfer two genes to cress plant Arabidopsis thaliana. Plants produced 20-100 mcg/gram of plant tissue, PHB. Production should be raised 100 fold to commercialize it. One potential problem is plants become sick. This is perhaps because new genes divert carbon away from the essential metabolites to PHB production. But if these genes are introduced in beet, potato plants which have too much excess of energy-storing carbon such as starch, such diversion of carbon to PHB is not damaging. However, USA only can think of PHB instead of starchy food. Genetically-engineered Arabidopsis plants so far have produced small amounts of pure PHB. But it is expected that co-polymers of PHB-HV can be made directly in crops.

Oilseed crops are most amenable targets for seed-specific PHA production.

Since both oil and PHA are derived from acetyl-CoA, it is diversion of acetyl CoA to PHB accumulation. Plastids of plants are targeted for PHB synthesis by engineered genes. Arabidopsis thaliana is closely related to oil producing crop rapeseed which in fact is a target crop for PHB production on agricultural scale. Rapeseed, Sunflower and soyabean are the crops which can be genetically transformed to produce PHA. ZENECA Seeds and Monsanto are working on it. Monsanto will commercialise it by 2003 AD. Monsanto is trying to improve the performance of this bioplastic. The bioplastic will be initially used to produce paper coating and a film for food packaging.

The Forest Genetics Research Institute (Suwon, Korea), a subsidiary of the South Korean Ministry of Agriculture and Forestry, has developed a process to produce PHB in the chloroplasts of genetically engineered aspen trees (poplars). Two genes from Alcaligenes eutrophus have been transferred to poplar plant cells. PHB present in chloroplasts can be obtained from leaves. Leaves are dried, crushed to fine powder and then bioplastic is extracted with chloroform.

The Massachusetts Institute of Technology (MIT) has been awarded two patents in USA covering the production of biodegradable plastics in plants. These are seventh and eighth in a series exclusively to Metabolix (Cambridge, Mass). Metabolix is developing a range of technologies for PHA production including enzyme catalysed polymerisation and fermentation routes.

It has also been shown by Metabolix Inc. that transfer of the proprietary PHA genes into plants results in accumulation of PHA polymers in the new host. Current research aims to optimize expression of these genes and target polymer synthesis to easily processable tissues like the seeds or tubers. Polymer extracted from these sources is expected to compete directly on a price-performance basis with current nonrenewable plastics. Full scale plant crop production is expected in four to ten years.

The polymers produced in plants are structurally identical to those from bacteria, so that processors can develop applications using fermented material in expectation of using plant-derived PHAs in the future. PHAs are also a useful industrial source of chiral building blocks for the chemical industry. Stereochemically pure monomers (R -3-hydroxyacids and their derivatives) are readily available by depolymerization of the PHA polymers. Monomer derivatives have been supplied to customers by Metabolix under research agreements.

Production of PHA in Genetically Engineered Bacteria

Now PHA can also be produced efficiently and with novel properties in genetically engineered bacteria. Recombinant E. coli harboring multicopy plasmid carrying A. eutrophus PHA biosynthesis genes is developed. PHB concentration of 80 g.l-1 and 2.0 g.l-1 h-1 is obtained.

Metabolix researchers have transferred proprietary genes for PHA production into the safe and widely used industrial microbial strain, Escherichia coli, where the new genes result in rapid production of high levels of polymer. This strain, E. coli K12, is already the source of a number of FDA approved food additives and medical products. The raw materials for the transgenic fermentations are widely available sugars like glucose, which are derived from renewable plant crops.

Transgenic fermentation systems have several advantages over the non-engineered systems. As a result, the economics of PHA production are now more favourable than at any previous time. Better polymer yields, easier recovery, and production of new copolymers are all enabled by the use of E.coli—the tried and true workhorse of the biotechnology industry. Faster growth means that engineered E.coli can produce PHA in just 24 hours, compared to three days or more for non-engineered strains. High levels of polymer, up to 90 per cent of cell weight, are also achievable, so less carbon substrate is wasted to make extraneous biomass, and polymer isolation becomes more straightforward. Finally, E.coli has the best understood genetics and biochemistry of any organism in the world, so that further metabolic engineering (e.g. to produce new polymer compositions) is possible.

Advantages of using recombinant E.coli for production of PHA are :

(a) Wide range of substrates (lactose, xylose, sucrose) can be used. Therefore whey, agricultural wastes and byproducts and molasses which are cheaper raw materials can be used.

(b) Engineered E.coli can produce PHB in 24 hours while non-engineered producers take 3 days.

(c) It is easier and less costly to purify polymers from recombinant E.coli than from A.eutrophus, since E.coli becomes fragile due to accumulation of biopolymer.

(d) Recombinant E.coli does not possess PHA depolymerase as what natural PHA producers possess. Hence synthesized PHA is not degraded by producer recombinant E.coli.

(e) Molecular weight of PHB produced by fermentation of recombinant E.coli can be controlled by modulating activity of PHA synthase enzyme.

Average Molecular weight of PHA produced by recombinant E.coli is 4 X 103 kDa.

Average Molecular weight of PHA produced by A.eutrophus is 6 X 102 to 1.2 X 103 kDa. Molecular weight decreases by increase in enzyme activity.

(f) Polydispersity index can be controlled by modulation of synthase activity.

Factors (e) and (f) thus affect polymer properties and processibility and well defined characteristics can be obtained. Newer applications such as stronger biodegradable fibre will be possible due to higher molecular weight of PHAs. Commercial production with recombinant bacteria will be very soon demonstrated.

Price Factor

Commercial applications and wide use of PHAs is hampered due to its price. Today price with natural producer like A.eutrophus is US $16 per kg. This is about 18 time more expensive than polypropylene. For PHA to be commercially viable price should come to US $3-5 per kg. With recombinant E.coli as producer of PHA, price can be reduced to US $4 per kg. which is close to other biodegradable plastic materials such as polylactides and aliphatic polyesters. With transgenic plants producing PHA price comes to only US 20 cents per kg. Most important aspect is its biodegradable nature helps to overcome the problem of environmental hazard. And the value of clean environment will otherwise also outweigh conventional plastics on price factor in favour of more and more use of PHA.

Apart from environmental advantages (better waste management), Biopol is made from renewable sources- sugars refined from crops rather than fossil fuels. Biopol need not be thrown away and can be recycled or burned cleanly to provide energy in an incinerator.

Possible Applications of PHAs

Early investigations of PHA granules by electron microscopy after freeze-etching showed that the polymer in the granule underwent a cold drawing process indicating the plastic nature of polyester and suggesting that it can be processed as a conventional thermoplastic. In addition to its potential as plastic material PHA is useful source of stereoregular compounds which can serve as chiral precursors for the chemical synthesis of optically active compounds, particularly in synthesis of some drugs or insect pheromones. These substances are biologically active only in the correct stereochemical configeration.

PHAs can be easily depolymerised to a rich source of optically pure bifunctional hydroxy acids. PHB, for example can be readily hydrolysed to R-3-hydroxybutyric acid and is used in the synthesis of Merck's anti-glaucoma drug Truspot. Along with R-1,3-butanediol, it is also used to synthesise b-lactams.

PHBV received European approval for food contact use in 1996. This opens opportunities in food service and packaging industry.

Plant-derived PHAs may in future be depolymerised and used for bulk chemical manufacture. Western Europe, for example produces several million tonnes of butanols, half of which are used directly, or following esterification, as solvents. Besides replacing existing solvents b-hydroxy acid esters and related derivatives are likely to find growing use as green solvents similar to lactic acid esters. b-hydroxy acids are more resistant to hydrolysis and are therefore better suited to certain applications. Hydroxy acids may also be converted into crotonic acids, 1,3-butanediol, lactones etc. all of which have existing markets of thousands of tonnes.

PHB, (i) is 100 per cent biodegradable, (ii) can be processed like thermoplastic and (iii) is 100 per cent water resistant, so that it could be used for similar applications as conventional commodity plastics.

Practical Applications of PHA

(1) Packaging films (for food packages), bags, containers, paper coatings.
(2) Biodegradable carrier for long term dosage of drugs, medicines, insecticides, herbicides, insecticides or fertilizers.
(3) Disposable items such as razors, utensils, diapers, feminine hygiene products, cosmetics containers, shampoo bottles, cups etc.
(4) Starting material for chiral compounds.
(5) Medical applications - Surgical pins, sutures, staples, swabs, wound dressings, bone replacements & plates and blood vessel replacements, Stimulation of bone growth by piezoelectric properties.

Industrial Production of PHAs and Other Biodegradable Plastics

Company Areas of interest

(1) Berlin Packaging Corp. (USA) Marketing
(2) Bioscience Ltd. (Finland) Medical applications
(3) Bio Ventures Alberta Inc. (Canada) Production in recombinant E.coli.
(4) Metabolix Inc. (USA) Production in transgenic plants
(5) Monsanto (USA)
(6) Polyferm Inc. (Canada) Production from cheap substrates
(7) ZENECA Bio-Products (UK) Production by A.eutrophus (former ICI, UK)
(8) ZENECA Seeds (UK) Production in transgenic plants
(9) Petrochemia Danubia (PCD) Production using A.lactus

It was shown that PBH-HV, 1 mm molding was completely degraded after 6, 75, 350 weeks in anaerobic sewage, soil & sea water respectively.

Biodegradability of PHAs

One of the properties that distinguishes PHAs from petroleum-based plastics is their biodegradability. Produced naturally by soil bacteria, the PHAs are degraded upon subsequent exposure to soil, compost, or marine sediment. Despite their biodegradability the PHAs still have good resistance to water and moisture vapor, and are stable under normal storage conditions and during use.

Biodegradation of PHAs is dependent upon a number of factors such as the microbial activity of the environment and the exposed surface area. In addition, temperature, pH, molecular weight and crystallinity are important factors. Biodegradation starts when micro-organisms begin growing on the surface of the plastic and secrete enzymes that break down the polymer into its molecular building blocks, called hydroxyacids. The hydroxyacids are then taken up by the micro-organisms and used as carbon sources for growth. In aerobic environments the polymers are degraded to carbon dioxide and water, whereas in anaerobic environments the degradation products are carbon dioxide and methane.

A number of reports have demonstrated that PHAs are compostable over a wide range of environmental conditions. In one report, the maximum biodegradation rates were observed at moisture levels of 55 per cent and temperatures of around 60°C-conditions similar to those used in most large-scale composting plants. Up to 85 per cent of the samples degraded within 7 weeks, and PHA coated paper was rapidly degraded and incorporated into the compost. In another study, the quality of PHA compost was determined by measuring seedling growth relative to a control. Seedling growth of around 125 per cent of the control was found for a 25 per cent PHB copolymer compost indicating that the compost can support a relatively high level of growth.

Biodegradation of PHAs has also been tested in various aquatic environments. In one study in Lake Lugano, Switzerland, items were placed at different depths of water as well as on the sediment surface. A life span of 5-10 years was calculated for bottles under these conditions (assuming no increase in surface area), while PHA films were completely degraded in the top 20 cm of sediment within 254 days at temperatures not exceeding 6°C.

Biolac

Other bioplastic which is easily biodegradable are polylactides (PLA) and polyglycolides (PGA). The earlier efforts in this field are by American Cyanamid Corporation which developed the first synthetic absorbable suture material. The product, 'Dexon' was a polyglycolic acid homopolymer. 'Vicryl' was developed then by Dupont which has 92:8 glycolic acid : lactic acid as copolymer. PLA and PGA are thermoplastics and are biodegradable polyesters. Low Mr polylactic acid and polyglycolic acid are made by direct polymerization of respective acids. The high Mr, PLA and PGA are made by ring opening polymerization of lactide and glycolide which are cyclic diesters of respective acids. Polyglycolic acid and polylactic aid have degradation time in few days and few weeks respectively while polylactides and polyglycolides have degradation time in few months to years.

Researchers at the University of Wisconsin, USA, have produced biodegradable and photodegradable polymers from l-lactic acid. The researchers are working on a project to produce l-lactic acid from whey permeate. The technology to ferment whey to lactic acid is in fact quite old but the end product is mixture of l-lactic acid and d-lactic acid. The pure l-lactic acid is worth nearly 3 times as much the mixture of l- and d- lactic acids. Pure l-lactic acid has several uses. It can be used to manufacture polylactides (plastics made from lactic acid that are bio- and photodegradable), as food preservative, as flavour enhancer and as acidulant and in pharmaceutical industry (in IV solution and drug delivery).

The Ecological Chemical Products Co. (ECOCHEM) has recently opened a $ 20 million commercial plant in Adel, Wisconsin to produce high-pure natural lactic acid and polylactide polymers for food and pharmaceutical applications. ECOCHEM is a joint venture of DUPONT and Con. Agra. Inc. ECOCHEM's manufacturing process has environmental advantages such as raw materials derived from natural byproducts of cheese industry, no new wastes generated during its production and recycled side streams producing either useful or environmentally acceptable co-products.

O O O O CH2
C CH2 C CH

H2C C HC C
O O CH2 O O

Glycolide Lactide

Production of lactic acid-based plastics from starchy food wastes and by-products is also on the way of commercialization. The Argonne National Laboratory has licensed key steps in its Biolac process to Kyowa Hakko USA

Inc., a US subsidiary of Japan's Kyowa Hakko, which deals in fermentation products. Kyowa Hakko plans to carry out further research and development aimed to commercialization.

PLA and PGA are having mainly medical applications, as sutures, as ligament replacements, for resorbable plates and screws in fracture fixation (i.e. in orthopedic repairs), for controlled drug release, for arterial grafts. The Biolac plastic has commercial applications for compost bags, coatings for paper, seeds, pesticides, fertilizers and agricultural mulch films for timed release of pesticides and fertilizers. In two key steps that have been licensed, one converts glucose of starchy wastes to lactic acid and the other converts lactic acid to polylactic acid.

Potential market for PLA plastics and coatings as forcasted by Argonne National Laboratory, Illinois :

- ❖ Controlled fertilizer and pesticide application - > 500 000 tonnes
- ❖ marine plastic applications - 250 000 tonnes
- ❖ degradable conditioner coatings for paperback stock - > 100 000 tonnes
- ❖ compost waste bags, sacs etc. - tens of thousands of tonnes
- ❖ agricultural mulch film - 75 000 tonnes
- ❖ Polylactic acid can be produced in US since the US produces 5 million tonnes of food wastes in the manufacture of fried potatoes and 1-2 million tonnes wastes in the cheese industry. Kyowa Hakko USA Inc. wants to use this fact for producing polylactic acid by the Biolac process.

The 5000 tonnes/year capacity plant at a cost of $ 8 million will be operational at Port Cargill on Minnesota river near Minneapolis. It will make polylactic acid from lactic acid which is obtained by bacterial fermentation of sugars from corn, potato, milk, sugarbeet.

Archer Daniels Midland Co. and Warner Lambert Co. are among the companies gearing up to produce lactic acid from corn or starch that could be used in next generation biopolymer plastic.

Two Japanese companies Kobe steel Limited (Kobe) and Shimadzu Corp. (Kyoto) have perfected a low-cost continuous process for manufacture of biodegradable polymer - poly-2-lactic acid (PLLA). It can be produced at tens of thousands of capacity. PLLA melts at 170-1800Cand has vicat softening point of 580C, a tensile strength of 700kg/cm and a transparency of 94 per cent. It can be processed into a film of 10 to 500 Mm thickness and can be injection moulded. It can be produced at $2.54 to $4.23 per kg. It can be used for food containers, soil retention sheetings and agriculture film.

The physical properties of polylactic acid are similar to polystyrene. It can also be modified to make it similar to PE (Polyethylene) or PP (Polypropylene). It has performance benefits similar to petrochemical-based plastics but is biodegradable by composting. Applications of polylactic acid as plastic can be - disposable fast food, dairy and delivery containers, food service ware, medical garments, waste bags.

Asahi Chemicals and Institute of Physical and Chemical Research in Japan have jointly discovered a new species of bacteria which synthesizes biopolyesters. The bacteria accumulate large quantities of polymers. The bacteria can use wide variety of compounds with 2-22 carbon atoms as a source of carbon. In experiments, inexpensive oils and fats were used as carbon sources which produced polyester at a maximum yield of 45 per cent. The yield will be increased to 60-70 per cent when recovery process is optimized. High efficiency plastic with low production cost is the aim of research work. The use of wide range (2-22 carbon atoms) compounds is advantageous. Product type and yield differ with type of feed used. The compound with 18 carbon atoms produces the highest recovery yields.

Biodegradable "enviroplastic" is developed by Planet Technologies, USA. It will be used for cosmetic, medical disposables, food service, personal hygiene product industries. It will dissolve in sludge compost or water without leaving any environmentally harmful residues.

Researchers at University of Iowa are investigating the synthesis of biodegradable plastic using enzymes in organic solvents. Sucrose and adipic acid are the substrates used and action of lipases and proteases is sought for to link sugar and diacid into copolymer chain.

Polyglutamic acid (PGLU) is a water soluble polymer produced by Bacillus species. Bacillus subtilis releases polyglutamate in growth medium. Fermentations to produce PGLU have been patented some years ago but commercial production is not reported. Yields to the tune of 40 g/l of PGLU in 5 days are reported.

O COOH

|| |

--------O - C - C - C - C - NH ----------

n

Japanese researchers at National Food Research Institute have developed water-resistant, biodegradable plastic films from corn protein. This newly developed plastic is expected to have wider applications such as food trays, because of low material cost and fabricability. The process is based on a protein called Thujene. It remains in corn after removal of starch. Thujene is spread into thin transparent film after being dissolved in acetone. When it has thickness of around 70 μm the film is as strong as commercial wrapping film with respect to boring. The film rarely permeates water. It will be enzymatically decomposed in one month in ordinary soil.

Conclusion

Thus, while tackling the issues related to environmental protection and cleanliness, we started with attack on symptoms rather than causes of pollution (Measurement of pollution and Treatment technologies). Then subsequently we gave stress on Environmental Impact Assessment (EIA) and could work on better planning and better control. And today we have started to attack the root cause of pollution - prevention of pollution. We are talking about "clean technologies". We are aiming for biodegradable and ecofriendly products and processes. Bioplastics is only a part of the large efforts that we are determined to make. Bioplastics is a reality and is a practical truth. Our willingness and improvement in technologies will give it a wider success.

Many opportunities exist for the application of **synthetic biodegradable polymers** in the biomedical area particularly in the fields of tissue engineering and controlled drug delivery. Degradation is important in the biomedical area for many reasons. Degradation of the polymer implant means surgical intervention is not required for removal, eliminating the need for a second surgery. In the role of tissue engineering biodegradable polymers can be designed such to approximate soft tissues, providing a polymer scaffold that can withstand resistance, provide a suitable surface for cell attachment and growth and degrade at a rate that allows the load to be transferred to the new tissue[1]. In the field of controlled drug delivery biodegradable polymers offer tremendous potential as a basis for drug delivery, either as a drug delivery system alone or in conjunction to functioning as a medical device.

This article will focus on the development of such applications, review the chemistry of some polymers including synthesis and degradation, describe how properties can be controlled by proper synthetic controls such as copolymer composition, highlight special requirements for processing and handling, and discuss some of the commercial devices based on these materials.

Polymer Chemistry and Material Selection

Some biodegradable polymers, their properties and degradation times can be found in Table 2 in this document.

When investigating the selection of the polymer for biomedical applications, important criteria to consider are;

- ❖ The mechanical properties must match the application and remain sufficiently strong until the surrounding tissue has healed.
- ❖ The degradation time must match the time required.
- ❖ It does not invoke a toxic response.
- ❖ It is metabolized in the body after fulfilling its purpose.

- It is easily processable in the final product form with an acceptable shelf life and easily sterilized.

Mechanical performance of a biodegradable polymer depends on various factors which include monomer selection, initiator selection, process conditions and the presence of additives. These factors influence the polymers crystallinity, melt and glass transition temperatures and molecular weight. Each of these factors needs to be assessed on how they affect the biodegradation of the polymer[2]. Biodegradation can be accomplished by synthesizing polymers with hydrolytically unstable linkages in the backbone. This is commonly achieved by the use of chemical functional groups such as esters, anhydrides, orthoesters and amides. Most biodegradable polymers are synthesized by ring opening polymerization.

Processing

Biodegradable polymers can be melt processed by conventional means such as compression or injection molding. Special consideration must be given to the need to exclude moisture from the material. Care must be taken to dry the polymers before processing to exclude humidity. As most biodegradable polymers have been synthesized by ring opening polymerization a thermodynamic equilibrium exists between the forward polymerization reaction and the reverse reaction that results in monomer formation. Care needs to be taken to avoid an excessively high processing temperature that may result in monomer formation during the molding and extrusion process.

Degradation

Once implanted a biodegradable device should maintain its mechanical properties until it is no longer needed and then be absorbed by the body leaving no trace. The backbone of the polymer is hydrolytically unstable. That is, the polymer is unstable in a water based environment. This is the prevailing mechanism for the polymers degradation. This occurs in two stages.

1. Water penetrates the bulk of the device, attacking the chemical bonds in the amorphous phase and converting long polymer chains into shorter water-soluble fragments. This causes a reduction in molecular weight without the loss of physical properties as the polymer is still held together by the crystalline regions. Water penetrates the device leading to metabolization of the fragments and bulk erosion.
2. Surface erosion of the polymer occurs when the rate at which the water penetrating the device is slower than the rate of conversion of the polymer into water soluble materials.

Biomedical engineers can tailor a polymer to slowly degrade and transfer stress at the appropriate rate to surrounding tissues as they heal by balancing the chemical stability of the polymer backbone, the geometry of the device and the presence of catalysts, additives or plasticisers.

An example of the structure of some of the types of polymer degradation can be viewed in Figure one in this article.

Applications

As previously mentioned biodegradable polymers are used commercially in both the tissue engineering and drug delivery field of biomedicine. Specific applications include.

- Sutures
- Dental devices
- Orthopedic fixation devices
- Tissue engineering scaffolds.

11.10 Chemically Assisted Degradation of Polymers

Chemically Assisted Degradation of Polymers is a type of polymer degradation that involves a change of the polymer properties due to a chemical reaction with the polymer's surroundings. There are many different types of possible chemical reactions causing degradation however most of these reactions result in the breaking of double bonds within the polymer structure.

Examples of Chemically Assisted Degradation

Degradation of Rubber by Ozone

One common example of chemically assisted degradation is the degradation of rubber by ozone particles. Ozone is a naturally occurring atmospheric molecule that is produced by electric discharge or through a reaction of Oxygen with solar radiation. Ozone is also produced with atmospheric pollutants reacted with UltraViolet Radiation. For a reaction to occur, ozone concentrations only have to be as low as 3-5 parts per hundred million (pphm) and when these concentrations are reached, a reaction occurs with a thin surface layer (5 x10-7 metres) of the material. The ozone molecules react with the rubber which in most cases is unsaturated (contains double bonds), however a reaction will still occur in saturated polymers (those containing only single bonds). When reaction occurs, scission of the polymer chain (breaking of double covalent bonds) takes place forming decomposition products:

Chain scission increases with the presence of active Hydrogen molecules (for example, in water) as well as acids and alcohols. Along with this type of reaction, cross linking and side branch formations also occur by an activation of the double bond and these make the rubber material more brittle. Due to the increase in brittleness due to the chemical reactions, cracks form in areas of high stress. As propagation of these cracks increases, new surfaces are opened for degradation to occur.

Degradation of Poly (vinyl) Chloride (PVC)

Degradation can also occur as a result of the formation, and then breakage of double bonds, such as solvolysis in PVC (Peacock). Solvolysis occurs when a Carbon-X bond, with X representing a halogen, is broken. This occurs in PVC in the presence of an acid species. Active Hydrogen atoms will remove a Chlorine atom from the polymer molecule, forming Hydrochloric acid (HCl). The HCl produced may then cause dechlorination of adjacent Carbon atoms. The dechlorinated Carbon atoms then tend to form double bonds, which can be attacked and broken by ozone, just like the degradation of rubbers described above.

Degradation of Polyester

Degradation of polyester may occur without the presence of the acidic catalyst that causes degradation of PVC. During hydrolysis water acts as the reactive catalyst instead of the acid. It causes degradation mainly at high temperature and pressure during processing.

In this process the water molecule will attack the C-O ester bond, splitting the polymer in half. The water molecule will then dissociate, with one Hydrogen atom forming a carboxylic acid group on the Carbon atom with the double bonded Oxygen, while the remaining atoms form an alcohol on the other chain end. These reactive products may also cause further degradation of the polymer chain. This chain scission lowers average molecular weight of the polymer, decreasing the number and strength of intermolecular bonds as well as the degree of entanglement. This will increase chain mobility, decreasing strength of the polymer and increasing deformation at low stresses.

Protection against Chemically Assisted Degradation

Both physical and chemical barriers can be used to protect a polymer from chemically assisted degradation. A physical barrier must provide continuous protection, must not react with the polymer's environment, must be flexible so that stretching may occur and must also be able to regenerate (after wear processes). A chemical barrier must be highly reactive with the polymer's surroundings so that the barrier reacts with the environmental conditions rather than the polymer itself. This barrier involves addition of a material into the polymer blend during fabrication of the polymer. Due to this, the barrier addition must have a suitable solubility, must be economically feasible and must not hinder the production process. For the barrier to be activated, the addition must diffuse to the surface and so a suitable diffusivity is also required. There are four theories on how these types of barriers protect the polymer material:

- *Scavenger Theory*: the protective layer reacts with the ozone rather than the polymer.
- *Protective Film Theory*: the protective layer reacts with the polymer producing a thin film on the polymer surface which is inert and cant be penetrated.
- *Re-linking Theory*: the protective layer causes broken double bonds to be reformed.
- *Self Healing Theory*: the protective layer reacts with degraded polymer chains to form low molecular weight material which forms an inert film on the surface.

Of these theories, the Scavenger Theory is the most common and most important. However, more than one theory can act at the same time and the theory that takes place depends on the protective materials, the polymer and surrounding environment.

11.11 Thermal Degradation of Polymers

Thermal degradation of polymers is molecular deterioration as a result of overheating. At high temperatures the components of the long chain backbone of the polymer can begin to separate (molecular scission) and react with one another to change the properties of the polymer. Thermal degradation can present an upper limit to the service temperature of plastics as much as the possibility of mechanical property loss. Indeed unless correctly prevented, significant thermal degradation can occur at temperatures much lower than those at which mechanical failure is likely to occur. The chemical reactions involved in thermal degradation lead to physical and optical property changes relative to the initially specified properties. Thermal degradation generally involves changes to the molecular weight (and molecular weight distribution) of the polymer and typical property changes include reduced ductility and embrittlement, chalking, colour changes, cracking, general reduction in most other desirable physical properties.

The Mechanism of Thermal Degradation

Most types of degradation follow a similar basic pattern. The conventional model for thermal degradation is that of an autoxidation process which involves the major steps of initiation, propagation, branching, and termination.

Initiation

The initiation of thermal degradation involves the loss of a hydrogen atom from the polymer chain (shown below as R.H.) as a result of energy input from heat or light. This creates a highly reactive and unstable polymer 'free radical' (R*) and a hydrogen atom with an unpaired electron (H*). The strength of the C-F bond means that there it is much harder for thermal degradation to initiate.

Propagation

The propagation of thermal degradation can involve a variety of reactions and one of these is where the free radical (R*) reacts with an oxygen (O_2) molecule to form a peroxy radical (ROO*) which can then remove a hydrogen atom from another polymer chain to form a hydroperoxide (ROOH) and so regenerate the free radical (R*). The hydroperoxide can then split into two new free radicals, (RO*) + (*OH), which will continue to propagate the reaction to other polymer molecules. The process can therefore accelerate depending on how easy it is to remove the hydrogen from the polymer chain.

Termination

The termination of thermal degradation is achieved by 'mopping up' the free radicals to create inert products. This can occur naturally by combining free radicals or it can be assisted by using stabilizers in the plastic.

The Research Methods of Thermal Degradation of Polymers

TGA

(Thermogravimetric Analysis) (TGA) refers to the techniques where a sample is heated in a controlled atmosphere at a defined heating rate whilst the samples mass is measured. When a polymer sample degrades, it mass decreases due to the production of gaseous products like carbon monoxide, water vapour and carbon dioxide.

DTA and DSC

(Differential thermal analysis) (DTA) and (differential scanning calorimetry) (DSC): Analyzing the heating effect of polymer during the physical changes in terms of glass transition, melting, and so on. These techniques measure the heat flow associated with oxidation.

Ways of Polymer Thermal Degradation

Depolymerisation

Under thermal effect, the end of polymer chain departs, and forms low free radical which has low activity. Then according to the chain reaction mechanism, the polymer loses the monomer one by one. However, the molecular chain doesn't change a lot in a short time. The reaction is shown below. This process is common for polymethymethacrylate (perspex).

$$CH_2\text{-}C(CH_3)COOCH_3\text{-}CH_2\text{-}C^*(CH_3)COOCH_3 \rightarrow CH_2\text{-}C^*(CH_3)COOCH_3 + CH_2{=}C(CH_3)COOCH_3$$

Random Chain Scission

The backbone will be break down randomly, could be occurred at any position of the backbone. The molecular weight decreases rapidly, and cannot get monomer in this reaction, this is because it forms new free radical which has high activity can occurs intermolecular chain transfer and disproportion termination with the CH2'group.

$$CH_2\text{-}CH_2\text{-}CH_2\text{-}CH_2\text{-}CH_2\text{-}CH_2\text{-}CH_2' \rightarrow CH_2\text{-}CH_2\text{-}CH{=}CH_2 + CH_3\text{-}CH_2\text{-}CH_2' \text{ or } CH_2' + CH_2{=}CH\text{-}CH_2\text{-}CH_2\text{-}CH_2\text{-}CH_3$$

Side-Group Elimination

Groups that are attached to the side of the backbone are held by bonds which are weaker than the bonds connecting the chain. When the polymer was being heated, the side groups are stripped off from the chain before it is broken into smaller pieces. For example the PVC eliminates HCL, under 100-120°C.

$$CH_2(Cl)CHCH_2CH(Cl) \rightarrow CH{=}CH\text{-}CH{=}CH + 2HCl$$

Oxidation of the Polymer

Polyphenylene oxide is well known for oxidation.

12

Synthetic Butadiene Rubber (SBR)

12.1 Rubber Polymers

Natural Rubber

Rubber is an example of an elastomer type polymer, where the polymer has the ability to return to its original shape after being stretched or deformed. The rubber polymer is coiled when in the resting state. The elastic properties arise from the its ability to stretch the chains apart, but when the tension is released the chains snap back to the original position.

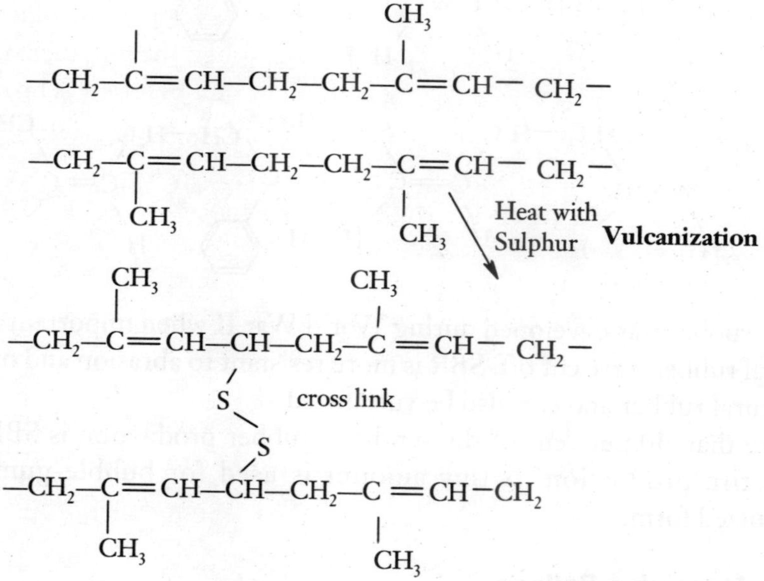

$$
\begin{array}{c}
\qquad\qquad\qquad\qquad CH_3 \\
| \qquad\qquad\qquad\qquad | \\
-CH_2-C{=}CH-CH_2-CH_2-C{=}CH-CH_2- \\
-CH_2-C{=}CH-CH_2-CH_2-C{=}CH-CH_2- \\
| \qquad\qquad\qquad\qquad | \\
CH_3 \qquad\qquad\qquad\qquad CH_3
\end{array}
$$

Heat with Sulphur **Vulcanization**

$$
\begin{array}{c}
CH_3 \qquad\qquad\qquad CH_3 \\
| \qquad\qquad\qquad\qquad | \\
-CH_2-C{=}CH-CH-CH_2-C{=}CH-CH_2- \\
/ \\
S \quad \text{cross link} \\
\backslash \\
S \\
/ \\
-CH_2-C{=}CH-CH-CH_2-C{=}CH-CH_2 \\
| \qquad\qquad\qquad\qquad | \\
CH_3 \qquad\qquad\qquad CH_3
\end{array}
$$

Natural rubber is an addition polymer that is obtained as a milky white fluid known as latex from a tropical rubber tree. Natural rubber is from the monomer isoprene (2-methyl-1,3-butadiene). Since isoprene has two double bonds, it still retains one of them after the polymerization reaction. Natural rubber has the cis configuration for the methyl groups.

Charles Goodyear accidentally discovered that by mixing sulfur and rubber, the properties of the rubber improved in being tougher, resistant to heat and cold, and increased in elasticity. This process was later called vulcanization after the Roman god of fire. Vulcanization causes shorter chains to cross link through the sulfur to longer chains.

The development of vulcanized rubber for automobile tires greatly aided this industry.

Synthetic Rubber - Styrene-Butadiene (SBR)

Some of the most commercially important addition polymers are the copolymers. These are polymers made by polymerizing a mixture of two or more monomers. An example is styrene-butadiene rubber (SBR) - which is a copolymer of 1,3-butadiene and styrene which is mixed in a 3 to 1 ratio, respectively.

Styrene-Butadiene Rubber

```
      H    H                 H    H
       \  /                   \  /
  3n  C=C        H  +    n  C=C   styrene
      /   \     /      |    /    \
     H     C=C         |   H     C6H5
          /   \        |
         H     H       v

 -  - H2C       H2C—H2C       CH2—H2C       CH2—H2C        CH2-
        \      /       \     /       \   /        \       /
         C=C            C=C            C            C=C
        /    \         /    \        /  \          /    \
       H      H       H      H      H   C6H5      H
```

SBR rubber was developed during World War II when important supplies of natural rubber were cut off. SBR is more resistant to abrasion and oxidation than natural rubber and can also be vulcanized.

More than 40 per cent of the synthetic rubber production is SBR and is used in tire production. A tiny amount is used for bubble-gum in the unvulcanized form.

Needle Through a Balloon

The polymer rubber chains exist in random loose clumps in the unstretched state. At the nipple end of the balloon, there is lots of rubber and therefore many, many polymer chains - still loosely coiled. These chains can be pierced without popping the balloon because the the chains can still be stretched. This is because they allow the skewer in between the chains without breaking the chains or the bonds that connect them. But on the sides of the balloon, these chains are stretched almost to their limit and very far apart. The piercing is too much for the stretched chains and they break apart., and the balloon pops.

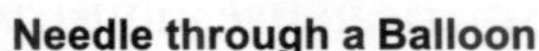

Needle through a Balloon

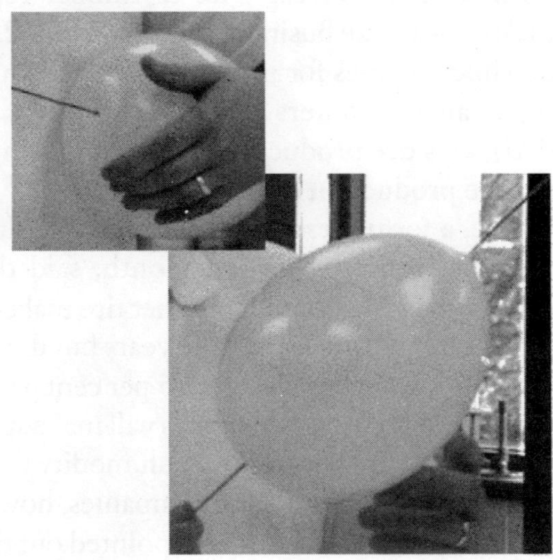

12.2 Synthetic Rubber

The rubber industry has been suffering from a poor economy and rising feedstock costs just like the rest of the chemical industry. Yet, in some sectors, its problems go even further. Styrene-butadiene rubber (SBR) is in such bad shape that some producers are shutting capacity and putting their businesses up for sale; one company even declared bankruptcy. Ethylene-propylene rubber (EPDM) makers are faring a little better, but a decline in the auto industry and additional capacity may dampen this sector as well.

According to the International Institute of Synthetic Rubber Producers (IISRP), the 2.2 million-metric-ton-per-year North American synthetic rubber industry showed no growth in 2002. Demand for SBR—the largest segment—fell by 1.8 per cent during the year. Polybutadiene and isoprene rubber showed some growth in 2002, EPDM and polychloroprene declined, and acrylonitrile-butadiene rubber was flat.

The rubber industry group forecasts a moderate 1.6 per cent annual growth rate for synthetic rubber through 2007, with 1.3 per cent growth for SBR, 2.2 per cent for polybutadiene, and 2.9 per cent for EPDM.

Timothy Rae, Bayer Polymers' head of regional product management for polybutadiene rubber and butyls, says slow growth has combined with a 1990s capacity buildup, particularly in Asia, to create intense oversupply in synthetic rubber, and producers are cutting back production to adjust. "Overall, the rubber industry is going through a couple of years of difficult times, and production has pulled back significantly to reflect the demand," he says.

Observers say the North American tire industry is in a state of decline. For example, earlier this month, Goodyear Tire & Rubber reported that sales volume for its North American tire business decreased by 7.2 per cent in 2002, to 104 million units, while volumes for most other regions increased.

Stephen J. Zinger, an elastomers consultant with Chemical Market Associates Inc. (CMAI), says tire production has been moving to Asia. "Labor is a big component in the production of tires," he says.

But Thomas J. Reese, a former executive with Continental Tire, speaking at a CMAI conference in Houston late last month, said the tire industry's problems run deeper than that. He pointed out that tire makers improved their warranties from 20,000 to 80,000 miles over the years but didn't cut capacity to compensate. Whitewall tires, which used to get a 5 per cent premium in Detroit, are gone, as are snow tires. He predicts cell phones will make spare tires obsolete.

Moreover, Reese said, tires are becoming a commodity that consumers buy on the basis of price and warranty alone. Tire companies, however, are pouring millions of dollars into their brand names. Reese pointed out that "40,000 miles is 40,000 miles, but majors are still trying to get a premium for their 40,000."

Bob Nelson, general sales manager at Goodyear Chemicals, says tire production is the biggest consumer of synthetic and natural rubber. The business uses the vast majority of the emulsion and solution SBR, polybutadiene rubber, and isoprene rubber that is made. "The rubber industry goes as the tire industry goes," he says.

Nelson says operating rates for plants making emulsion SBR, a commodity tire rubber, are between 60 and 65 per cent—and this is up from about 55 per cent a couple of years ago. "There is a lot of old capacity out there because most of the emulsion SBR plants in North America and Europe were built in the early 1940s," he says.

Part of the problem for emulsion SBR, Nelson says, is that it is being replaced by solution SBR in many applications requiring higher performance. He notes, for example, that tire makers looking to decrease rolling resistance without sacrificing traction are turning to solution SBR, or even isoprene terpolymers or other copolymers. As a result of this substitution, demand for emulsion SBR decreased by 4.3 per cent last year, while solution SBR demand rose 3.8 per cent, according to IISRP.

Another issue for synthetic rubber, particularly for polybutadiene and emulsion SBR, is competition with natural rubber, which can be substituted for polybutadiene in tire applications by 3 to 5 per cent and up to 20 per cent in other applications, according to Goodyear's Nelson. When natural rubber prices are low, tire makers change their formulations.

Jack McLaughlin—head of Bayer's marketing unit for tires and high-impact polystyrene—says that five years ago, when natural rubber prices were very low, tire producers added more natural rubber to their formulations. Recently, natural rubber prices have climbed significantly, although tire makers have not yet switched back to synthetic rubber.

On the supply side, the synthetic rubber industry is being squeezed by high petroleum costs spurred by the run-up to the war in Iraq and an oil workers' strike in Venezuela. Also, the natural gas supply crisis that has hit North America has been hurting tire makers and other rubber fabricators.

On top of the energy price problem, the industry's single biggest chemical raw material, butadiene, is in short supply. Observers explain that naphtha-based ethylene crackers—throttled back because of high naphtha prices and low demand for ethylene derivatives—are producing little coproduct butadiene. "You have fewer molecules going through the ethylene crackers and fewer by-products coming out because you are using lighter feeds," Goodyear's Nelson says.

CMAI's Zinger agrees. "The price of butadiene is moving up significantly, and that is putting a lot of pressure on rubber producers," he says. CMAI says butadiene prices will increase by nearly 50 per cent this year, having increased by 100 per cent since the beginning of 2002.

McLaughlin says tight butadiene supplies impact more than just prices. "Butadiene is not easily available," he explains. "Just getting the raw materials to produce is causing problems throughout the industry."

All these factors have led to trouble for synthetic rubber producers. Three—Goodyear, Ameripol Synpol, and DSM—are considering selling their businesses.

Goodyear says it is putting its chemical business up for sale to improve its finances; the company has also restructured nearly $3 billion in debt to the same end. The chemical business had sales in 2002 of nearly $940 million and operating income of about $70 million. About 65 per cent of the chemical business' output goes to Goodyear's own tire business.

International Specialty Products is in discussions to buy Ameripol Synpol, which filed for bankruptcy in December. Last year, Ameripol shelved its Odessa, Texas, SBR plant; it laid off about 80 workers at its remaining Port Neches, Texas, facility earlier this year.

Observers say DSM's elastomers operations, which contain SBR and EPDM businesses, are up for sale. The company, which is focusing on fine and specialty chemicals, sold its European petrochemicals business to Saudi Basic Industries Corp. last year, and rubber is now seen as nonstrategic. "Ever since SABIC's acquisition of their petrochemical assets, this business has kind of stood alone," Zinger says.

Zinger says financial industry buyers—and he puts ISP, which has no direct connection to rubber, in this category—are the most probable purchasers of these businesses. "Because profit margins have been poor, there hasn't been a lot of interest. The companies that would be interested do not have a lot of cash flow," he explains.

In the Meantime, rubber producers are responding to poor profit margins through cost cutting. "Everything is being impacted," Bayer's McLaughlin says. "How many visits to a customer do you need during a year? How much tech service do you supply? What custom work do you do? Or do you strip down to having one product in a wooden crate that goes out the door at a certain rate?"

Bayer has closed a 36-year-old polybutadiene plant in Sarnia, Ontario. Bayer's Rae says the company has moved production to its plant in Orange, Texas, which was expanded in 1999 and can make both solution SBR and polybutadiene rubber. "It gives us the flexibility to react to changing market requirements," Rae says.

Goodyear is using the Six Sigma efficiency program to improve operations. Nelson credits Six Sigma with lowering costs by $40 million in the firm's chemical business. The company also closed 48,000 metric tons of capacity at its Houston SBR plant in May 2001. The company had expanded its Beaumont, Texas, plant by 110,000 metric tons a year earlier to make, like Bayer, both solution SBR and polybutadiene.

EPDM has performed better than most other synthetic rubbers, but its growth has slowed from the 1980s and 1990s, when automotive and commercial roofing markets were expanding at a 6 to 8 per cent annual clip, according to Joe Gatto, who heads Crompton Corp.'s EPDM business.

The slowdown came to a head in 2001 when, thanks to the recession, EPDM consumption declined by 14 per cent in North America and 7 per cent globally, Gatto says. Last year was flat, and he expects 3 per cent growth this year. "I think there has been a solidification in the downstream markets, so I would expect some growth," he says.

Torkel Rhenman, global business director for Nordel EPDM at the DuPont Dow Elastomers joint venture, contends that the industry is in a process of renewal. "As modern capacity is being brought on-line, older, higher cost units are being mothballed," he says.

Gatto is concerned about a decline in sales to the automotive industry, which comprises about 50 per cent of the EPDM market, either directly through applications such as seals or indirectly as an impact modifier for automotive plastics. Ford and General Motors recently announced a reduction in car output for this year. However, Rhenman expects stronger growth in Asia to offset the North American results.

Overall car output aside, Gatto says more and more EPDM is being used per vehicle. He says EPDM's high heat resistance is winning it more applications under the hood, where it performs better than natural rubber. It is also replacing polychloroprene in belt applications.

In addition, use of EPDM/polypropylene thermoplastic vulcanizates (TPVs) is growing in sealing applications and in dense parts. "TPV is probably the fastest growth market for EPDM, although it is partly replacing thermoset EPDM," Rhenman says.

But TPVs are facing stiff competition from thermoplastic olefins, particularly in automotive interior applications, Gatto acknowledges. EPDM is also under pressure as an impact modifier for polymers because new catalyst systems allow rubberlike properties to be incorporated directly into the polymers, which eliminates the need to compound them with EPDM.

Rhenman is excited about the impact that the metallocene-catalyst EPDM is having on the market. DuPont Dow makes metallocene-based EPDM at a plant that it is leasing from Dow Chemical in Seadrift, Texas. Dow obtained the Unipol-technology plant in the Union Carbide acquisition. "Metallocene EPDM is taking significant share from the older Ziegler-Natta technology due to greater product cleanliness and significant advantages in manufacturing efficiencies," he maintains. The company is expanding its Plaquemine, La., nonmetallocene EPDM unit by 35,000 metric tons next year.

ExxonMobil is building a plant in Baton Rouge, La., that will make metallocene EPDM and other ethylene elastomers. The 90,000-metric-ton plant is expected to be completed during the third quarter of this year.

Crompton doesn't make metallocene EPDM, but Gatto claims he isn't concerned about its arrival, saying his company can achieve the flexibility it needs with Ziegler-Natta catalyst systems. "We feel that the products we can make with our systems are meeting and exceeding the market needs, and that is where we ought to be," he says.

Gatto is concerned, however, by all the capacity coming onstream. It may take a while, he says, for demand to catch up to the new supply.

But as competitive as the EPDM business gets, it's not likely to approach the difficulties of the rest of the synthetic rubber industry. For Goodyear's Nelson, there's only one thing to do when the market gets as bad as it's been.

"You try to control the things you can control and let the other things put a knot in your stomach," he says.

12.3 Synthetic Rubber Output

India's synthetic rubber production rose 3.2 per cent in April-May, but rising demand from auto tyre makers led to higher imports, the Rubber Board said.

Total synthetic rubber output was 16,997 tonnes during the period, compared with 16,475 tonnes a year ago, it said in a statement on Monday.

Consumption in first two months of fiscal 2008/09 rose 6.7 percent to 49,275 tonnes, while demand from the automotive tyre sector increased 9.2 percent to 32,468 tonnes.

Tyre makers account for more than 60 percent of the domestic consumption.

Higher consumption and non-availability of some variety of synthetic rubber led to more imports, which jumped 26.7 percent to 37,495 tonnes, the Rubber Board said.

Synthetic rubber production is dominated by a clutch of players like Reliance Industries Ltd (RELI.BO: Quote, Profile, Research) and Apar Industries Ltd (APAR.BO: Quote, Profile, Research).

Reliance produces about 73,000 tonnes of polybutadiene (PBR) variety annually, whereas Apar produces about 23,600 tonnes of nitrile rubber. India entirely imports its butyl and styrene-butadiene rubber (SBR) requirements.

The country's consumption in 2007/08 stood at 297,155 tonnes.

12.4 SBR (Styrene Butadiene Rubber) Sheet

Description

SBR (Styrene Butadiene Rubber) Sheet Emulsion SBR contains 23.5 per cent styrene and 76.5 per cent butadiene. SBR is the most widely used synthetic rubber. Adhesives and chewing gum have been identified as two growing markets for BR applications.

Application

- SBR is used in tyre and tyre products
- Commercial blend SBR for industrial gaskets
- Used in tread rubber (accounts for 76 % of global consumption)
- Mechanical goods(15 % of global consumption)
- Automotive (5 % of global consumption)
- Adhesives, floor tile and shoe soles, (4 % of global consumption)

Capabilities

- ❖ Excellent tensile strength
- ❖ Abrasion resistance
- ❖ Flexibility

12.5 Raw Materials

The tyre industry is raw material intensive, which accounts for more than 60 per cent of the production cost. Therefore, prices of raw materials directly affect the profitability of tyre companies. Since most of these raw materials are petroleum based, their prices fluctuate with the international prices of petroleum products. The main raw materials for tyre are rubber (natural or synthetic), carbon black, nylon tyre cord and rubber chemicals. Except natural rubber, the costs of all other raw materials in tyre production are related to crude oil prices. Table 12.1 shows the proportion of each raw material in terms of their value and weight.

Table 12.1: Raw materials

Raw Materials	*By Value (%)*	*By Weight (%)*
Rubber	52	49
Carbon Black	23	10
Nylon tyre cord	8	24
Chemicals	15	12
Others	2	5
Total	**100**	**100**

Natural Rubber

Natural rubber accounts for 52 per cent of the value of the tyre. In India mixture of both natural as well as synthetic rubber is used for making tyres. However the consumption ratio is towards higher usage of natural rubber due to Indian climatic conditions, over loading of vehicles and poor road condition. In India the consumption of natural to synthetic rubber is 80:20 which is in stark contrast to international ratio. The industry uses RSS—4 grade rubber.

India's 90 per cent of the rubber production comes from Kerala. Domestic rubber production has increased at a compounded growth rate of 9 per cent annually from 1991 to 1997 after which the production slowed down. However In FY 2001, rubber production soared and crossed double-digit mark at 10.2 per cent.

Synthetic Rubber

Synthetic rubber is generally of two types—poly-butadiene rubber (PBR) which forms 40 per cent of the synthetic rubber used in tyres. The other variety is Styrene Butadiene Rubber (SBR) primarily used in passenger car radials to give the grip to the tyres. At present, IPCL is the only domestic producer of PBR. However it able to meet only 44 per cent of the tyre industry's requirement. Thus India is a significant importer of synthetic rubber. There is an urgent need to increase production capacity of SBR to supplement natural rubber.

Carbon Black

Carbon black is a key raw material used in the manufacture of automotive tyres. More than 70 per cent of the demand for carbon black is from the tyres segment. Carbon black feed stock (CBFS) is the key raw material used to manufacture carbon black. Roughly 2.2 tonnes of CBFS is required to produce one tonne of carbon black. Its main use is as a reinforcing agent in tyres.

Though there are more than twenty types of CB, the ones used for tyre production are mainly of three types, N220, N330 and N660. N660 is mainly used in the carcass of the tyres, N330 is used for the tread and N220 is used for the tread of heavy-duty tyres. On an average, about 45 per cent of the CB consumed by the tyre industry is of the N660 variety, 28 per cent of N220 and 27 per cent of N330 variety.

Truck tyres consume 20 kgs of CB per tyre, while smaller tyres like Maruti consume 1.5 kgs. Overall approximately 60-65 per cent of the CB produced in India is consumed by the tyre industry. Indian market is dominated by the top three players in the industry—Philips Carbon Black, Hitech Carbon (unit of Indian Rayon) and Cabot India (a subsidiary of Cabot Corporation, US).

Nylon Tyre Cord

This is mainly a reinforcing material and lends strength and tenacity to the to a tyre. It is placed below the tyre tread, in contact with the road. Almost 90 per cent of nylon cord manufactured in India is consumed by the industry. The tyre cord fluctuates in consonance with the prices of caprolactum its main input.

Rubber industry in Post GATT Era

With the lifting of physical barriers on imports of all commodities by April 2001, as also phasing out of various subsidies for exports, the rubber industry is in for a very rough tide. With the slowdown in economy compounding the problem, the automobile majors are in for a major shake-out.

Table 12.2: Consumption Patterns of Major Raw Materials

(All Figures in Tonnes)

Raw Materials	*Consump.*	*Tyre-Sector*	*Non-Tyre Sector*	*Import*	*Tyre Imports**	*Non-Tyre Imports*
Natural Rubber	628,000	50%	50%	16,400	85%	15%
SBR	53,800	49%	51%	33,200	77%	23%
PBR	49,200	81%	19%	11,400	85%	15%
Carbon Black	245,000	69%	31%	30,000	66%	34%
Nylon	66,000	95%	5%	25,000	100%	Nil
Rubber Chemicals	24,000	60%	40%	3,000	85%	15%
Steel Tyre Cord	1,800	100%	Nil	1,800	100%	Nil
Butyl Rubber	37,900	64%	36%	37,900	64%	36%

* Mainly duty free imports against export of tyres.

12.6 Synthetic Rubbers

Name of Product	*Poly Butadiene Rubber (Grade 1220)*
Description	This grade is equal to BR-01 and produced by solution polymerization of butadiene 1, 3 monomer in presence of cobalt Octoate + DEAC catalyst. Stabilizer/Antioxidant: A mixtures of TMP oil and Irgnox power in Benzene solvent. This product is nominated:"High Cis" with (>97 % Cis).
Application	In Tire Computing, Rubber Products, Adhesive, Shoe sole, and Coating resin.
Licensor	Nippon Zeon.
Packing	40700 Kg net Box-Pallet each Box contains 20 bale of PE-wrapped PBR each bale is 35Kg 19600MT/40 feet container.

Name of Product	*Styrene Butadiene Rubber (SBR 1500)*
Description	These products are produced by polymerization in a solvent media by adding some chemicals and additives. (a copolymer contains 23.5 per cent styrene and 76.5 per cent butadiene) Stabilizer: DTO in 1500 grade production. STP in 1502 grade production these rubbers are applicable in internal Rubber Mixing.
Application	In Tire Computing, Rubber Products, Adhesive, Shoe sole, and Coating resin.
Licensor	Japan Synthetic Rubber.
Packing	910 Kg net Box-Pallet each Box contains 26 bale of PE-wrapped SBR each bale is 35Kg up to 23.66 MT/40 feet container.

Polybutadiene Rubber

Quality Statement

Property	*Test*	*Unit*	*Value*	*Specification*
Mooney viscosity (ML1+4) 100i/EC	ASTM D1416		45	45+/-4
Volatile Matter	ASTM D1416	Wt %	0.04	Max 0.5
Ash	ASTM D1416	Wt %	0.10	Max 0.5
CIS Content	NMR	Wt %	97	Min 97
Antioxidant	—	—	—	Non-Staining
Organic Acid	—	Wt %	Less than 0.1	—
Cure characteristics Formulation ISO 2322			45+/-4	
MH-Max Torque		dNm	28.1	
ML-Min Torque		dNm	4.35	
Ts1		min	2.4	
Tc50		Min	6.3	
Tc90		Min	8.4	
N300%; 35 °C cure at 145 °C		16	MPa	
Tensile		MPa	17.6	
Elongation at Break		%	320	

Styrene Butadiene Rubber Grade 1502

Quality Statement

Property	*Test*	*Unit*	*Value*	*Specification*
Mooney Vicosity (ML1+4) 100C	ASTM D1416	WT%	52	46-58
Volatile Matter	ASTM D1416	Wt %	0.21	Max 0.75
Ash	ASTM D1416	Wt %	0.27	Max 1.50
Bould Styrene	NMR	Wt%	23.8	Max 24.5
Antioxidants				Non-Staining
Organic Acid		Wt %	3.3	Max 7.0
Soap		WT%	Less than 0.1	
Cure characteristics		**Formulation ISO 2322**		
MH-Max Torque		dNm	21.8	
ML-Min Torque		dNm	2.89	
Ts1		Min	3.8	
Tc50		Min	8.7	
Tc90		min	15.3	
M300%; cure at 145 °C		MPa	18.9	
Tensile		MPa	27.8	
Elongation at Break		%	400	

Test Formulation: ISO 2322

Polymer 100: HAF N330 50: Zinc Oxide 3: Stearic Acid 1: Sulphur 1.75: TBBS 1 (ISO 2322)

SBR 1500

Property	*Units*	*Test Method*	*Value*
Volatile Matter	wt %	ASTM D-1416	0.75 max.
Ash	wt %	ASTM D-1416	1.5 max.
Organic Acid	wt %	ASTM D-1416	5.00-7.25
Soap	Wt %	ASTM D-1416	0.50 max.
Bound Styrene	Wt %	ASTM D-1416	22.5-24.5
Raw Viscosity (ML 1+4 @ 100 °C)	—	ASTM D-1646	46.8-58.0
Compound Viscosity (ML 1+4 @ 100 °C)	—	ASTM D-1646	84 max.
Tensile Strength (35 Minutes Cure)	Kg/cm^2	ASTM D-412	250 min.
Ultimate Elongation (35 Minutes Cure)	%	ASTM D-412	470 min.
300% Modulus (35 Minutes Cure)	Kg/cm^2	ASTM D-412	119-159

SBR 1712

Test	*Value*
Mooney viscosity, MB 1÷4 (100°C) Rubber laminated M	46-58
Conventional tensile strength, Mpa, min	21,5(220)
Relative elongation at break, %	550-750
Permanent strain after break, %, max	20
Rebound elasticity, % min	28
Viscosity spread within a batch, max	7
ETA extract, %	32.5+/-2.5
Volatile, %	0.35
Mass fraction of organic acids, %	4-5.6
Mass fraction of organic soaps, %, max	0.20
Mass fraction of antioxidant, % VS-1 VTS-150 Agidol-2	0.15-0.35 1-1.4 0.8-1.5
Mass fraction of oil, %	26.0-29.0
Mass fraction of metals, %, max: Copper Ferrum	0.0002 0.005
Mass fraction of bound second monomer, %, styrene	22-25
Mass fraction of ash, soluble in water, %, max	0.6

EPDM

KEP grades	*Oil (phr)*	*Mooney Viscosity (ML1+4@125?)*	*ENB Content (wt.%)*	*Ethylene Content (wt.%)*	*Product Form *1)*
		Copolymers			
KEP-020P		25 *2)	-	71.0	P
KEP-070P		69 *2)	-	71.0	P
		Terpolymers			
Low Diene					
KEP-435		34	2.3	57.0	B
		Medium Diene			
KEP-210		23	5.7	65.0	B
KEP-240		42	4.5	57.0	B
KEP-270		71	4.5	57.0	B
KEP-510		23	5.7	71.0	B
KEP-570F		59	4.5	70.0	FB
KEP-570P		53	4.5	70.0	P
KEP-7141		27	4.5	52.0	B
KEP-281F		83 *3)	5.7	67.0	FB
KEP-1030F		86 *3)	4.5	62.0	FB
KEP-901	100	50 *3)	4.8	70.0	B
KEP-960(F)	50	49 *3)	5.7	70.0	B, FB
KEP-980	75	58 *3)	4.5	70.5	B
		High Diene			
KEP-321		30	9.5	52.0	B
KEP-330		28	7.9	57.0	B
KEP-350		56	7.9	57.0	B
KEP-650		49	8.7	59.0	B
KEP-650L		41	7.9	57.0	B
KEP-650L		41	7.9	57.0	B
VISTALON SERIES	**Oil (phr)**	**Mooney Viscosity (ML1+4, 125?)**	**ENB Content (wt %)**	**Ethylene Content (wt %)**	**Product Form *1)**
Vistalon 503K	-	34.0 *2)	-	55.0	B
Vistalon 606	-	65.0 *2)	-	54.0	B
		Terpolymers			
Vistalon 2504	-	25.0	4.7	58.0	B
Vistalon 7500	-	82.0 *3)	5.7	56.0	SFB
Vistalon 8600	-	81.0 *3)	8.9	58.0	SFB
Vistalon 5730	30.0	36.0	7.9	59.0	B

1) P : Pellet, B : Dense Bale, FB : Friable Bale
2) ML1+4 @100°C
3) ML1+8 @125°C
4) The above values in the table are typical data, Not specifications. Contact your regional sales representative to determine grade availability in your region.
1) B : Bale, SFB : Semi-friable bale

2) ML1+4@100?
3) ML1+8@125?
4) The above values in the table are typical data, Not specifications. Contact your regional sales representative to determine grade availability in your region.Synthetic Rubbers

12.7 Styrene-Butadiene Rubber SBR 1712

Characteristic	*Target*	*Tolerance*	*Test method*
Appearance	Dark brown to black bales		
Mooney Viscosity	50	+/-5	ML (1+4) @100 C
Oil %	27.5	+/-1.5	
Bound Styrene %	23.0	+/- 1.0	ASTM D 5775
Loss on heating	0.40%	max	105 C
Ash content	0.60%	max	ASTM D 4574

12.8 Styrene Butadiene Rubber (SBR) Revisited

Styrene butadiene rubber (SBR) is the most widely used synthetic rubber. Emulsion SBR contains 23.5 per cent styrene and 76.5 per cent butadiene. SBR is used in tyre and tyre products, including tread rubber, which accounts for 76 per cent of global consumption. Other uses include mechanical goods, 15 per cent; automotive, 5 per cent; miscellaneous.

12.9 SBR Rubber Molding.

Chemical Designation: Styrene Butadiene, Buna S, GRS

Molded SBR rubber is an economical substitution for Natural Rubber in many applications because it provides significant cost savings. SBR is a good choice for applications where finished product exposure includes acids, organic salts and alkalis. In addition, SBR Rubber is soft and has good flexing characteristics at low temperatures, like Natural Rubber. It is wear resistant and non-toxic.

Do not use SBR Rubber for applications that involve petroleum derivatives or exposure to chemicals. Special formulations of SBR can provide sunlight and heat resistance.

Red SBR Rubber is popular for use as a gasket in low-pressure applications, such as washers and gaskets for heating and plumbing. Black SBR is popular for use in applications where abrasion is a factor.

Chemical Designation: Styrene Butadiene, Buna S, GRS

Molded SBR rubber is an economical substitution for Natural Rubber in many applications because it provides significant cost savings. SBR is a good choice for applications where finished product exposure includes acids, organic salts and alkalis. In addition, SBR Rubber is soft and has good flexing characteristics at low temperatures, like Natural Rubber. It is wear resistant and non-toxic.

Do not use SBR Rubber for applications that involve petroleum derivatives or exposure to chemicals. Special formulations of SBR can provide sunlight and heat resistance.

Red SBR Rubber is popular for use as a gasket in low-pressure applications, such as washers and gaskets for heating and plumbing. Black SBR is popular for use in applications where abrasion is a factor.

12.10 Rubber Injection Molding

Rubber Injection Molding is an ideal process for forming high volume production, large quantities of small to medium size parts, complex inserts, close dimensional tolerances, insert molding and components that require uniformity. Rubber injection molding is particularly suited for products similar to the grommets shown above.

Door Stop

Manufactured by rubber injection molding utilizing colour compounds, molded in multiple colours including green, yellow, beige, red, blue and brown.

The process of rapidly forcing an exact amount of rubber from a tube/cylinder into a closed, heated mold. This process provides economical advantages including:

- Short molding cycle
- Lower unit cost
- High dimensional tolerances
- Absence of flash
- Little scrap/waste

12.11 Rubber Compression Molding

Compression Molding is the original production method for molded rubber. It is ideal for low to medium volumes and a useful molding process for forming bulky parts, gaskets, seals and O-rings. It is a widely used, efficient, economical production method for many products particularly low production volumes of medium to large parts, materials with a high cost and applications that demand extreme hardness. The process involves compressing preformed rubber in a mold with a press. During the compression process, rubber is forced into the mold cavity to form the final product.

Component for Agricultural Equipment Rubber to metal compression molding

Large Press Capability

Transfer and compression molding equipment with the ability to run 42" x 42" and 48" × 56" molds to mold your largest part requirements.

Large Shot Capability

Rubber Injection Molding 10lb-15lb-20lb shot size

Rubber Compression Molding offers advantages over other methods by providing:

- ❖ Tooling savings
- ❖ Short setup time (saves on short production runs)
- ❖ The capacity to process stiff, high durometer materials
- ❖ Least amount of waste

12.12 Rubber Transfer Molding

Rubber Transfer Molding combines the advantages of injection molding with the ease of compression molding. Rubber transfer molding is the ideal process for forming parts that require:

- ❖ Exact positioning
- ❖ Bonding rubber to fragile metal parts - such as wire
- ❖ Mold designs that contain multiple cavities and can trap air
- ❖ Intricate parts with lower volume requirements

Dental Equipment Component

Silicone rubber created with transfer molding, molded in a variety of colours; mint, lavender, sky blue, peach and light gray

The process of transfer molding involves pushing unvulcanized rubber through sprues into a heated mold. The rubber is placed in a chamber at the top of the mold then placed in a press. During the transfer process, rubber is forced through the mold to prevent trapped air while forming the final product.

Transfer molding offers several advantages over other methods by providing:

- ❖ Shorter production cycles
- ❖ Maintains closer dimensional tolerances than compression molding
- ❖ Provides uniformity
- ❖ Fast mold setup

Large Press Capability

Transfer and compression molding equipment with the ability to run 42" 5 42" and 48" × 56" molds to mold your largest part requirements.

Large Shot Capability

Rubber Injection Molding
10lb-15lb-20lb shot size

12.13 SBR - Styrene Butadiene Rubber

Styrene Butadiene Rubber (SBR) is an elastomeric copolymer that consists of styrene and butadiene. SBR has good abrasion resistance and good stability. SBR is resistant to mineral oils, fats, aliphatic, aromatic and chlorinated hydrocarbons.

- ❖ SBR has a glass transition temperature of approximately -55°C
- ❖ SBR has a possible temperature range of - 40 to +212 °F

- ❖ Other names include GRS, Buna S
- ❖ SBR has an elongation percentage of 450 - 500
- ❖ SBR has a useful temperature of -75 to 250°F
- ❖ SBR is not oil, ozone or weather resistant. Not recommended for electrical applications

Styrene Butadiene Rubber (SBR) is commonly used in pneumatic tires or tubes, heels and soles, and gasketing applications.

Da/Pro Rubber has chemists available to assist you in determining if SBR is ideal for your application. We will evaluate the specifications of your component and make recommendation on the most effective and efficient solution.

12.14 Notification

Ministry Of Commerce
(Directorate General of Anti Dumping-and Allied Duties)
NOTIFICATION
New Delhi, the 21st January, 1999
PRELIMINARY FINDINGS

Subject: Anti-Dumping investigation concerning imports of Styrene Butadiene Rubber (SBR) from Japan, Korea, Turkey, Taiwan, USA, Germany and France.

30/1/97-ADD.- Having regard to the Customs Tariff Act 1975 as amended in 1995 are- the Customs Tariff (Identification, Assessment and Collection of Anti-Dumping Duty on Dumped Articles and for Determination of Injury) Rules, 1995, thereof

A. Procedure

1. The procedure described below has been followed with regard to the investigation:
   (i) The Designated Authority (hereinafter also referred to as Authority), under the above Rules, received a written application from Synthetics and Chemicals Ltd., Oriental House, 7, J Tata Road, Churchgate, Bombay-400020 (also referred to SCI hereinafter) on behalf of the domestic industry, alleging dumping of Styrene Butadiene Rubber (SBR) (hereinafter also referred to as subject goods) originating in or exported from Japan, Korea, Turkey, Taiwan, USA, Germany and France.
   (ii) Preliminary scrutiny of the application filed by petitioner revealed

certain deficiencies, which were subsequently rectified by the petitioner. The petition was, therefore, considered as properly documented.

(iii) The Authority, on the basis of sufficient evidence submitted by the petitioner decided to initiate the investigation against imports of SBR from Japan, Korea, Turkey, Taiwan, USA, Germany and France. The Authority notified the Embassy of Japan, Korea, Turkey, Taiwan, USA, Germany and France about the receipt of dumping allegation before proceeding to initiate the investigation in accordance with sub-rule 5(5) of the Rule.

(iv) The Authority issued a public noticed dated 7th April, 1998 published in the Gazette of India, Extraordinary, initiating anti-dumping investigations concerning Imports of SBR classified under custom code 4002.19 of Schedule 1 of the Customs Tariff Act, 1975 originating in or exported from Japan, Korea, Turkey, Taiwan, USA, Germany and France (hereinafter also referred to as the subject countries).

(v) The Authority forwarded a copy of the public notice to all the known exporters (whose details were made available by petitioner) and industry associations and gave them an opportunity to make their views known in writing in accordance with the rule 6(2):

(vi) The Authority forwarded a copy of the public notice to all the known importers (whose details were made available by petitioner) of SBR in India and advised them to make their views known in writing within forty days from the date of issue of the letter;

(vii) Request was made to the Central Board of Excise and Customs (CBEC) to arrange details of imports of SBR made in India during the past three year, including the period of investigation.

(viii) The Authority provided a copy of the petition to the known exporters and the Embassy of the subject countries in accordance with rules 6(3) supra. A copy of the petition was also provided to other interested parties, wherever requested;

(ix) The Authority sent a questionnaire to elicit relevant information to the following known exporters, in accordance with the rule 6(4):

(a) Japan

- ❖ Asahi Chemical Industry Co,
- ❖ Japan Elastomer Co. Ltd.
- ❖ Japan Synthetic Rubber Co.
- ❖ Mitsubishi Kasei Corporation
- ❖ Nippon Zeon Co. Ltd,
- ❖ Sumitomo Chemical Co.

(b) *Korea*

❖ Korea Kumho Petrochemical

(c) *Taiwan*

❖ Taiwan Synthetic Rubber Corpn.

(d) *France*

❖ Bayer AG
❖ Goodyear Chemical Europe
❖ Michelinet Cie
❖ Shell Chimie SSA.

(e) Germany

❖ Kombinal VEB Chemische Work

(f) *Turkey*

❖ Petkim Petrokimya AS

(g) *USA*

❖ Ameripol Synpol Corpn.
❖ DSM Copolymers Inc.
❖ Firestone Synthetic Rubber & Ltx Co.
❖ General Tire Inc.
❖ Goodyear Tire and Rubber Co.

A number of parties requested for extension of time, which was allowed by the Authority by two weeks.

The following exporters responded:

(a) *Japan*

❖ Mitsubishi Chemical
❖ JSR Corpn
❖ JTC Corpn
❖ Mitsui & Co.
❖ Fuji Chemical Development Co.

(b) *Korea*

❖ Korea Kumho Petrochemical Co. Ltd.

(c) *Taiwan*

❖ Taiwan Synthetic Rubber Corpn.

(d) *France*

❖ Michelin
❖ Bayer

(e) *Germany*

❖ Buna Sow Leuna Olefinverbund Gmbh (BSL) Germany

(f) *Turkey*

❖ Petkim Petrokimya Holdings

(g) *USA*

❖ The Goodyear Tire and Rubber Co.
❖ Ameripol Synpol Corpn.

(x) The Embassy of the subject countries in New Delhi was informed about the initiation of the investigation in accordance with rule 6(2) with a request to advise the exporters/producers from their country to respond to the questionnaire within the prescribed time. A copy of the letter, petition and questionnaire sent to the exporters was also sent to the Embassy, alongwith a list of known exporters/producers.

(xi) A questionnaire was sent to the following known importers of SBR calling for necessary information in accordance with rule 6(4):

❖ Apollo Tyre Ltd, Kochi
❖ Modistone Ltd. New Delhi
❖ Birla Tyres Ltd., Calcutta
❖ Ceat Ltd., Mumbai
❖ Govind Rubber Ltd., Alwar
❖ J K Industries Ltd., New Delhi
❖ Modi Rubber Ltd.; New Delhi
❖ MRF Ltd., Madras,

A number of parties requested for extension of time, which was allowed by the Authority by two weeks. Response to the questionnaire was filed by the following;

❖ Automotive Tyre Manufacture Association, New Delhi (ATMA).
❖ All India Rubber Industries Association, New Delhi
❖ Vimsons Rubber (P) Ltd., Kerala
❖ South Asia Tyres Ltd., Aurangabad

❖ Thejo Engineering Services (P) Ltd., Madras
❖ NARCO Ltd, Thane
❖ Birla Tyres, Calcutta
❖ MRF Ltd., Chennai
❖ Ceat Ltd, Mumbai
❖ Rishiroop Polymers (P) Ltd." Mumbai
❖ Apollo Tyres Ltd., New Delhi
❖ Dow Chemical International Ltd., Mumbai.
❖ J K Industries Ltd., New Delhi
❖ Vamshi Rubber Ltd., Hyderabad Chemical & Allied Products Exports Promotion Council, Calcutta
❖ All India Federation of Rubber Footwear Manufacturers.
❖ Kanara Small Industries Association, Mangalore

(xii) Additional information regarding injury was sought from the petitioner, which was also furnished;

(xiii) The Authority conducted on-the-spot investigation at the premises of petitioner to the extent considered necessary;

(xiv) The Authority kept available non-confidential version of the evidence presented by various interested parties in the form of a public file maintained by the Authority and kept open for inspection by the interested parties;

(xv) Cost investigations were also conducted to work out optimum cost of production and cost to make and sell the subject goods in India on the basis of Generally Accepted Accounting Principles (GAAP) and the information furnished by the petitioners also as to ascertain if anti-dumping duty lower than dumping margin would be sufficient to remove injury to the domestic industry.

(xvi) ****in this notification represents information furnished by an interested party on confidential basis and so considered by the Authority under the Rules;

(xvii) Investigation was carried out for the period starting from 1 st April 1996 to 31st August 1997.

(xviii) The Authority provided an opportunity to all interested parties to present their views orally on 10th November 1998. All parties presenting views orally were requested to file written submissions of the views expressed orally. The parties were advised to collect copies of the views expressed by the opposing parties and offer rebuttals, if any.

B. Petitioner's Views

2. The petitioner has raised the following major issues in their petition and subsequent submissions.

(a) Voluminous imports of SBR, which is showing an upward trend, are entering into India at a low price. Exporters/producers from the subject countries are dumping SBR into India at a price lower than their normal value. Inspite of increase in cost, these exporters have reduced their prices. Due to this, it is claimed that petitioner is forced to reduce their prices besides providing heavy discounts and enhanced credit facilities. It is further stated all these factors are causing heavy losses to petitioner and they are not able to recover even the cash cost. Inspite of reduction in price, their market share has gone down by 11 per cent.

(b) It is claimed that petitioner is suffering injury on account of the fact that their production as well as capacity utilisation has affected adversely, stocks of SBR has increased and man power is reduced by more than 5 per cent. The petitioner has also claimed threat of material injury, if dumped imports are not checked. Probability of further steep increase in imports and permanent injury to the domestic industry can not be ruled out. Dumped imports have retarded the growth of SBR industry and expansion. Financial constraint due to dumping has delayed the commencement of the projects.

(c) International Institute of Synthetic Rubber Producers adopts the following numbering system for various grades of styrene-butadiene rubber produced by emulsion polymerization system:

1000 series - Hot non-pigmented rubbers

1500 series - Cold non-pigmented rubbers

1600 series - cold back masterbatch with 14 or less parts oil per 100 parts SBR

1700 series - cold oil masterbatch

1800 series - cold oil black masterbatch with more than 14 parts oil per 100 parts SBR

1900 series - Emulsion Resin Rubber Masterbatch

In each of the above series various grades of SBR are characterised by level of copolymer (% styrene), emulsifier type, nominal Mooney Viscosity, Coagulation, specific gravity and product stain. Thus each grade of SBR differs from the other due to variation in one or more of these parameters. However, under each series the basic polymerisation technology remains the same and as such, the basic physical nature of the rubber also remains the same. In view of this various grades of SBR under each series barring few specific grades meant for specialized applications are interchangeable and can be substituted with each other in most of the applications by making suitable modifications in the compounding recipe. In this-connection, a careful look of the various tables given in the Synthetic Rubber Manual corroborates this fact. For example, various related polymers listed under Table 12.3-2 under 1500 grade has varying parameters such as emulsifier type, nominal Mooney Viscosity, coagulation system etc.

Further the interchangeability of various grades of SBR is possible even between grades falling under two different series. For example, 1712 grade of SBR can be substituted with 1500 grade and many other grades of rubber falling under 1500 & 1700 series even in Tyre tread application. Similarly, 1502 grade of SBR can be substituted with 1500 grade and many other grades of rubber falling under 1500 and 1700 series in various applications.

(d) Synthetic rubber has been meeting the increasing demand arising due to gap between demand and supply of natural rubber. Since the rubber plantation area is limited, the natural rubber is being supplemented by synthetic rubber consistently. SBR is a type of synthetic rubber and is produced in different grades keeping in view the specific needs of customers.

(e) The petitioner has expressed their opinion that there may not be any major difference in the manufacturing process of SBR being produced by them and produced by other Indian producers or foreign producers who are dumping goods to India.

(f) The petitioner has claimed the normal value on the basis of inquiries made by them from the Embassy of India in the case of Japan, Turkey and Korea. In the case of Germany and France, it is claimed on the basis of quotation from Spain. In the case of USA and China Taipei, it is arrived at on the basis of export to third country. The petitioner has claimed export price on the basis of DGCIS statistics. While arriving at export price to India, they have shown adjustment on account of ocean fright, insurance, dealer's commission and terms of sales.

(g) EU is a trade/customs union and there is free movement -of commodities among its member nation. EU countries are collectively invoking anti-dumping action under the WTO rules. Therefore anti-dumping action should be taken against all the member countries of EU for dumping in India. Else, on the event of anti-dumping measures on a particular nation of EU, it is very likely that the material shall be dumped through other member nations of EU. This is apprehended more so because some of the multi-national manufactures have their plants spread over different EU countries.

(h) Production Process as well as the production capacities overlap in producing different grades of SBR. The petitioner has claimed injury to domestic industry due to dumping for the SBR, which includes S 1502, S 1712 and S 1958. Wherever necessary/feasible, separate information on the specific grades are provided.

(i) The petitioner has requested to assess injury cumulatively from all the countries found dumping the product. The impact of dumping started more severe from January 1997 onward.

(j) The petitioner has claimed that the injury suffered by them is due to dumping and thus causal link between dumping and injury is established.

C3. Views Of Exporters, Importers And Other Interested Parties

1. Importer view

(a) There is a severe recession and inflation in the industry, which caused crisis particularly in the rubber industry, which cannot sustain burden of any unjustified antidumping duty.

(b) The supply from the petitioner (SCI) is irregular and quality is not satisfactory. They are not in a position to cater the full requirements of their consumers. They are misusing the monopoly in the manufacturing by exploiting the consumers. The import of SBR is under compulsion, otherwise due to non-availability of SBR, the consumer units are to be closed.

(c) The petitioner is having erratic supplies and frequent price increases. The petitioner shut down their plant without prior intimation to end user which results in their severe production loss.

(d) The quality of product of the petitioner is slightly inferior, compared to the imported products, landed cost of which is cheaper than that of indigenous product from petitioner. The dislocation in their production operation is another factor contributing not to rely on this source. The imposition of anti-dumping duty will not only make import of SBR more costly and to a certain extent prohibitive, but also paralyse the industry for want of quality product in the required volumes to meet the projected demand from the rubber based industry. The claim of the petitioner that they have sufficient capacity to cater to the complete requirement both in quality and quantity to the consumers is contrary to the fact. Anti dumping on SBR should never be imposed in the interest of rubber based industry.

(e) Tyre companies are forced to depend more and more on imports largely on account- of petitioner's mismanagement in vendor relations. Many of the tyre companies had resorted to imports due to erratic supplies, not adhering to delivery schedules, frequent quality complaints and price increases unrelated to cost increases. Even those companies, which still lift large qualities of SBR from petitioner, are unhappy with their dealings with them.

(f) Increase in imports cannot be looked at in isolation but has to be examined in relations to growth in exports. Some tyre companies have increased their level of exports significantly and, therefore, when increase in imports are commented upon, imports, "net of duty free imports against advance licenses" must be examined.

(g) The petitioner has themselves admitted in their published accounts that their performance is affected by raw material constraint due to congestion at Kandla Port, credit squeeze imposed by banking system and delayed payment by customers. In spite of this, their sales were higher and, therefore, there does not seems to be any impact of imports of SBR on petitioner. Commencing from December 1997, the petitioner has already announced four price increases.

(h) Radialisation of passenger tyres has resulted in a f1irly large shift from SBR to Natural Rubber (NR). Also the increase in iriports is due to that new generation vehicle manufacturers require "solution based" SBIZ tyre whereas the petitioner is manufacturing SBR based on `emulsion based' technology.

(i) The petitioner has not been operating efficiently and has not made efforts to improve their inteiial and external performance. In an environment where the tyre industry is exposed to international competition, if inefficiency on part of raw material supplier get passed on to it as costs, it will indeed be a sad commentary on the future prospects of tyre industry. If anti-dumping duty is imposed, it will contribute in raising the local price further by petitioner and this will be a convenient way of passing on their systems/operational inefficiencies to the user industry.

(j) The petitioner is offering much lower credit as compared to credit given by international vendors. The tyre industry is in serious liquidity crisis, and the request for longer credit has not got any favourable response from petitioner.

(k) For a manufacturer from South East Asia, the fall in currency value in relation to US dollar has been unprecedented. The effect of this fall was that to get the same domestic currency realisation from exports of this product as from a domestic sale, the exporter needed to quote a dollar price that reflected the percentage depreciation of his currency. This did not reflect an "intent to dump" which would have been an unfair practice. There must be a "normalisation" process of comparing domestic price and export price from that country for establishing dumping. If so, the margin will then be seen to be lower than de minimus.

(1) There is no causal link between the import and the alleged erosion to petitioner's health. The petitioner had raised an application for imposing anti-dumping duty in 1993 even when the duty levels were high and it was summarily rejected by the Govt. The current application is raised on similar flimsy grounds.

(m) Certain relevant data/inputs have been withheld by the petitioner ar marked as confidential 'and have not been disclosed. Thus the proceedings are not transparent. Petition is defective to this extent and the same may be dismissed forthwith.

(n) The pattern of consumption NR: SR: RECLAIM in India is 73:18:9 as compared to 35:65:0 in Western world. The reason for this consumption pattern in high prices of SBR over NR in India.

(o) The petition first started manufacture of SBR with a second hand plimported from USA in 1963. Such plant has lost past it prime. In every sense it has outlined its utility. The present expansion is a replacement of old plant.

(p) The main types of SBR are as specific, not "like articles" and not interchangeable. These are :

1) Emulsion Polymerised SBR (ESBR)
11) Hot Emulsified
III) Cold Emulsified
IV) Solution SBR
V) Thermo Plastic SBR

It is pointed out that all the above types are covered in customs heading 4002.19. Amongst the four different tyres, India produces only 2nd types. The 2nd type can be further classified into the following grades:

I) Oil extended grades: 1712, 1778
II) Pure SBR: Grades 1500, 1502
III) High styrene SBR: Grades 1958 with styrene contents upto 50 per cent.

Thus in view of above, it is stated, that SBR is a general term and cover a very wide spectrum of polymers most of which are neither manufactured by petitioner nor they have the capacity to do so. The inquiry is initiated for a general family of polymers known as SBR having various grades and types most of them not being manufactured by petitioner nor they have the capability or capacity to manufacture.

(q) The petitioner has admitted that their plant is a multipurpose plant capable of producing SBR, NBR etc. Even after such enlightened realisation and also having the necessary technology to produce NBR, the petitioner did not initiate steps to produce NBR, which would give better realisation, attracts anti-dumping duty and for which India is facing a shortfall.

(r) The claim of the petitioner that they are capable of producing total requirement of SBR in India is far from the truth. It is claimed that the production of petitioner alongwith other producers have been much lower than the required consumption. It is further stated that though the consumption of SBR has shown a steady growth, the petitioner production has stagnated from 1988. Thus on the one hand, the petitioner is not capable of producing the requirements of industry and on the other hand they want to block the necessary import of deficit quantity as well even in'the liberalised trade regime.

(s) In the petition, the alleged importer of dumped imports, are eight tyre companies. This implies that the entire non-tyre sector remains outside the ambit of anti-dumping investigation. The anti-dumping investigation shows each grade/commodity under investigation and the matier can not be generalised.

(t) The imports in 96-97 as compared to 95-96 have increased but as compared to 94-95 it has gone down. In 95-96, there was a worldwide shortage of SBR.

(u). The rupee dollar conversion during 1996-97 was US$ 1=Rs. 35.50 which now stands at US$ 1=Rs. 42.76. Hence rupee has depreciated by Rs. 7.26 per dollar which is 20.4 per cent. Therefore, the value of imports has gone up by over 20 per cent which is more than or substantially offset the incidence of injury in terms of loss of Rs. 5 to Rs. 20/- as stated by petitioner. Hence there is no injury to the petitioner.

(v) The onus of evidence of normal price in the exporting countries has not been satisfied by the petitioner and no proper procedure followed. The normal prices submitted are unsubstantiative, unauthentic and irrelevant.

(w) The causal link between injury and dumping is not properly established by the petitioner. The alleged case of injury can not be perceived to have been caused by imports.

(x) It is claimed that import price of SBR has been increased substantially at present due to changes in rate of duty and exchange rate. The landed price of imeorts have gone up by about 33 per cent in June 98 as compared to May 97.

(y) The single factor for reduction in the international price of SBR is the reduction in the prices of raw material for SBR, which are styrene and butadiene. The petitioner has not passed the benefit of reduction in cost of input to consumers and the petitioner had earned profit.

(z) It is stated that no causal link between the alleged dumped imports and injury as the company has attributed reasons to other factors in their published accounts of 199596 and 1996-97. Also there is no injury due to the followings

- The production of domestic industry has increased
- sales have increased
- capacity utilisation has increased
- there is a total loss of 7 per cent market share by the petitioner out of which 5 per cent is taken by other than dumped imports and 2 per cent loss of market share due to alleged dumping cannot be considered material.

II. Exporter's View

(a) The Designated Authority has no jurisdiction in the matter as the condition of serious injury to domestic industry is missing. There is no causal relationship between the dumped goods and injury, which is a pre-requisite for imposing anti-dumping duty. Lack of profitability to petitioner is due to uneconomic size and certain managerial deficiencies.' By imposing anti-dumping duties, no public interest will be served. Levy of anti-dumping duty would be contrary to the provisions of MRTP Act 1969 as it would tantamount to encouraging monopolist who would then exploit his dominant position.

(b) Levy of anti-dumping duty will be against the interest of tyre industry, which employ more than 7.5 lacs people with an annual sales turnover of US$ 2250 million. The total exports are 30 per cent of annual turnover. As against this the petitioner is employing not more than 300 persons in business and export sales is virtually zero.

(c) SBR is having various grades. Some of the grades fall under custom tariff entry no. 4002.19 and some other fall under custom tariff entry no. 3903.90 which are copolymers of styrene and butadiene but are neither 'like articles' to the grades of SBR produced by the petitioner nor are such type of co-polymers of styrene and butadiene produced by the petitioner in India. There are some types of SBR, which are not capable of being produced in the same plant such as that of petitioner and any or all of them are not substitutable with one another for the end use. It would therefore, be incorrect and vague to undertake to conduct an investigation against a class of product from one domestic source with another class of product from international source wherein the products grouped together may have different physical forms or technical characteristics and which are used in entirely different end use applications. It is the duty of Authority to identify the article liable for anti-dumping duty which has not been carried out since the term SBR is a broad generic term and it represents a class of products or articles.

(d) ESBR grades of 1502, 1712 and HSR grade 1958 are neither like articles nor interchangeable. It is contended that investigation for anti-dumping has been initiated for a wide class of articles rather than a singular product or articles and as such no clear and proper comparison of statistical data can be done to evaluate whether there is any increased quantity of import necessitating any action for imposition of anti-dumping measures:

(e) Investigation does not cover several countries which individually accounts for more than 3 per cent of imports or if individually accounts for less than 3 per cent collectively accounts for more than 7 per cent and thus it is discriminatory. The Authority should act in non-discriminatory manner and also investigate import from countries such as Belgium, Brazil, Italy, Switzerland etc..

(f) The current rules framed for the investigation are unlawful and liable to be struck down. The exporter has objected to the proceedings conducted without full transparency.

(g) There is another producer of solution type of co-polymers of styrene and butadiene commonly known as SBS type of synthetic rubber and falls under custom tariff heading 4002.19. This producer has not been named in the petition.

(h) Since the petitioner's facility and that of Gujarat Apar Polymers Ltd. is a multipurpose plant, the combined output of all grades of ESBR, NBR and SBR lattices should be considered if at all, for the purpose of combined investigation of the antidumping measures for SBR and NBR. If the investigation pertaining to a class of unlike articles or elastomers is proceeded, then it is appropriate to club the production of SBR latexes of M/s Aptotex Latices Ltd. (APL) and that of Thermoplastic rubbers by Ws ATV Projects Ltd. since the import statistical data is mixed up for these type of copolymers of styrene and butadiene and since the existing production facility of the petitioner was and can also be utilised for the production of NBR grades for which antidumping duty protection has already been given by the Govt.

(i) The petitioner is charging different rates for different grades. In order to undertake an objective analysis of the petition, the Authority must have grade wise data of production, sales, raw material consumption, cost of production, import volume and price and type of SBR, which is not possible in the present case.

(j) Korean exporter has represented that there is no basis whatsoever produced by the petitioner to consider export price to Mexico of unknown type of grades of SBR as the normal value for Korea RP for all grades of SBR.

(k) The Petitioner has not submitted the data on normal value from all countries correctly. There is no consistent trend in prices. In respect of adjustments in normal value and export price, no evidence has been submitted. At some places, the price of SBR 1712 has been shown higher as compared to SBR 1502, which is not possible. It is stated that data presented by petitioner is fabricated and misleading and, therefore, should not be relied upon.

(1) It is evident that there is no injury caused to the petitioner by the import of SBR during the POI if volume effect and price effect is taken into account. The import data of DGCI&S should not be relied upon, as they are inaccurate and unreliable. The reliable data can be procured from Rubber Board.

(m) The petitioner has not suffered injury as

- The volume of import has not increased.
- There is no decline in production if all grades of .ESBR are combined and data are corrected though there is a decline in the production of SBR grade 1502 and 1712.
- There is a gap in the demand and supply position and petitioner has failed to meet the demand since long time.
- The market share of petitioner has been stable between 56-61 per cent over the years.
- The increase in imports if any is due to increase in exports which necessiates the imports under advance licensing scheme.
- Fiscal year 1995-96 should not be considered to assess injury in Period of

investigation, as this year was an abnormal year in SBR case:

- ❖ The landed prices of imports are higher as compared to the domestic prices charged by the petitioner.
- ❖ There is no evidence of price undercutting or suppression. Price changes if any is due to lower price of raw material i.e. styrene and butadiene.
- ❖ The inventory represents equivalent to 15 days production, which is normal.
- ❖ Though the production capacity of petitioner is 41000 tonnes p.a. they have never achieved this in last 5-6 years even when the SBR demand was on peak. The petitioner can not produce SBR equal to its installed capacity, as the plant is old and obsolete.
- ❖ Reduction of work force by 5 per cent is normal.
- ❖ The reason for delay in expansion of project is not dumping.
- ❖ There is no threat of injury.

(n) The petitioner is losing its share in domestic market to other producers of SBR who are operating smaller capacity plants more efficiently and profitability.

(o) There are certain other factors which are causing injury to petitioner these are:

- ❖ Improper location of plant site at Bareilly, which is far from the source of raw material.
- ❖ Long time taken to complete the expansion project of ESBR.
- ❖ Improper arrangement for procurement of raw material and inventory management as evident from their published accounts.
- ❖ Unexplained delay in the commencement of production of NBR.
- ❖ there is a reduced proftability, but the SBR business continues to earn profit.

In view of above, there is no causal link between dumping and injury if any.

Some of the submissions of importers and exporters overlap and hence not repeated for the sake of brevity.

(D) Examination Of the Issues Raised

4. The submissions made by exporters, importers, petitioner and other interested parties have been examined and considered and have been dealt at appropriate places in these findings.

(E) Product Under Consideration

5. The product involved in the present investigation is Styrene Butadiene Rubber (SBR) originating in or exported from the said countries classified under custom subheading 4002.19 of the customs Tariff Act. The classification is, however, indicative only and in no way binding on the scope of the present

investigation. SBR is a type of synthetic rubber. SBR is one of the major elastomers and is mixed with natural rubber for manufacture of Auto tyres and tubes, bicycle tyres and tubes, footwear, belts and hoses etc. Styrene and' butadiene are the main raw material to manufacture SBR.

Some variation/type of SBR exists depending upon the extent of Styrene content and the other physical/chemical properties. Content of Styrene in SBR may vary depending upon the application for which it is used. SBR Polymers may be hot type (have code in 1000 series), cold type SBR (have code in 1500 series), cold type oil extended SBR (have code in 1700series) and high Styrene resin master batches (have code in 1900 series) etc. It is claimed by petitioner that they are manufacturing and marketing SBR grades S 1502, S 1712 and S 1958 and these grades are mostly being imported. It is fiuther claimed that these grades have specific application but these are produced with common equipment, facilities and production process.

It is brought to the notice of Authority that there are various type of SBR. Some types of SBR fall under custom Tariff No. 3903.90 and there are certain types of solution polymerised SBR termed as SSBR which fall under custom Tariff Heading No. 4002.19. There are some emulsion polymerised SBR which are graded in 1000 series, 1500 series, 1700 series and 1900 series.

The Authority observes that items falling under 3903.90 are not the product under consideration and SBR in 1000 series is not produced by petitioner and thus are not product under consideration.

The petitioner is producing SBR grade 1502, 1712 and 1958. Though there are different grade in 1500, 1700 and 1900 series, they can be produced in the same plant and process of manufacturing is same. The Authority therefore, hold that product under consideration is SBR of 1500 series, 1700 series and 1900 series under customs sub heading 4002.19 of the customs Tariff Act..

(F) Like Articles

Petitioner has claimed that products being imported and produced by domestic industry are of similar quality and all tyre companies and other major non-tyre companies who have foreign technical collaboration have their name in their approved list of supplier s of SBR. It is further claimed that in view of this fact and the fact that the same end users of SBR are using petitioner's product and imported product interchangeably, the product being imported and produced by the domestic industry are similar and thus like products under the anti-dumping rules. It is observed by the Authority that all SBR are copolymers of Styrene and Butadiene and serve the same general purpose though the same have different specific application. The Authority finds that the manufacturing process, equipment and other facilities needed for producing different grades of SBR are common and does not involve any special equipment to produce different grade of SBR. The various grades merely depict the level of co-polymers and other parameters. The Authority observe that the each type/grade of SBR produced by the domestic industry has a characteristics closely resembling to each respective type/grade of SBR imported from the subject countries and these are technically and commercially substitutable and, therefore, are like Articles within the said rules. For instance; S 1502 produced by domestic industry has characteristics closely resembling to S 1502 imported from subject countries and therefore is a like article.

(G) *Domestic Industry*

7. The petitioner has been filed by M/s Synthetic and Chemicals Ltd. Oriental House, 7 J, Tata Road, Churchgate, Mumbai-400020. Ws Apar, M/s Gujrat Apar and M/s Apcotex Lattices are stated to be other producer of SBR. According to the statistics published by "Rubber Statistical news" of Rubber Board, India, the total production of SBR during the POI was 56980 MT. The petitioner produced 43861 tonnes of all grades of SBR. Thus the petitioner accounts for more than 25 per cent of domestic production of SBR and therefore, has a standing to file a petition on behalf of domestic industry under the rules.

(H) *DUMPING*

8. Under Section 9A(1)(c), normal value in relation to an article means:

(i) the comparable price, in the ordinary course of trade, for the like article when meant for consumption in the exporting country or territory a determined in accordance with the rules made under sub-section (6); or

(ii) when there are no sales of the like article in the ordinary course of trade in the domestic market of the exporting country or territory, or when because of the particular market situation or low volume of the sales in the domestic market of the exporting country or territory, such sales do not permit a proper comparison, the normal' value shall be either:

(a) comparable representative price of the like article when exported from the exporting country or territory or an appropriate third country as determined in accordance with the rules made under sub-section (6); or

(b) the cost of production of the said article in the country of origin along with reasonable addition for administrative, selling and general costs and for profits, as determined in accordance with the rules made under sub-section (6):

Provided that in the case of import of the article from a country other than the country of origin and where the article has been merely transshipped through the country of export or such article is not produced in the country of export or there is no comparable price in the country of export, the normal value shall be determined with reference to its price in the country of origin.

9. The Authority sent questionnaire to the exporters from the subject countries in

terms of the section cited above. The claims made by exporters with regard to normal value and export price are as under.

I. Claims Of Exporters

Country: USA
Exporter: Ameripol Synpol Corpn. (ASC)

They have claimed-that there is no basis for including ASC in dumping allegation due to the reasons that the product they export to India are unavailable locally. The numbers of grades manufactured in India are limited and it is further claimed that they are the sole producer in the world for one of the grades sold into India. It is further claimed that there is no big difference between export price to India, their domestic price and price of export to third countries. The sales volumes of India have been quite steady for, years and the volumes on a percentage of total imports, into India are miniscule. It is further claimed that on these accounts they should be excluded from the dumping allegations. They have. stated that no incentive is offered by Govt. of USA to them on export to India.

It is further claimed that during the POI they exported SBR grade 8401 and plant cleans up (PCU). Grade 8401 is a special grade and they are only producer in the world of this material. 8401 is a special TRAXOL(R) oil extended (37.5 phr), SBR polymer in the 39.5 per cent bound styrene. It features high damping characteristics and low air permissibility. It finds application in tire liners, tire treads, automotive mounts, racing tires and heaving extended good where it provides improved noise damping plus better traction and faster stopping than conventional SBR. PCU is a term used to denote plant cleans up which is comprised of scraps of all kind of SBR grades collected from equipment cleans up floor sweeping etc. and it as not even an off specification of SBR material.

The exporter has submitted the information in respect of grade 8401 and PCU in relation to domestic price and export price to India. It is claimed that export price were either FOB factory or FOB ports of Houston. The exporter has claimed adjustment on account of inland freight when prices were quoted FOB port. In respect of domestic sales, it is claimed that list price are ex-works. They have claimed adjustment on account of discount/commission. Thus no dumping margin has been claimed in respect of above grades. No costing information submitted.

Exporter: The Good Year Tire & Rubber company

It is claimed by the exporter that after reviewing their, records and discussion with the sales and marketing people, they have not participated actively in the Indian market for SBR during the relevant period of inquiry. Instead they had made a series of transfer to Goodyear's affiliated company in India, when the petitioner failed to meet certain supply commitments made to Goodyear India. It is also claimed that they have not sold any

SBR to any other customer in India during the relevant period nor have any plans to do so.

They have not submitted any other information.

Country : France
Exporter: Bayer

It is claimed that they structure their business into sales to dealers in Europe and India and to end users in Europe. This price structure is very similar for dealers in both India and Europe. A system of rebates exist within Europe especially for large quantity contracts which is equivalent to effective price reduction of between 3 per cent to 7 per cent. This system does not exist for them to India, as it is large quantity net prices. It is further claimed that they sold SBR to India at prices equivalent to European price during the period of investigation. They have submitted information in respect of Krylene HS 260 grade. It is stated that selling prices are negotiated case by case. No list prices exist for they serve no useful purpose in highly competitive market.

They have claimed adjustments on account of Inland freight, handling, overseas freight, overseas insurance while calculating export price to India. The ex-factory price claimed for export is *** US$ per kg.

They have claimed price adjustment on account of Inland freight in their domestic sales. They have not submitted the data on domestic sales at ex-factory level. They have also not submitted average domestic price and cost data. It is stated that raw material cost and direct labour cost are identical for all product sold. Overhead costs are allocated equally on all the production volume of krylene HS 260. It is also claimed that they do not benefit from subsidies by regional or national govt. or authorities.

ii) Exporter: Michelin, France

It is claimed by the exporter that they had not exported SBR into India and, therefore, zero anti-dumping be levied. It is further stated that if they receive any orders for supply of SBR to India in future, they reserve their right to make a request for a new shipper review.

Country: Germany
Exporter: Bung Sow Leuna Olefinvefbund Gmbh (BSL). Germanx through Dow Chemicals International Ltd., Mumbai.

It is claimed that the name of the company is changed from Kombinat VEB Chemische Worke Bune to Buna Sow Leuna Olefinverbund Gmbh. It is claimed that during the period of investigation, they exported one .consignment to India of ***MT in July, 1997. This consignment is negligible in view of total consumption of SBR in India. Further this was the very first trial order which usually incurs additional expenses for the customer for testing. Thus they are justified in charging a price slightly below market price level in order to share these testing expenses with the customer. It is further claimed that the export price charged was higher as compared to domestic price and hence no dumping has been admitted. It is further stated that as they have exported one single consignment of small quantity, they are not desirous of submitting a detailed report on the questionnaire sent by Authority.

Country: Taiwan
Exporter: Taiwan Synthetic Rubber Corpn (TSRC)

The exporter has submitted information in relation to grade 1502 and grade 1712 of SBR.

Normal value: In respect of normal value they have submitted the domestic price structure. They have claimed the price adjustment on account of commission (which is *** per cent average of selling price) and inland freight..

Export Price- They have claimed the price per unit on the basis of weighted average price. They have claimed the price adjustments on account of discount/commission, inland freight, handling, harbor construction fee and promotion charges, overseas freight (including shipping charges) and overseas insurance.

The dumping margin is thus claimed as under:
Grade margin as per cent of export price
SBR 1502 7.14 per cent
SBR 1712 6.57 per cent

Country: Turkey
Exporter: Petkim Petrtokimya Holding AS

They have claimed that their annual production and sales has been to the extent of 14,00,000 MT and they export about 20 per cent of their production. Their export to "the European communities" is over 50 per cent of their export sales and till date they have not faced any investigation of anti-dumping, safeguard etc. It is further claimed that they produce SBR 1502 and SBR 1712 type and they do not produce S 1958 type. It is further stated that in no way their export threaten Indian domestic producer and they do not foresee any rise in their exports to India.

Normal value: They have claimed weighted average domestic price of US$ *** PMT in respect of SBR 1502 and US$ * * * PMT in respect of SBR 1712 at ex-factory level. They have not claimed any adjustments in respect of domestic prices.

Export price: They have claimed price adjustments on account of interest, commission, ocean freight and ocean insurance. The ex-factory price in respect of export to India is US# * * * for SBR 1502 and SBR 1712 grades. The dumping margins are thus claimed as under

Grade Dumping margin as per cent of Export price

SBR 1502 28.82 per cent

SBR 1712 13.66 per cent

They have claimed to earn profit on both grade of SBR during the period of investigation..

Country: Korea

M/s Korea Kumho Petrochemicals Co. Korea

They have submitted information in respect of grades 1502, 1712, 1778 and KHS 58. It is stated that in Korea, there are two SBR makers and other producer is M/s Hyundai Petrochemical Co and they are not aware of their exports to India.

Normal value: They have claimed the normal value on the basis of prices prevailing in the domestic prices which are based on weighted average: They have claimed price adjustments on account of discounts/credit interest, packing, inland freight, storage, handling, taxes, technical services, promotion travelling expenses, bad debts expenses and advertising expenses. It is claimed that domestic packing is different from export packing because it is designed to fit for robot and average circulating is I.3 times. Regarding tax it is claimed that import duty for raw material on export sales is to be refunded whereas for domestic sales, it is to be absorbed by the company.

Export Price: They have claimed the CIF export price on the basis of weighted average during the period of investigation: They have claimed the price adjustment on account of discount/commission, packing, inland freight, inspection fee, overseas freight and overseas insurance. It is further stated that other adjustments on account of rebate; in custom duty on raw material used for export production, donation, R &D work etc. are not yet ascertained and they will submit relevant data during verification.

As per the costing information, they are making profit on 1502 and 1712 grades.

The comparison of normal price and export prices shows the dumping margin as under:

Grade, Dumping margin as per cent of export price

SBR 1502. 9.43 per cent

SBR 1712 2.85 per cent

Country: Japan

Exporter: Mitsubishi Chemical

They have claimed that they had not made exports to India during POI and they do not intend to initiate any export of SBR into the Indian market under the current circumstances. Thus, it is stated that they do need recognize a necessity to respond to the questionnaire.

Exporter: SHIPPER - Fuji Chemical Development Co. Ltd.

Fuji Chemical Development Co. Ltd. has claimed that they are the Shipper of Mitsui & Co. and the data given by Mitsui & Co. will apply mutatis Mutandis to them and therefore no separate response to the questionnaire is being filed.

Exporter: Mitsui & Co.

It is claimed by them that they are one of the authorised trading houses of JSR company and Fuji Development Co. Ltd., is their shipper. It is further claimed that the CIF or CFR price offered by JSR is submitted to Mitsui from JSR. The price is calculated by JSR using the certain freight rates, commission rates etc that is given to JSR. For shipment, Mitsui shall instruct Fuji to ship JSR material on a certain vessel. Regarding normal value, they have not submitted separate data and stated that it is as per JSR rate.

Exporter: JTC Corpn.

It is stated that M/s. JTC Corpn. Japan is an exporter _of SBR bought from JSR. They have not claimed separate normal value.

JSR Corpn. (formerly known as Japan Synthetic Rubber Co. Ltd.)

It is claimed by the exporter that they have exported three grades of SBR during the POI to India. These grades are S 1500, S 1712, and S 1502. It is also stated that JSR also sells within its domestic market these 3 grades of SBR, apart from other grades. The grades sold in Japan are similar to the respective grades exported to India.

Normal value: They have claimed the normal value on the basis of weighted average price. They have claimed adjustment on account of inland freight, Insurance, interest and storage while arriving at the ex-factory price. They have also claimed deduction of U$ **** PMT on account of extra grades produced for the domestic market which results in lower production and capacity utilisation and consequently the fixed cost per unit are higher for SBR produced and sold in domestic market.

Export Price: It is claimed that no list price exists. They have claimed deduction on account of commission to Japanese trading house, Indian trading house, Inland freight, Insurance, overseas freight, overseas insurance, interest and commission while arriving at the export price to India.

The dumping margin thus claimed is as under:

Grade	Dumping margin as per cent of Export price
S 1502	10.32
S 1712	No Dumping
S 1500	No Dumping

J. 11. Examination of the claims of the exporters by the Authority

As indicated earlier the Authority is considering the product under consideration as SBR 1500 series, SBR 1700 series and SBR 1900. series. The Authority also notes that the some grade may be numbered differently in various countries. Wherever the information are made available grade-wise, the Authority has calculated separate dumping margin as the end uses and prices varies in different grades of SBR.

The claims of various exporters are examined as under:

Country: USA
Exporter: Ameripol Synpol Corpn (ASC)

The Authority notes that the exporter has not submitted information in respect of SBR 1500 series, SBR 1700 series and SBR 1900 series. They have submitted information in respect of SBR grade 8401 grade and PCU. No costing information has been submitted. It is not explained that whether grade 8401 and PCU are like article to SBR 1500 series, SBR 1700 series and SBR 1900 series. SBR 8401 is claimed to be different grade of SBR altogether. Since no information has been provided by the exporter in respect of product under consideration, the Authority considers the information provided by petitioner as best available information and determines the normal value and export price as claimed by the petitioner at U$ 1286 PMT and at U$ 985 PMT respectively. The dumping margin is calculated as U$ 301 PMT which is 30.55 per cent of export price.

Country: France
Exporter: Bayer

The Authorities notes that the exporter has not submitted information in respect of SBR 1500 series, SBR 1700 series and SBR 1900 series. They have submitted some information in respect of Krylene HS 260 grade. No costing information has been Dubmitted. It is not substantiated that grade Krylene HS 260 are like article to SBR 1500 series, SBR 1700 series and SBR 1900 series. Since no information is provided by the exporter, the Authority considers the information provided by the petitioner as the best available information and determines the normal value and export price as claimed by the petitioner at U$ 1100 PMT and at U$ 858 PMT respectively. The dumping margin is calculated as U$ 242 PMT which is 28.20 per cent of export price.

Country: Germany
Exporter: BSL

The Authority notes that the exporter has not submitted information in respect of their exports to India. In view of this, the Authority considers the information provided by the Petitioner as the best available information and determines the normal value and export price as claimed by the petitioner at U$ 1100 MT and at U$ 881 PMT respectively. The dumping margin is calculated at U$ 219 PMT which is 24.86 per cent of export price.

Country: Taiwan
Exporter: Taiwan synthetic Rubber Corpn.

Normal value: The Authority notes that the normal value is claimed on the basis of weighted average domestic price and adjustment on account of commission and inland freight has been claimed. The Authority determines the normal value as claimed at U$ ***in respect of SBR 1502 and at U$ ***in respect of SBR 1712 for the purpose of preliminary determination subject of verification.

Export Price: The export price is on the basis of weighted average export price to India. They have claimed the adjustment on account of discount and commission, inland freight, handling, harbour construction fee and promotion charges, overseas freight (including shipping charges) and overseas insurance. These expenses as claimed except harbour construction fee & promotion charges are allowed for preliminary determination subject to verification. Harbour construction fee and promotion charges are not allowed, as its nature is not clear. The export price is thus determined at U$ *** PMT in respect of SBR 1502 grade and are U$ *** PMT in respect of SBR 1712 grade subject to verification. The dumping margin is calculated at U$ *** PMT in respect of SBR 1502 grade which is 6.68 per cent of export price and at U$ *** PMT in respect of SBR 1712 grade which is 6.07 per cent of export price.

Country: Turkey
Exporter: Petkin Petrokinys ho1ding AS

Normal value: The Authority notes that the normal value is on weighted average of domestic price and no adjustment is, claimed. It is allowed subject to verification and is determined at U$ ***PMT in respect of SBR 1502 grade and at U$ ***PMT in respect of SBR 1712 grade.

Export Price: The export price is calculated on the basis of weighted average of export price. The claim on account of interest, commission,' ocean freight and ocean insurance are allowed for preliminary determination subject to verification. The export price is determined at U$ *** PMT in respect of SBR 1502 grade and at U$ *** in respect of SBR,1712 grade. The dumping margin is arrived at U$ *** PMT in respect of SBR 1502 grade which is 28.82 per cent of export price and at U$ *** PMT in respect of SBR 1712 grade which is 13.66 per cent of export price.

Country: Korea
Exporter: Korea Kumho Petrochemicals Company Korea.

Normal value : The normal value is calculated on the basis of weighted average of domestic price. The Authority allows the adjustments on account of discount/credit interest, packing, inland freight, handling and taxes as claimed for the purpose of preliminary determination subject to verification. The adjustments on account of storage, technical services, promotional travelling expenses, bad debts expenses and advertisement expenses are not allowed. Thus the normal value is determined at U$ *** PMT in respect of SBR 1502 grade and at U$ *** in respect of SBR 1712 grade.

Export price : The export price is on the basis of weighted average of export sales to India. The adjustments as claimed on account of discount/commission, packing, inland freight, inspection fee, overseas freight and overseas insurance are allowed for the preliminary determination subject to verification. The export price is determined at U$ *** PMT in respect of SBR 1502 grade and at U$ *** PMT in respect of SBR 1712 grade. The dumping margin is calculated at U$ *** PMT in respect of SBR 1502 grade which is 11.66 per cent of export price and at U$ ***PMT in respect of SBR 1712 grade which is 5.29 per cent of export price.

Country: Japan

Normal value : The normal price is calculated on the weighted average price. The adjustment on account of inland freight, insurance and interest are allowed as claimed subject to verification for preliminary determination. The claim on account of storage cost is not allowed. The claim on account of extra grade produced for the domestic market is based on estimates and is not actually incurred by the exporter and hence not allowed.

Export price: The export price is calculated on the basis of weighted average price. The claim on account of commission to Japanese Trading house, Indian Trading house, Inland freight, insurance, overseas freight, overseas insurance, interest and commission are allowed subject to verification for the purpose of preliminary determination -

The dumping margins are calculated as under:

Grade Dumping margin as per cent of export price

S 1502 17.32 per cent

S 1712 7.64 per cent

S 1500 3.36 per cent

In respect of non-co-operating exporters from subject countries who has not responded to questionnaire the Authority determine the dumping margin at the highest dumping margin determined for the co-operating exporters from each country. Thus the dumping margin as percentage of ex factory export price for non-cooperating exporter from Taiwan is ***PMT 6.68 per cent, from Turkey is *** PMT 28.82 per cent from Korea is U$ ***PMT 11.66 per cent and from Japan is U$ ***PMT 17.32 per cent.

The dumping marsin in resneet of subject countries & exporters are determined as under:

Country	*Exporter*	*Grade Dumping*	*Margin as Per cent of Export Price*
1. USA	All exporter	All grade	30.55 per cent
2. France	All exporter	All grade	28.20 per cent
3. Germany	All exporter	All grade	24.86 per cent
4. Taiwan	Taiwan synthetic	S 1502	6.68 per cent
	Rubber Corpn.	S 1712	6.07 per cent
	Other exporters	All grade	6.68 per cent
5. Turkey	Petkin Petrokinya	S 1502	28.82 per cent
	Holding	AS S 1712	13.66 per cent
	Other exporter	All grades	28.82 per cent

Country	*Exporter*	*Grade Dumping*	*Margin as Per cent of Export Price*
6. Korea	Korea Kumho Petrochemical Co.	S 1502	11.66 per cent
		S 1712	5.29 per cent
	Other exporters	All grade	11.66 per cent
7. Japan	JSR Corpn.	S 1502	17.32 per cent
		S 1712	7.64 per cent
		S 1712	3.36 per cent
	Other exporters	All grades	17.32 per cent

K. 12. Injury

Under Rule 11 Supra, Annexure-II when a finding of injury is arrived at, such findings shall involve determination of the injury to the domestic industry, '...taking into account all relevant facts, including the volume of dumped imports, their effect on prices in the domestic market for like articles and the consequent effect of such imports on domestic producers of such articles...". In considering the effect of the dumped imports on prices, it is considered necessary to examine whether there has been a significant price undercutting by the dumped imports as compared with the price of the like product in India, or whether the effect of such imports is otherwise to depress prices to a significant degree or prevent price increase, which otherwise would have occurred, to a significant degree.

Annexure II (iii) under rule 11 supra further provides that in case where imports of a product from more than one country are being simultaneously subjected to Anti dumping investigation, the Designated Authority will cumulatively assess the effect of such imports, only when it determines that the margin of dumping established in relation to the imports from each country is more than two percent expressed as percentage of export price and the volume of the imports from each country is three percent of the imports of the like article or where the export of the individual countries less than three percent, the imports cumulatively accounts for more than seven percent of the imports of like article, and cumulative assessment of the effect of imports is appropriate in light of the conditions of competition between the imported article and the like domestic articles.

The Authority notes that the margin of dumping and quantum of imports from subject country are more than the limits prescribed above. Cumulative assessment of the effects of imports is appropriate since the export prices from the subject country were directly competing with the prices offered by the domestic industry in the Indian market.

For the examination of the impact of imports on the domestic industry in India, the Authority may consider such further indices having a bearing on the state of the industry as production, capacity utilisation, sales quantum, stock, profitability, net sales realisation, the magnitude and margin of dumping etc. in accordance with Annexure II(iv) of the rules supra.

(I) The effect of the dumped imports shall be assessed, in accordance with para (vi) of the Annexure II to the Rules, in relation to the domestic production of the like article when available data permit separate identification of that product on the basis of such criteria as the production process, producers' sales and profits. If such separate identification of that product is not possible, the effect of the dumped imports shall be assessed by the examination of the product of the narrowest group or range of products, which includes the like product, for which the necessary information can be provided.

(ii) It is observed that though the different types of SBR have different characteristics, usage etc., there are a number of processes which use common equipment and facilities. In view of the fact that the production processes as well as the production capacities overlap each other in varying proportions, it would not be appropriate nor feasible to assess the injury to the domestic industry for each individual type of SBR. The Authority therefore, in accordance with the para 6 of Annexure II to these Rules, considers it appropriate to assess the injury for all types of SBR cumulatively.

(a) *Quantum of Import*: To analyse the trend of quantum of import from the subject countries over the years, the Authority observed the statistics published by DGCIS, Calcutta. However, the petitioner and some of the exporters had pointed over some factual errors in the data published by DGCIS and pleaded not to rely upon the data of DGCIS. It is also observed that the petitioner is producing 1502, 1712 and 1958 grades of SBR whereas as per DGCIS statistics the import data are classified as under:

Code:

40021901 Oil Extended SBR

40021902 SBR with Styrene contents of over 50 per cent

400021909 Others

The petitioner has represented that under code 40021901, the imports are mainly of SBR 1712 grade, under code no. 40021902 the imports are mainly of SBR 1958 grade and under 40021909, the imports are mainly of SBR 1502 grade. The Authority enquired from Rubber Board regarding the availability of statistics of SBR grade-wise. It was stated by Rubber Board that it is collecting details of SBR from the publication 'foreign trade statistics of India' published by DGCI&S Calcutta. The publication does not contain details of break up of import of SBR under different grades. None of the importers/exporters/petitioner has submitted statistics relating to imports of SBR grade-wise which can be relied upon. Therefore, the Authority considers the statistics published by DGCIS as 'best available information' to analyse the injury and other parameters.

As per the information published by DGCI&S, the quantum of imports is as under:

94-95 95-96 POI Annualised

From subject countries12602 9593 17206 12145

Others 2572 1795 6305 4450

Total 15174 11388 23511 16595

It is observed that the total imports of SBR was 15174 MT, 11388 MT and 23511 tonnes (16595 MT on annualised basis) in 1994-95, 95-96 and in POI. Thus there was a decrease of 24.94 per cent in imports in 95-96 over 1994-95 and increase of 45.71 per cent in POI (on annualised basis) over 95-96. The volume of imports from the subject countries was 12602 MT, 9593 mt and 12145 tonnes in 94-95, 95-96 and in POI (on annualised basis). Thus the volume of imports from subject countries was 83 per cent, 84 per cent and 73 per cent in total imports. The volume of imports from other countries from which dumping is not alleged was 2572 MT, 1795 MT, and 6305 MT in POI (on annualised basis) which were 17 per cent, 16 per cent and 27 per cent of total imports. In absolute terms, the quantum of imports from subject countries decreased by 23.86 per cent in 95-96 over 94-95 and increased by 26.60 per cent in POI (on annualised basis) over 95-96. The share of other countries shows a decrease of 30.2 per cent in 95-96 and shows an increase of 147.91 per cent in POI (on annualised basis) over 95-96. The volume of imports from subject countries were lower by 3.62 per cent in POI (on annualised basis) over 1994-95 and from other countries it was higher by 73.02 per cent in POI over 94-95, Thus it is observed that:

❖ Quantum of total imports were lower in 95-96 over 94-95, but higher in POI (on annualised basis) significantly when compared to 95-96.

❖The share of imports from the subject countries were lower in total imports in POI (On annualised basis) when compared to 94-95 and 95-96 and the share of other countries in total imports were higher in POI when compared: to 94-95 and 95-96.

❖ In absolute terms, there was increase in imports from the subject countries in POI (on annualised basis) when compared to 95-96 but there was a marginal decrease when compared to 94-95.

(b) *Production capacity and capacity utilisation*: It is stated by the petitioner that total installed capacity of domestic industry is 49700 MT p.a. out of which the petitioner is having an installed capacity of 41000 MT. There is no increase in the installed capacity of the petitioner over last 3 years though they have undertaken to expand the capacity from 41000 MT to 70000 MT per annum Production of SBR by the petitioner was 29132 MT, 29139 MT and 42327 MT (29878 on annualised basis) in 94-95 and 95-96 and in POI. Thus is observed that there is no significant change in production over the years. The capacity utilisation was 71.05 per cent, 71.07 per cent and 72.87 per cent in 94-95, 95-96 and in POI (on annualised basis).

(c) *Closing stock*: The closing stock of SBR with the petitioner was 230 MT, 1462 MT and 1348 MT as on 31.3.95, 31.3.96 and 31.8.97 respectively. This shows that closing stock of the petitioner has increased in 95-96 over 94-95 but has marginally decreased as on 31.8.97 over 31.3.96. The stock as on 31.8.97 represents equivalent to 16 days production..

(d) *Employment*: It is observed that the number of employees with petitioner was 1674, 1607 and 1604 as on 31.3.95, 31.3.96 and on 31.8.97, which does not show any significant change.

(e) *Volume of sales and market share*: It is observed that the petitioner's volume of turnover of SBR was 32415 MT, 28940 MT and 41906 MT in 94-95, 95-96 and in POI (29581 MT on annualised basis). This shows that there was decline of about 10.72 per cent in 95-96 over 94-95. There was an increase of about 2.21 per cent in sales in POI (on annualised basis) over 95-96. It is thus observed that whereas production increased by about 2.56 per cent in POI over 95-96, sales increased by 2.21 per cent. It is also observed that the total demand of SBR was 48850 MT, 51920 MT and 77994 MT in 94-95, 95-96 and in POI (55055 MT on annualised basis. The share of petitioner in total demand was 66.35 per cent, 55.74 per cent and 53.73 per cent in 94-95, 95-96 and in Period of investigation. Thus the share of the petitioner in total demand is declining over the years.

(f) *Expansion of Project*: The petitioner has claimed that due to dumping by the exporters from the subject countries at lower prices has put pressure on their financial position and the commercialization of the project has been further delayed. It is claimed that they are increasing the capacity from 41000 TPA to 70000 TPA and after the expansion they are capable of meeting the demand of entire country.

I. Conclusion on Iniury

13 In view of foregoing paragraphs it is observed that:

a) Quantum of imports from subject countries and total imports have increased in POI as compared to 95-96. The percentage increase was more from countries not allegedly dumping as compared to subject countries. The imports have marginally decreased in POI as compared to 94-95 from subject countries.

(b) The production and capacity utilization have been stable over the years.

(c) The closing stock as on 31.8.97 shows a significant increase as compared to 31.3.95 but shows a decrease as compared to 31.3.96.

(d) The number of employees with the petitioner does not show any significant changes.

(e) The quantum of sales was marginally higher in POI as compared to 1995-96. However the market share of petitioner is declining over the years due to increase in imports.

(f) There is a decline in average selling price of all grade of SBR, period of investigation as compared to 95-96 effecting profits in 1996-97. The petitioner has incurred losses in April 97 to August 97 period.

(g) There is a delay in the commercialization of project which increases the installed capacity from 41000 TPA to 70000 TPA of the petitioner.

The increase in quantum of imports from subject countries, higher closing stocks, decline in the market share of the petitioner over the years, decline in unit sales realization, decrease in profitability/incurring of losses shows that domestic industry has suffered material injury.

The Authority holds that domestic industry has suffered material injury.

M. Causal Link

14. In establishing that the material injury to the domestic industry has been caused by the imports from subject countries, the Authority holds that increase in quantum of imports from subject countries resulted in decline in the market share of petitioner, were undercutting the prices of domestic product forcing the domestic industry to sell below its fair prices. Resultantly the domestic industry was not in a position to recover fair selling price. The material injury to the domestic industry was, therefore, caused by the dumped imports from the subject countries.

N. Indian Industry's Interest and Other Issues

15. The purpose of anti dumping duties, in general, is to eliminate dumping which is causing injury to the domestic industry and to re-establish a situation of open and fair competition in the Indian market, which is in the general interest of the country.

16. It is recognized that the imposition of anti dumping duties might affect the price levels of the products manufactured using the subject goods and consequently might have some influence on relative competitiveness of these products. However, fair competition on the Indian market will not be reduced by the anti dumping measures, particularly if the levy of the anti dumping duty is restricted to an amount necessary to redress the injury to the domestic industry: On the contrary, imposition of anti dumping measures would remove the unfair advantages gained by dumping practices, would prevent the decline

of the domestic industry' and help maintain availability of wider choice to the consumers of SBR. Imposition of antidumping measures would not restrict imports from the subject countries in any way, and, therefore would not affect the availability of the product to the consumers.

17. To ascertain the extent of anti-dumping duty necessary to remove the injury to the domestic industry, the Authority rely upon reasonable selling price of SBR in India for the domestic industry, by considering the optimum cost of production at optimum level of capacity utilization for the domestic industry.

O. Landed-value

18. The landed value of imports is determined on the basis of export price of SBR, determined as detailed above in the para relating to dumping, after adding the prevailing level of customs duties and one percent landing and two percent handling charges.

P. Conclusions

19 It is seen after considering the foregoing, that

(a) SBR described under para 5 and originating in or exported from Japan, Taiwan, Turkey, France, USA, Germany & Korea has been exported to India below normal value, resulting in dumping;

(b) The Indian industry has suffered material injury

(c) The injury has been caused cumulatively by the imports from the subject countries.

20 It is considered necessary to impose anti dumping duty, provisionally, pending final determination, on all imports of SBR originating in or exported from the subject countries, pending investigations.

21. It was considered whether a duty lower than the dumping margin would be sufficient to remove the injury. Landed price of the imports, for the purpose, was compared with the fair selling price of the domestic industry, determined for the period of investigations. Wherever the difference was less than the dumping margin, a duty lower than the dumping margin is recommended. Accordingly, it is proposed that provisional anti dumping duties be imposed, from the date of notification to be issued in this regard by the Central Government, on SBR originating in or exported from Japan, Taiwan, Turkey, USA, & Korea falling under Customs sub-heading 4002.19 of the Customs Tariff Act, pending final determination.

22. The anti-dumping duty shall be the amounts mentioned in column 6, provided that the duty shall be the difference between the amounts mentioned

in column 5 and the landed price of imports per Kg. in case such difference is more than the amounts mentioned in column 6:

Sl. No.	*Country*	*Exporter*	*Grade*	*Amount (Rs per Kg)*	*Amount (Rs per Kg)*
1	2	3	4	5	6
1.	USA	All exporter	1700 series	48.20	3.48
			1900 series	62.16	6.77
2.	Taiwan	All exporter	1500 series	57.33	2.21
			1900 series	56.90	2.21
3.	Turkey	All exporter	1500 series	57.50	8.26
			1900 series	60.69	8.26
4.	Korea	All exporter	1500 series	54.48	3.38
			1700 series	48.20	0.98
			1900 series	55.28	3.38
5.	Japan	All exporter	1500 series	59.82	3.71
			1700 series	48.20	0.92
			1900 series	61.78	5.20

23. Landed value of imports for the purpose shall be the assessable value as determined by the Customs under the Customs Act, 1962 and all duties of customs except duties levied under Section 3, 3A, 8B, 9 and 9A of the Customs Tariff Act, 1975.

Q. Further Procedure

24. The following procedure would be followed subsequent to notifying the preliminary findings

(a) The Authority invites comments on these findings from all interested parties and same would be considered in the final findings;

(b) Exporters, importers, petitioners and other interested parties known to be concerned are being addressed separately by the Authority, who may make known their views, within forty days of the despatch of this notifications. Any other interested party may also make known its views within forty days from the date of publication of these findings.

(c) The Authority would disclose essential facts before announcing the final findings.

Bibliography

A. Deschamps *et al.*, *Transactions of the Sixth World Biomaterials Congress*, I (Minneapolis: Society for Biomaterials, 2000), 364.

A. Stemberger *et al.*, *Transactions of the Sixth World Biomaterials Congress*, I (Minneapolis: Society for Biomaterials, 2000), 369.

B. Wong *et al.*, *Transactions of the Sixth World Biomaterials Congress*, I (Minneapolis: Society for Biomaterials, 2000), 363.

B.D. Ratner *et al.*, *Biomaterials Science* (San Diego: Academic Press, 1996), 6.

Cheremisinoff, P. 1989, *Handbook of Polymer Science and Technology*, M Dekker, New York.

CM Vaz *et al.*, *Transactions of the Sixth World Biomaterials Congress*, I (Minneapolis: Society for Biomaterials, 2000), 429.

Daniels AU, Chang MKO, Andriano KP, et al., *J Appl Biomat*, 1(1):5778, 1990.

DF Williams, *The Williams Dictionary of Biomaterials* (Liverpool, UK: Liverpool University Press, 1999), 40.

DF Williams, *The Williams Dictionary of Biomaterials* (Liverpool, UK: Liverpool University Press, 1999), 42.

DF Williams, *The Williams Dictionary of Biomaterials* (Liverpool, UK: Liverpool University Press, 1999), 318.

Ezrin, Meyer, *Plastics Failure Guide: Cause and Prevention*, Hanser-SPE (1996).

Gilding DK, and Reed AM, "Biodegradable Polymers for Use in Surgery—Polyglycolic/Poly(lactic acid) Homo- and Copolymers," Polym, 20:1459—1484, 1979.

Gogolewski S, and Pennings AJ, "Biodegradable Materials of Polyactides, Porous Biomedical Materials Based on Mixtures of Polyactides and Polyurethanes," *Makromol Chem, Rapid Commun*, 3:839845, 1990.

Han YK, Edelman PG, and Huang SJ, "Synthesis and Characterization of Crosslinked Polymers for Biomedical Composites," *J Macromol Sci-Chem*, A25(57):847869, 1988.

HQ Mao et al., *Transactions of the Sixth World Biomaterials Congress*, I (Minneapolis: Society for Biomaterials, 2000), 252.

J Kohn, *Transactions of the Sixth World Biomaterials Congress*, I (Minneapolis: Society for Biomaterials, 2000), 84.

K Ishihara et al., *Polymer Journal* 31 (1999): 1231–1236.

Kohn J, and Langer R, "Bioresorbable and Bioerodible Materials," in Biomaterials Science: An Introduction to Materials in Medicine, Ratner BD, Hoffman AS, Schoen FJ, and Lemons JE (eds), New York, Academic Press, pp 64—72, 1996.

Lewis, Peter Rhys, Reynolds, K and Gagg, C, *Forensic Materials Engineering: Case studies*, CRC Press (2004)

M Wang et al., *Transactions of the Sixth World Biomaterials Congress*, I (Minneapolis: Society for Biomaterials, 2000), 81.

M. A. Villetti, J. S. Crespo,M. S. Soldi, A. T. N. Pires, R. Borsali and V. Soldi. Thermal degradation of natrural polymers. Journal of Thermal Analysis and Calorimetry, Vol. 67 (2002) 295~303

Mitra, S, Ghanbari-Siahkali, A, Amdal, K 2006, "A novel method for monitoring chemical degradation of crosslinked rubber by stress relaxation under tension", *Polymer Degradation and Stability*, Vol. 91, no. 10, pp 2520-2526

N Peppas, *Transactions of the Sixth World Biomaterials Congress*, I (Minneapolis: Society for Biomaterials, 2000), 554.

N Yiu, "Design of Polyrotaxanes Aiming at Intelligent Biomaterials," in *Supramolecular Approach to Biological Function* (Minneapolis: Society for Biomaterials, 2000).

Peacock, A, Calhoun, A 2006, *Polymer Chemistry Properties and Applications*, Hanser Gardner Publications Inc., Munich.

Pietrzak WS, Sarver DR, and Verstynen ML, "Bioabsorbable Fixation Devices: Status for the Craniomaxillofacial Surgeon," J Craniofacial Surg, 8(2):87, 1997.

Pietrzak WS, Verstynen ML, and Sarver DR, "Bioabsorbable Polymer Science for the Practicing Surgeon," J Craniofacial Surg, 8(2):92, 1997.

Pitt CG, Hendren RW, Schindler A, et al., "The Enzymatic Surface Erosion of Aliphatic Polyesters," *J Controlled Rel*, 1:314, 1984.

Q.U.V Accelerated Weather tester operation manual, the Q-Panle company, Cleveland, OH, USA.

RS Ward, "Thermoplastic Silicone-Urethane Copolymers: A New Class of Biomedical Elastomers," *Medical Device & Diagnostic Industry* 22, no. 4 (2000): 68–77.

S Brocchini et al., *Journal of Biomedical Materials Research* 42 (1998): 66–75.

S Halstenberg et al., *Transactions of the Sixth World Biomaterials Congress*, I (Minneapolis: Society for Biomaterials, 2000), 427.

Storey RF, Wiggins JS, Mauritz KA, et al., "Synthesis and Fabrication of Completely Absorbable Composites for Biomaterials," *ACS Div Polym, Chem Polym Preprs*, 30(2):492493, 1990.

SW Kim, *Transactions of the Sixth World Biomaterials Congress*, I (Minneapolis: Society for Biomaterials, 2000), 250.

T Yoneyama et al., *Artificial Organs* 24 (2000): 23–28.

UV Weathering and Related Test Methods, Cabot corporation, www.cabot-corp.com

Vacanti JP, Morse MA, Saltzman WM, et al., *J Ped Surg*, 23(1):39, 1988.

Weidner, S, Kuhn, G, Freiedrich, J, Schroder, H 1996, "Plasmaoxidative and Chemical Degradation of Poly(ethylene terepthalate) Studied by Matrix-assisted Laser Desorption/Ionisation Mass Spectrometry",*Rapid Communications in Mass Spectroscopy*, Vol. 10, No. 1, pp 40-46

Wright, David C., *Environmental Stress Cracking of Plastics* RAPRA (2001).

Index

Absorption qualities, 61-63
Accumulator, 207-208, 210, 214-215, 226-227, 229-230, 236
Acrylic acids, 117-118
Added-value products, 6, 32-33
Aerospace, 119, 121, 128, 133, 146, 151, 155, 159-160, 162, 165
Aliphatic carbonates, 198, 202
Alma-Ata, 274
Alternating copolymers, 26, 312-313
Amino acids, 182, 184
Amorphous
    materials, 138, 253, 255, 267, 331
    polymers, 96, 102, 252-254, 302-303
    thermoplastic, 94, 137, 251
Anionic polymerization, 14, 18-19, 27, 310, 312
Antibodies, 193
Anti-dumping duty, 382, 385-389, 407-408
Antimony, 280, 283, 284
Antioxidants, 30, 199, 202, 292, 327, 330-31
Applications
    aerospace, 119
    electrical, 378
    fixation, 321
    high temperature, 118-119
    industrial, 133, 339
    patent, 203, 343
    pharmaceutical, 177-178, 349
    plastic surgery, 172-173
    structural, 120
Aquatic systems, 339
Aqueous medium, 192, 274, 276, 279-281
Aramid fibres, 131
Aromatic
    carbonates, 198, 202
    groups, 328, 331
Arsenic, 279-280, 283-284
Artificial weathering, 293, 295, 297
Atactic polystyrene, 94
Auxiliary opacifer, 240

Backbone chain, 30
BAE Systems, 122-123
Bakelite, 317
Barrel, 205, 207-208, 210-214, 217-218, 220-21, 225-26, 231-32, 241
    diameters, 222
Benzoyl peroxide, 13, 202
Bicomponent polymer fibers, 134-135, 138-141
    ideal, 139
    single, 136-137
Biocompatibility, 174, 179
Biodegradability, 333, 335-336, 339, 348
Biodegradable
    plastics, 334-335, 338, 340, 342-344, 347, 351
    polymers, 180-181, 184, 334, 350, 352-354
Biological interfaces, 179
Biomass, 171, 341-342
Biomechanics, 175-176
Biomedical applications, 177-178, 194, 352
Biomedicine, 190, 194, 354
Bioplastics, 335-336, 339, 344, 348, 352
Biopolymers, 99, 170-171, 187, 335-336, 345
Birefringence Properties in Polymers, 91
Bisphenol, 9-10, 180, 198, 202, 316
Bleaching rate, 200-201
Block copolymers, 26-28, 30, 97-98, 107, 194
Blowing agent, 204-212, 214-216, 219-227,

229-232, 234-237, 240-241, 243-244
chemical, 204, 219
fluid mixture of non-nucleated, 210-211
gaseous, 205
physical, 204-205, 210, 219-220
residual chemical, 219
supercritical fluid, 222-223, 229, 235, 241, 243
Branched
macromolecules, 309
polymer, 16

Capacity utilisation, 383, 388, 403, 405-406
Capillary rheometer, 259, 262
Carboxylic acid groups, 355
Cellulose
chains, 302
fibers, 144, 301, 317
molecules, 300, 302
Centrifugation, 341
Chain branching, 302
Chain extenders, 58-59
Chain mobility, 355
Chain movement, 303
Chain propagation, 307
Chain reaction, 292
Chains
branched, 302
homopolymer, 28
mole fraction of, 299
polypeptide, 170, 313
rubber, 30, 290
Chewing gum, 366
Chromatography, 86-87, 89, 91, 285
C-MOLD analyses, 245-246, 253-254
Coalescence, 277, 279-280
Coalescent separation, 279, 282
Commodity plastics, 347
Comonomers, 29
Component foam, 68-69
Components
molten, 141
principle, 57
absorbable, 321-322, 324
properties of, 101, 159
Compression molding, 108, 158, 375-377
equipment, 375, 377
Condensation polymers, 313-314
Cooling rate, 246, 253, 255, 265, 267-268
Copolymer
chains, 187, 351
equation, 26-27
grades, 29, 45
structures, 26, 28
Copolymerization, 12, 23-24, 26, 28, 311
Co-polymers, 335, 340-341, 343, 393
Coupling agents, 30, 194, 195, 324
Covalent bonds, 324
Cross-exp model, 247
Cross-sectional dimension, 206, 209, 211-213, 224, 233-234, 236-238
Crude oil, 1, 5, 10, 31-32
Crystalline
arrangement, 101
domains, 94, 301-302
polymers, 252-254, 268
Crystallisation, 83, 98, 152, 327
Customs Tariff Act, 378-379, 408-409

Decomposition products, 354
Dispensable systems, 69
Dow's Polyurethane Systems, 81
Dowty Propellers, 122
Drug delivery, 173, 180, 190-193, 349, 352
Drying temperature, 113-114
Dumped imports, 383, 387-388, 403-404, 407
DuPont, 25

E.coli, 345-347
Elastomeric properties, 302
Elastomers, 55, 81-82, 95-96, 107, 133, 290, 303, 332, 390, 392
cast polyurethane, 58
thermoplastic polyurethane, 58
Emulsion polymerization of vinyl monomers, 17
Epoxide groups, 118
Epoxies, 9, 126, 128-129, 139, 160
Ethane, 2, 4, 6
Ethene, 171, 298
Ethylene, 1-4, 6-8, 10-11, 28-29, 298, 301, 304, 309-310, 327
Extensometer, 268-269, 271
Extruder
barrel, 212, 218, 221, 225, 231-232
screws, 208, 211, 232

Extrusion barrel, 205, 207, 212, 232

FDA, 177-178, 345
Fermentation, 171, 178, 341, 345, 351
Fiberglas™, 115
Fiberize thermoplastic materials, 138
Fibers
  bicomponent, 134-135, 138-141
  multicomponent, 135, 140-141
  natural, 142, 144
  tricomponent, 140
Filtration products, 135, 141
Flash devolatilization, 102-104
Flexural strength, 143
Fluid polymer, 206
Fluid polymeric stream, 219, 221, 224
Fluid
  carrier, 199-200, 203
  confining, 263-264
  supercritical, 208, 213, 220, 233
Foam pigs
  light density, 61
  medium density, 62
Formaldehyde, 274-276, 278-281
Free radicals, 13-15, 17, 291-292, 324, 329, 357
Functional groups, 21, 25, 170, 184, 307, 314
Functionality, 327

Glass transition, 147, 247, 303, 323-324, 357
Glass-reinforced plastic component, 125
Glassy amorphous domains, 95
Glycols, 6
Goodyear Tire and Rubber Co, 363-364, 380-381, 395

Handi-Foam, 68-69
Heart valves, 172-174, 179, 191
Heat resistance, 373-374
High-pressure dilatometry, 263
Homogeneous solution, 97, 216, 231
Homopolymers, 26-27, 105, 255, 268
HVEM, 111-112
Hydrocarbons, 2, 8-9, 220
Hydrochloric acid, 282-285, 287, 355
Hydrogen
  atoms, 15-16, 308, 328, 355, 357
  bonding, 314
Hydrolysis, 182, 290, 320, 347
Hydrophilic, 182, 191
Hydroxy butyrate, 336
Hydroxyl groups, 10, 118-119, 302
Hydroxy-terminated oligomers, 322-323

Injection
  molding, 35-36, 38-39, 45-46, 49-50, 58-59, 78, 84, 158, 205, 216-217, 233, 236, 238, 241, 341-342
  molding machine, 84, 157, 242-243
Isocyanates, 57
Isoprene, 310

Lactic acid, 99, 171, 349-350
Layers
  protective, 356
  single, 130
Length-to-thickness ratio, 209, 212, 235, 243
Light scattering, 302
Linear
  chains, 25, 184, 195
  polymers, 298
Liquid crystalline order, 101-102
Liquid moulding, 151-152

Marine plastic applications, 350
Matrix resin, 321, 324
Mechanical properties of polymers, 92, 116
Mechanical valves, 229-230
Medical polymers for biomaterials applications, 178
Melt density, 262
Melt processing techniques, 95
Melt spinning, 78, 80
Melting points, 134-135, 138
Merit index, 41-42
Metabolix, 344-345
Metering devices, 214, 220, 225
Methacrylates, 198, 202
Methyl groups, 360
Microcellular material, 204-206, 216-218, 224, 226, 235
  precursor of molten polymeric, 211
  producing injection-molded, 210
Microcellular nucleation, 209, 211, 216, 236-237
Micro-organisms, industrial polymer production, 340

Mold polypropylene, 242-243
Mold polystyrene, 242-243
Mold thickness reduction temperature, 242
Molded polymeric material, 203, 235
Molding chamber, 206-215, 224, 226-229, 231-233, 236-238, 240-241
  temperature, 209, 238
  movable wall, 227
  pressurized, 206
Molding machine, 241, 243
Molecular chains, 358
Molecular mass, 10, 12, 15-18, 20-21, 23-24, 29
  distribution, 17, 21
Molecular weight
  distribution, 92, 356
  final, 323
  low, 204, 219
Molecules
  large, 2
  polymeric, 311
  unstable, 14
Molten glass, 134
Molten polymers, 35-36, 134, 136-137, 205, 250
  separate streams of, 136-137
  spinner centrifuges, 135
Monomer
  common, 301
  constituent, 290
  cyanoacrylate, 19
  diene, 20
  formation, 353
  formulation, 198-199, 202-203
  isoprene, 301
  liquid, 19
  polymer, 334
  polymerized, 198
  single, 24
  vinyl chloride, 6
Morphology, 94, 97, 101-102, 104-109, 111-112, 301, 303, 309, 328, 330
  dual semicontinuous, 108-109
  final, 105-106, 316
  particulate, 108-109
Multi-functional acrylates, 198, 202

Naphtha, 1, 3, 8, 10-11
Natural polymers, 180-181, 191
Natural rubber, 95, 290, 302, 310, 360, 362-363, 365, 367-368, 373-374, 384, 386, 392
NIMRC, 122-123
NovaMatrix, 177-178
Nucleation, 205, 215-216, 219, 224, 226-227, 229, 231
  sites, 215-216
Nucleator, 216, 223-224, 226-230, 240
Nucleic acids, 170, 190

Organic chemistry, 156
Orientational order, 101
Orifices, 134-137, 140-141, 206-207, 210-212, 218, 221-222, 224, 227-228, 232, 241
O-rings, 303, 375
Ozone, 291, 354-356, 378

Peripheral wall, 135-137, 139, 141
Petrochemical processing, 1, 5, 9
Petroleum products, 367
Photochromic
  additives, 196-199, 201-203
  materials, 196-197
  molecules, 196, 200-201
  response, 197-198
Photographic film industry, 319
Photons, 329
Pickering, 121, 123
Piezoelectric properties, 347
Plasticator, 241, 243
Plasticizers, 6, 30, 303.330, 332, 335, 354
Plasticizing units, 241
Plastics
  based, 171, 318
  degradation, 319-320
  engineering, 56, 67
  fiber-reinforced, 125
  glass-reinforced, 120, 125, 127-128, 130
  graphite-reinforced, 128
  industry, 245
  materials, 320
  production of, 158
  reinforced, 126, 128-132, 144, 270
  waste, 318
Poisson's ratio, 257-258, 270
Polyacrylamide, 218
Polyacrylic acids, 191-192

Polyamides, 23, 139, 218, 315
Polyanhydrides, 180-181, 191
Polybutadiene, 27, 104, 108, 361, 363-364, 366
Polycaprolactones, 55, 57, 59, 336
Polycarbonates, 9, 32, 56, 67, 94, 139, 202
Polydioxanone, 180-181
Polydispersity, 327, 342
Polyethylene
  high molecular weight, 310
  high-density, 94, 298
Polyglycolic acid, 349
  surgical mesh, 321-322, 324
Polyhydroxyalkonates, 336-337
Polyhydroxybutyrate, 181-182
Polylactic acid, 170-171, 336, 349-351
Polymer Analysis, 86, 92
Polymer chains, 15, 18, 79-80, 98, 113, 170, 303-304, 315, 354-355, 357-358, 361
Polymer chains, degraded, 356
  growing, 308
Polymer degradation, 178, 180, 318-320, 354
Polymer fibers, 134, 315
  manufacture of, 134
Polymer films, 83, 316
Polymer flows, 256, 268
Polymer grades, 29
Polymer manufacturing, methods of, 1
Polymer matrix, 30, 157, 292, 303
Polymer molecules, 299, 301, 326, 328, 337, 355, 357
Polymer nomenclature, 245
Polymer pellets, 223, 263
Polymer synthesis and modification, methods of, 12
Polymeric macromolecules, 298
Polymeric matrix, 201, 203
Polymerization
  addition, 307
  degree of, 16, 18
  precipitation, 102, 106, 108
  radical chain-growth, 306-308
  rapid, 310
  step-growth, 21-22, 316
Polyolefins, 29, 32, 209, 218, 278, 291, 315-336
Polypeptide, 170
Polypropylene, 29, 32, 42, 49, 51, 56, 67, 84, 111, 125, 138-139, 149, 242-243, 291, 329-331
Polyrotaxanes, 183-184
Polysaccharides, 188, 335
Polystyrene, 15, 18, 23, 51, 104, 106, 139, 171, 194-195, 204, 218, 222-243, 303, 331-333
  high-impact, 106, 108, 363
Polytetrafluoroethylene, 277-278
Polyurethane products, 52, 57, 59
  manufacture, 70
  systems, 81-83
Prepolymers, 9-10, 25, 58-59
Prestressing material, 131
Propane, 2, 4, 6, 181
Propene, 310
Propylene, 1, 3, 6, 8, 11, 28, 304, 310

Radiation
  ionizing, 326-328, 331
  ultraviolet, 196-197, 295
Radical polymerization, 306-307, 309-311
Reactivity ratios, 26-27
Reciprocating screw, 219, 224-226, 232
Reflux condenser, 282, 285-288
Resin, vinyl ester, 117-118
Rheological properties, 101, 246-247
Rheology, 90, 92-93, 259
Rubber compression molding, 375-376
Rubber injection molding, 374-377
Rubber production, 367
Rubberlike properties, 365
Rubbers
  isoprene, 361-362
  polybutadiene, 362, 364, 370
  silicone, 377
  styrene-butadiene, 360-361, 366, 373, 383
  thermoplastic, 390

Schematic specific-heat diagram, 253
Schematic thermal conductivity diagram, 254
Semi-crystalline
  polyesters, 321
  polymers, 251, 253-254, 268
  thermoplastics, 94
Shear stress, 131, 247
Shear viscosity, 250, 260
Shift function, 257

Silicone-urethane copolymers, 186
Solid dispersion, 274-276, 279-288
Specialty polymers, 9, 29, 32
Specimen geometry, 270-271, 273
Spectroscopy, 86, 90
Spinodal decomposition, 102, 104-106, 108, 112
Stress relaxation, 257
Structure
  colloidal scale, 86
  hollow plastic, 51
  polymeric, 100, 298
  skin-foam-skin, 238, 239
Styrene
  butadiene rubber, 366, 368, 373, 377-378, 391
  monomers, 8, 16, 18
Sulphuric acids, 283-284, 288
Supramolecular chemistry, 180, 183
Syndiotactic polymers, 310
Synthetic polymers, 1, 181, 190, 333, 335
  most important primary sources of, 1

Taiwan Synthetic Rubber Corpn, 380-381, 396
Tensile modulus, 38, 41, 108-110
Tensile properties of plastics, 269
Terpolymers, 312
Thermal analysis, 86, 90, 267
Thermal conductivity, 245-246, 250, 253-254, 266
Thermal degradation, 15, 187, 266, 315, 356-357
Thermal expansion, 135, 138
Thermodynamic properties, 246
Thermoplastic
  composites, 147-148, 150-151, 153
  elastomers, 28, 95, 195, 218
  materials, 10, 135, 137-139, 141, 151
  polyesters, 23
  polymers, 39, 94, 147, 149, 160, 218, 316, 340
  systems, 59, 150
Thermoset polymers, 147-148
Tire makers, 362-363
Transmission electron microscopy, 110-112
Transversely-isotropic, 257
Trommel panels, 60, 73-74, 76
Twin Screw Extrusion, 78
Tyre products, 366, 373
Tyrosine-derived polycarbonate poly, 182

Ultrafiltration, 277-278, 280, 284
UV radiation, 290, 293, 295, 297

Vertical baffle, 136-137
Vinyl chloride, 304-305, 307
Vinyl groups, 118, 324
Vinyl monomers, 6, 13, 17
Viscoelastic properties, 86
Viscosity, 134, 136-137, 139, 141, 154, 178, 222-243, 247, 259-260
  high, 152
  nominal Mooney, 383-384

Waste-to-energy systems, 100, 248
Water-repelling properties, 371
Water-soluble polymers, 67, 151, 212

X-ray scattering, 30, 54

Ziegler-Natta catalysts, 20, 28, 310